MATH MATTERS

MATH MATTERS

JAMES V. RAUFF

Millikin University

JOHN WILEY & SONS
New York • Chichester • Brisbane • Toronto • Singapore

ACQUISITIONS EDITOR Ruth Baruth

MARKETING MANAGER Debra Riegert

PRODUCTION EDITOR Chris Curioli/Ken Santor

COVER DESIGNER Steve Jenkins

TEXT DESIGNER Lee Goldstein

MANUFACTURING MANAGER Susan Stetzer

ILLUSTRATORS Wellington Studios

ILLUSTRATION COORDINATOR Rosa Bryant

COVER PHOTO Tony Stone Images/David Muench

This book was set in 10/12 Times Roman by Bi-Comp, Inc., and printed and bound by Courier Stoughton. The cover was printed by New England Book Components.

Library of Congress Cataloging-in-Publication Data

Rauff, James V.

Math Matters / James V. Rauff

p. cm.

Includes index.

ISBN 0-471-30452-2 (pbk. : alk. paper)

1. Mathematics. I. Title

QA39.2.R38 1996

510--dc20 94-42054

CIP

Printed in the United States of America

10 9 8 7 6 5 4 3 2 1

Preface

PURPOSE

Math Matters is about some of the basic ideas in mathematics that people are likely to encounter in the normal course of their lives. It is about money, statistics, relations, probability, graphs, decision making, codes, logic, languages, and much more. *Math Matters* was written to help people develop their mathematical problem solving skills and to help them see how mathematics is a part of modern society.

Math Matters is for college students who are majors in areas other than mathematics, science, or engineering. It can be used as a text in the traditional liberal arts mathematics course or in a course for mathematical competency. *Math Matters* is also for students (not necessarily in school!) who wish to learn about some of the practical topics of modern mathematics. *Math Matters* assumes a basic knowledge of artihmetic and a year of high school algebra.

DEVELOPMENTAL PROCESS

Math Matters developed over 20 years of teaching fundamental mathematical concepts to 18–22 year old college students, to students who have returned to college after years of life experience, and to evening and weekend adult students who were trying to understand the mathematics around them. These students taught me that they felt more connected to mathematics when the mathematics was connected to their life. They understood that all of the examples and applications could not be completely "real," but they insisted that I offer some justification for the material that I wanted them to learn. They wanted straight answers. They wanted a text that asked them to pause and think as they read. They wanted a text that was concise, accurate, and inexpensive.

I have tried to make *Math Matters* the text that my students told me to write.

Several classes of students have used manuscript pages of *Math Matters* and have offered suggestions for improvement. Their suggestions have influenced the final version of this book.

FEATURES

1. Situations. Each chapter of *Math Matters* is devoted to the mathematics involved in one or more "real-life" situations. Some of the situations are simplified so that the mathematics does not become unwieldy, but the basic premise behind each situation is real.

Here is a sampling of the 36 situations in *Math Matters*. "Understanding Your Bank Card Bill" (Situation 3) motivates a discussion of installment buying and APR. "Lotteries" (Situation 12) leads to an investigation of permutations and combinations. Linear optimization is explored while seeking "The Hottest Chili in the West" (Situation 21). A tour of the metric system follows "Driving in England" (Situation 23). "Counting Hats in Ixtapa" (Situation 30) takes the reader on a tour of counting words in languages around the world. Logical reasoning is examined by "Following IRS Instructions" (Situation 31). "Thinking Machines and Imaginary Worlds" (Situation 36) takes a glimpse at artificial life and cellular automata.

2. Interactive text. Throughout the exposition, *Math Matters* asks the reader for responses to questions and gives the reader space to write her response. When the student writes in the book, she connects herself to the situations and topics at hand. This connection enables the student to begin to own the mathematics.

3. Practice Exercises. The Practice Exercises are designed to provide immediate feedback to the reader on his understanding of the text. The Practice Exercises test basic understanding at the point of impact with the material.

4. Essay Exercises. Essay exercises occur throughout the text. These exercises further develop the connection between the mathematics and the world. The essay exercises also probe the depth of understanding of the student.

In addition, the available *Instructor's Guide* contains writing activities for each chapter and suggestions on how to use writing to learn mathematics.

ORGANIZATION

The chapters of *Math Matters* are essentially independent and can be covered in any order. Exceptions are Chapters 6 and 9, which use the concepts of linear equations developed in Chapter 5. The order of topics presented is one which I have found most successful in my courses. Many students bring with them an interest in money (Chapter 1), statistics (Chapter 2), and gambling (Chapters 3 & 4). These topics are quite comfortably followed by a discussion of functions (Chapter 5) and optimization (Chapter 6).

A one-semester course could typically work through Chapters 1–5 along with an additional chapter from Chapters 6–12. A two-semester course is needed to cover the entire text.

Each major mathematical topic in *Math Matters* is organized according to the following scheme.

1. A real life "Situation" is described and a question arising from it is posed.
2. A discussion or exploration of possible solutions or approaches follows.
3. The formal mathematics needed to solve the problem is introduced. This introduction proceeds carefully with numerous examples with clearly discussed step-by-step instructions. "Practice Exercises" pause the development of the mathematics so that the reader can verify his or her understanding.
4. The formalism is applied to the Situation and to similar problems. "Practice Exercises" intervene so that the student can verify that he or she understands the solutions.
5. "What Do You Know?" chapter exercises provide cumulative reinforcement.

NOTE TO THE STUDENT

Math Matters is an interactive text. Your success in using *Math Matters* depends upon your participation in the story of each chapter. I will frequently ask you for a response to a question and give you space to write your response. It is important that you actually WRITE IN YOUR BOOK! By writing you are connecting yourself to the situations and topics at hand. When you establish this connection you begin to take control of the mathematics.

Another way of connecting with *Math Matters* is to work the Practice Exercises when you come to them. Don't read a chapter of *Math Matters* and then go back and work the Practice Exercises. Work these exercises immediately when they pop up. Check your answers. Go back to the text and review. Work the exercises again until you can see why the given answer is correct. Don't just accept it! In this way you will come to own the mathematics naturally.

Math Matters assumes a participatory reader. Read *Math Matters* with a pencil in hand.

Supplements

Math Matters is a self-contained text. However, there is an instructor's guide available that contains summaries of each chapter, the answers to even-numbered exercises, and an essay with examples on "Using Writing to Teach Mathematics."

Acknowledgments

Math Matters came about as the result of many students working hard to understand fundamental mathematical concepts and pressing me to explain and justify the mathematics I wanted them to learn. I thank each one of them for helping me to understand how they learned or didn't learn mathematics.

I also would like to single out for special mention three people who were more directly responsible for the creation of *Math Matters*:

Patrice M. Zaccagni, a sales representative, who suggested that I might write this book and established my first contact with the editors at John Wiley & Sons.

Ruth Baruth, the acquisitions editor for *Math Matters*, who worked tirelessly for its production and was unwavering in her support for the project form its inception.

Rebecca Rauff, a free-lance editor and writer, who helped me formulate my ideas into a prospectus and an initial chapter, and who provided personal love and support throughout the creation of *Math Matters*.

I would also like to thank Ken Santor, Chris Curioli, Eileen Navagh and the rest of the editors, artists, and designers at John Wiley & Sons for their excellent work on the production of *Math Matters*.

Finally, I wish to thank the following colleagues who reviewed the manuscript of *Math Matters*, providing many useful and insightful suggestions and corrections:

Sharon Rose Butler
Pikes Peak Community College

David Capaldi
Toll Gate High School

Beatrice Eastman
William Patterson College

Gilberto Gaza
El Paso Community College

Peter Georgakis
Santa Barbara City College

Herbert Gindler
San Diego State University

John Karlof
University of North Carolina at Wilmington

Jean Krichbaum
Broome Community College

Anne Landry
Dutchess Community College

Charles Nelson
University of Florida, Gainesville

Gayle Palka
Broome Community College

John Peterson
Bringham Young University

Randall Swift
Western Kentucky University

Steve Sworder
Saddleback College

Lynn Wolfmeyer
Western Illinois University

To Ed, Yvonne, Becky, and Andy

Contents

Chapter 1. MATHEMATICS OF THE MARKETPLACE 1

Chapter 2. MATHEMATICS OF DESCRIPTION 37

Chapter 3. MATHEMATICS OF COLLECTIONS 75

Chapter 4. MATHEMATICS OF PREDICTION 111

Chapter 5. MATHEMATICS OF RELATIONSHIPS 169

Chapter 6. MATHEMATICS OF OPTIMIZATION 213

Chapter 7. MATHEMATICS OF SPACE 249

Chapter 8. MATHEMATICS OF NUMBERS 283

Chapter 9. MATHEMATICS OF GAMES 311

Chapter 10. MATHEMATICS OF OTHER CULTURES AND OTHER TIMES 343

Chapter 11. MATHEMATICS OF REASONING 369

Chapter 12. MATHEMATICS OF COMPUTERS 409

Chapter One

Mathematics of the Marketplace

What's the Point? You will encounter mathematics most frequently in the marketplace. Sale prices, discounts, interest rates, car loans, home mortgages, and bank cards are likely to be a part of your life throughout your life. The activities in this chapter will help you to understand the terminology and mathematics of the marketplace.

Situation 1

Is That a Good Price?

You are shopping for a new CD player. Fred's Electronics Warehouse is selling the player you want at Fred's "everyday low price" of $249.99. The CD Partment Store normally sells the same CD player for $280, but this week the player has been marked down 14%. Marion's Music Mansion is selling the same player at a 10% markup on the dealer cost of $220. Which store offers the lowest price for the CD player? Which store offers the highest price?

If you said that the CD Partment Store has the lowest price and Fred's has the highest, then you were right! The mathematics involved here has to do mostly with percentages. Percentages are a special kind of ratio, and percentage problems are a special kind of proportion. So, we'll begin our discussion of the solution to Situation 1 with ratio and proportion.

1.1 RATIO AND PROPORTION

The **ratio** of two quantities is the fraction formed by the two quantities. For example, if there are two men and six women on a bus, then the ratio of men to women on the bus is 2/6. The ratio of women to men on the bus is 6/2. Notice that the quantity following the word *to* appears as the denominator (bottom) of the fraction. Ratios can be written using the fraction bar, the word *to,* or with a colon.

	Notation		
Ratio	**Fraction Bar**	**"To"**	**Colon**
Men to women	2/6	2 to 6	2:6
Women to men	6/2	6 to 2	6:2

PRACTICE

There are 7 elephants, 8 zebras, and 14 gazelles at a watering hole.

1. Find the ratio of elephants to zebras.

2. Find the ratio of elephants to gazelles.

3. Find the ratio of gazelles to zebras.

4. Find the ratio of gazelles to elephants.

5. Find the ratio of zebras to gazelles.

6. Find the ratio of zebras to elephants.

Ratios, because they are fractions, should be reduced to lowest terms. Thus, the ratio of men to women on the bus becomes $2/6 = 1/3$, or 1 to 3, or $1:3$. A 1 in the denominator is retained in ratios. Thus, the ratio of women to men on the bus becomes $6/2 = 3/1$, or 3 to 1, or $3:1$.

PRACTICE

7. Reduce each of the ratios in Practice Exercises 1–6 to lowest terms.

A BASKETBALL TEAM WON 80 GAMES AND LOST 6 GAMES.

8. Find the ratio of the number of games lost to the number of games won.

9. Find the ratio of the number of games won to the number of games lost.

10. Find the ratio of the number of games lost to the number of games played.

11. Find the ratio of the number of games played to the number of games won.

A **proportion** is an equality between two ratios. Each of the following equalities are proportions.

$$2/5 = 6/15$$

$$3:4 = 9:12$$

$$x:6 = 12:7$$

The proportion $3:4 = 9:12$ is read "Three is to four as nine is to twelve."

Complete the table matching the symbolic proportion with the English proportion.

English	Symbols
Two is to three as four is to six.	$2/3 = 4/6$
Five is to seven as ten is to fourteen.	________
Three is to one as nine is to three.	$3/1 = 9/3$

Usually, we know three of the quantities of a proportion and we need to find the fourth. For example, suppose a car travels 200 miles on 15 gallons of gasoline. How many miles can the car go on 25 gallons of gasoline?

To answer this question, we first write the proportion "200 miles is to 15 gallons as x miles is to 25 gallons." In symbols,

$$200/15 = x/25$$

Now we can reduce the left ratio 200/15 to 40/3 to make the numbers a little easier to work with. This gives us

$$40/3 = x/25$$

Now we **cross multiply,** a technique of algebra. Multiply the quantities as shown by the arrows.

$$\frac{40}{3} \times \frac{x}{25}$$

Now we have

$$40 \cdot 25 = 3 \cdot x$$

(In this book, the $\cdot$ will be used to indicate multiplication.) Next, rewrite the equality so that the expression involving x is on the left.

$$3 \cdot x = 40 \cdot 25$$

Now multiply the two numbers on the right-hand side of the equation.

$$3 \cdot x = 1000$$

Finally, divide both sides by 3.

$$3 \cdot x/3 = 1000/3$$
$$x = 333.33 \text{ miles}$$

The car can go 333.33 miles on 25 gallons of gas.

Let's look at another example. Notice that the method of solution is always the same!

Problem 1 Fern can type 60 words per minute. How long will it take Fern to type 1200 words?

Solution

1. Write the proportion: "60 words is to 1 minute as 1200 words is to x minutes" ($60/1 = 1200/x$).
2. Cross multiply: $60 \cdot x = 1 \cdot 1200$
3. Multiply: $60 \cdot x = 1200$
4. Divide by 60: $60 \cdot x/60 = 1200/60$
   $x = 20$
5. Write the answer: Fern can type 1200 words in 20 minutes. ■

Problem 2 A bullfrog that weighs 12 ounces can jump 20 feet. How far can a 2-pound bullfrog jump?

Solution

1. Write the proportion: "12 ounces is to 20 feet as 32 ounces is to x feet." *It is important that the two ratios of a proportion be ratios involving the same units.* It would be incorrect to write "12 ounces is to 20 feet as 2 pounds is to x feet." ($12/20 = 32/x$).
2. Reduce the left ratio: $3/5 = 32/x$
3. Cross multiply: $3 \cdot x = 5 \cdot 32$
4. Multiply: $3 \cdot x = 160$
5. Divide by 3: $3 \cdot x/3 = 160/3$
   $x = 53.33$
6. Write the answer: A 2-pound bullfrog can jump 53.33 feet. ■

PRACTICE

12. A car goes 300 miles on 12 gallons of gasoline. How far can it go on 15 gallons of gasoline?

13. Twelve apples cost $2. How much will 17 apples cost?

14. If it takes Myrna 3 days to make 5 dresses for her antique doll, how many days will it take her to make 24 dresses?

15. There are 24 hours in a day. How many hours are there in 4 years?

1.2 PERCENT

Percent is a ratio of some number to 100 (*percent* = *to one hundred*). Thus, 45 percent, or 45%, means 45/100. Fill in the blanks.

35% means 35/100

27% means __/100

88% means 88/___

178% means 178/___

0.7% means 0.7/___

1.5% means __/100

(Answers: 27, 100, 100, 100, 1.5)

Percents are often expressed as decimals. So, 45%, which means 45/100, can also be written as 0.45. To write a percent as a decimal, simply divide the number before the % sign by 100. To write a decimal as a percent, multiply by 100.

Thus,

$$40\% = 40/100 = 0.40$$

$$1.35\% = 1.35/100 = 0.0135$$

$$200\% = 200/100 = 2$$

$$0.54 = (0.54) \cdot 100 = 54\%$$

$$1.56 = (1.56) \cdot 100 = 156\%$$

$$0.0003 = 0.0003 \cdot 100 = 0.03\%$$

PRACTICE

Fill in the blanks.

	Percent	Ratio	Decimal
1.	35%	_____	0.35
2.	_____	23/100	0.23
3.	23.4%	23.4/100	_____
4.	_____	4.5/100	0.045
5.	36%	36/100	_____

Percentage problems are proportions involving percents. Here are some typical percentage problems.

WHAT IS 14% OF 200?

FORTY-FIVE IS WHAT PERCENT OF 120?

SEVENTY IS 30% OF WHAT?

Each percentage problem has the form "A is B percent of C?", where A, B, and C are numbers or the word *what*. In the three examples above, we have

WHAT IS 14% OF 200?

A = WHAT, $B = 14$, $C = 200$

FORTY-FIVE IS WHAT PERCENT OF 120?

$A = 45$, B = WHAT, $C = 120$

SEVENTY IS 30% OF WHAT?

$A = 70$, $B = 30$, C = WHAT

PRACTICE

Fill in the blanks.

	Percentage Problem	A	B	C
6.	45 is what percent of 60?	45	what	_____

Percentage Problem	A	B	C
7. What is 30% of 120?	____	30	120
8. 75 is 40% of what?	75	____	what
9. What is 40% of 800?	____	____	____
10. 90 is what percent of 300?	____	____	____
11. 56 is 45% of what?	____	____	____

Each percentage problem "*A* is *B* percent of *C*" is actually the proportion "*A* is to *C* as *B* is to 100" or $A/C = B/100$. Percentage problems can therefore be solved just like proportions.

Problem 1 What is 40% of 70?

Solution

1. Identify A, B, C: A = what, $B = 40$, $C = 70$
2. Write as the proportion $A/C = B/100$, using x for the word *what*: $x/70 = 40/100$
3. Solve the proportion:

$$100 \cdot x = 70 \cdot 40$$
$$100 \cdot x = 2800$$
$$x = 28$$

4. Write the answer: 28 is 40% of 70. ■

Problem 2 75 is what percent of 90?

Solution

1. Identify A, B, C: $A = 75$, B = what, $C = 90$
2. Write as the proportion $A/C = B/100$, using x for the word *what*: $75/90 = x/100$
3. Solve the proportion:

$$100 \cdot 75 = 90 \cdot x$$
$$7500 = 90 \cdot x$$
$$90 \cdot x = 7500$$
$$x = 83.33$$

4. Write the answer: 75 is 83.33% of 90. ■

PRACTICE

12. What is 15% of 60?

13. 78 is what percent of 100?

14. 58 is 50% of what?

15. What is 400% of 12?

16. 300 is what percent of 40?

17. 70 is 0.4% of what?

As you work more percentage problems, you will begin to notice patterns in the solutions and begin to develop shortcuts. That's fine. However, if you stick to the "A, B, C's of percentage problems," you will always have one method that works for every problem.

Later we will use the techniques for working percentage problems to answer the questions of Situation 1.

1.3 MARKDOWNS AND MARKUPS

A retailer is in business to make a profit. As a result, the retailer will charge you more for an item than she paid for it. This additional charge for the item is called its **markup.** For example, if the retailer bought a TV for \$200 from the manufacturer and then sold that TV to you for \$250, the retailer's markup is \$250 − \$200 = \$50. We can express the relationship between what the retailer paid for an item (the dealer's cost), the markup, and what the retailer sells the item for (the selling price) by the following verbal formulas:

Markup = Selling price − Dealer's cost

Selling price = Dealer's cost + Markup

If a retailer decides to lower the regular price of an item in order to increase her chances of selling it, the difference between the regular price and the new price is called a **markdown.** For example, if a TV usually sells for \$250 and the retailer puts it on sale for \$230, then the markdown is \$250 − \$230 = \$20. A verbal formula relating the regular price, the markdown, and the sale price is

Markdown = Regular price − Sale price

Sale price = Regular price − Markdown

PRACTICE

Fill in the blanks.

1. Dealer's cost = \$200, selling price = \$300, markup = _____

2. Dealer's cost = \$200, selling price = _____, markup = \$45

3. Dealer's cost = _____, selling price = \$300, markup = \$60

4. Regular price = \$400, sale price = \$350, markdown = _____

5. Regular price = \$400, sale price = _____, markdown = \$60

6. Regular price = _____, sale price = \$350, markdown = \$100

Markdowns and markups are usually expressed as percentages. We hear or read sentences like

MARKED DOWN 50%!

TAKE AN ADDITIONAL 30% OFF!

HE MARKS UP HIS INVENTORY 30%

Markdown percentages are percentages of the regular price. So, when an advertisement claims that a shirt has been marked down 40%, this means that the markdown was 40% of the regular price.

Suppose that the regular price for a shirt was \$25. If the shirt's price is marked down 40%, what is the markdown?

____

____

____

If you said \$10, then you are right! We needed to answer the questions, "What is 40% of 25?" Make sure that you can see that 10 is 40% of 25. If you can't, go back and review Section 1.2.

What was the sale price of the shirt?

____

____

____

(Answer: Sale price = Regular price − Markdown

= \$25 − \$10

= \$15)

PRACTICE

7. A dress that usually sells for \$200 is marked down 30%. What is the sale price of this dress?

8. A refrigerator is marked down 50% from its regular price of $350. What is the sale price of the refrigerator?

9. A chair is on sale for $200. The regular price of the chair was $280. What is the markdown on the chair?

10. What percent of the regular price of the chair in Practice Exercise 9 is the markdown?

If we know the percent markdown and sale price, it is possible to compute the regular price. Suppose that a set of four automobile tires is on sale for $250. The tire salesman tells you that this is 35% off the regular price. What was the regular price of the tires?

Notice that if the sale price is the regular price minus 35% of the regular price, then the sale price must be 100% − 35% = 65% of the regular price. Thus, we need to answer the question, "$250 is 65% of what?" Or,

$$250/x = 65/100$$

$$250 \cdot 100 = 65 \cdot x$$

$$25{,}000 = 65 \cdot x$$

$$25{,}000/65 = x$$

$$x = \$384.62$$

The regular price of the set of tires was $384.62.

PRACTICE

11. A pair of running shoes has been marked down 20% to \$129.99. What was the original price of this pair of running shoes?

12. A sofa is on sale for \$519.50. This price is a 40% markdown from the regular price. What was the regular price of this sofa?

Markups are usually figured as percentages of the dealer's cost. Suppose that an electronics dealer customarily marks up all his merchandise 38% on dealer's cost. If the dealer bought a VCR for \$200, what will be his markup on this VCR?

Did you get \$76? You needed to answer the question, "What is 38% of 200?" If you didn't get \$76, go back and rework Section 1.2!

What is the selling price of this VCR?

Right! Selling price = Dealer's cost + Markup = \$200 + \$76 = \$276.

PRACTICE

13. A carpet dealer marks up his merchandise 20% on dealer's cost. If a carpet costs the dealer $500, for how much will he sell the carpet?

14. An auto dealer marks up each automobile 5% on her cost. If she buys a car from the manufacturer for $15,000, for how much will she sell the car?

If you know what a dealer paid for an item, then you can compute the markup on the dealer's cost. Suppose that you know a TV retailer paid the manufacturer of the TV $140 for the TV. The retailer is selling the TV for $187.60. What is the dealer's percent markup on cost?

The markup is $187.60 − $140 = $47.60. Thus, all we need to know is the answer to the question, "$47.60 is what percent of $140?" Or,

$$47.60/140 = x/100$$

$$0.34 = x/100$$

$$x = 34$$

The percent markup on cost is 34%.

Now we can answer the questions posed in Situation 1.

The CD player would cost $249.99 at Fred's Electronic Warehouse.

The markdown at the CD Partment Store is 0.14 · $280 = $39.20. So we would pay $280 − $39.20 = $240.80 for the CD player at the CD Partment Store.

Finally, the markup at Marion's Music Mansion is 0.10 · $220 = $220. So we would pay $220 + $22 = $242 for the CD player at Marion's.

At $240.80, the CD Partment Store offers the lowest price for the CD player!

PRACTICE

15. A sofa is selling for $500. The sofa retailer bought the sofa from the manufacturer for $260. What is the retailer's percent markup on cost?

16. A book is being sold for $29.95. The bookseller paid $20 for the book. What is the bookseller's markup on cost?

Situation 2 Investing Your Money

You have just received a holiday bonus totaling $2000 and you have decided to put it in a bank for 6 months and then use it to take a vacation to Luxembourg. There are three banks in your town. The First National Bank will pay you 5% simple interest on your money. The Ravine Bank and Trust will pay you 4% compounded daily. The Apple Tree Savings Bank will pay you 4.98% compounded quarterly. Which bank will pay you the most money after 6 months?

Situation 2 involves a lot of technical financial terms and some fairly complicated calculations. The three banks fall into two main categories. The First National Bank pays simple interest and the other two pay compound interest. We'll begin with simple interest.

1.4 SIMPLE INTEREST

Interest is money paid for the use of money. When a bank pays you interest on your savings accounts, the bank is paying you for the use of your money. When you borrow money from a bank, then you pay interest for the use of the bank's money. The amount of money that you put into a savings account or the amount that you borrow from a bank is called the **principal.**

Simple interest is computed as a percentage of the principal according to the **simple interest formula:**

$$\text{Interest} = \text{Principal} \cdot \text{rate} \cdot \text{time}$$

or

$$I = PRT$$

In the simple interest formula, I stands for the interest earned and P for the principal.

The interest rate R is expressed as a percentage over some time interval. We could have R given as 5% per year, 7% per month, 2% per day, 3% per week and so on. If no time period is specified, then the time period is assumed to be 1 year. The First National Bank in Situation 2 pays 5% *per year* because no time interval was specified.

T in the simple interest formula is the time over which the interest is computed. T must be written in the same units of time as R. Suppose that we are calculating the interest over 6 months. If R is 5% per year, then T is 0.50 because 6 months is half a year. If R is 7% per month, then T is 6 (6 months). If R is 3% per week, then T is 26 (26 weeks in 6 months).

The simple interest formula is easy to use. Let's apply it first to the First National Bank in Situation 2. In this case, we have $2000 to save, so our principal P is $2000. The interest rate is 5% per year, so $R = 5\% = 0.05$. The time period is 6 months, so $T = 0.50$. Thus, the interest earned in 6 months will be

$$\begin{aligned} I &= PRT \\ &= \$2000 \cdot 0.05 \cdot 0.50 \\ &= \$50 \end{aligned}$$

The First National Bank will pay you \$50 in 6 months for the use of your \$2000. This means that at the end of the 6 months, you will have \$2000 + \$50 = \$2050 for your trip to Luxembourg.

Complete the following table of simple interest calculations:

Principal	Interest Rate	Time	Interest
1. \$2000	4%	2 years	$2000 \cdot 0.04 \cdot 2 = \$160$
2. \$2000	3%	6 months	$2000 \cdot 0.03 \cdot 0.50 = \$30$
3. \$2000	2% per month	5 months	$2000 \cdot 0.02 \cdot 5 = \$200$
4. \$2000	________	2 years	$2000 \cdot 0.06 \cdot 2 = \$240$
5. \$2000	1% per week	3 months	$2000 \cdot 0.1 \cdot$ _____ $= \$260$
6. \$2000	________	1 year	$2000 \cdot 0.04 \cdot 12 = \$960$
7. \$2000	3%	4 years	____________ = \$240
8. \$1000	5%	2 months	____________ = \$8.33

(Answers: 6%, 13, 4% per month, $2000 \cdot 0.03 \cdot 4$, $1000 \cdot 0.05 \cdot (2/12)$)

PRACTICE

1. If you put \$500 into a bank account paying 4% simple interest, how much interest will you earn in 3 years? How much money (principal and interest) will you have in 3 years?

2. If you put \$600 into a bank account paying 2% simple interest per month, how much interest will you earn in 2 years? How much money will you have in 2 years?

3. If you put \$500 into a bank account paying 4% simple interest semiannually, how much interest will you earn in 3 years? How much money will you have in 3 years?

4. If you put \$600 into a bank account paying 2% simple interest, how much interest will you earn in 2 years? How much money will you have in 2 years?

Which of the simple interest rates, 5%, 4.5% per month, or 1% per week, pays the most interest in 1 year?

____

____

____

To decide which rate pays the most interest, we see how much interest is paid on a principal of \$100. If we let $P = \$100$, then the 5% rate pays $\$100 \cdot 0.05 \cdot 1 = \$5$ in interest, the 4.5% per month rate pays $\$100 \cdot 0.045 \cdot 12 = \$54$ in interest, and the 1% per week rate pays $\$100 \cdot 0.01 \cdot 52 = \$52$ in interest. The 4.5% per month rate is the best.

Another way to compare the three rates is to compare their annual yields. The **annual yield** (also known as **effective rate**) of an investment is the simple annual interest rate that would earn the same amount of interest in 1 year. We can find the annual yield for each rate by answering the percentage question, "The interest earned is what percent of the principal?" The annual yield of the 5% rate is 5% because \$5 is 5% of \$100. The annual yield of the 4.5% per month rate is 54% because \$54 is 54% of \$100. The annual yield of the \$1 per week rate is 52% because \$52 is 52% of \$100.

Notice that the numerical value of the annual yield matches that of the interest earned on \$100 for 1 year. In general, to find the annual yield, we need only figure the interest earned on \$100 for 1 year.

Suppose we have an investment paying a simple interest rate of 3.4% per month. The interest earned on \$100 in 1 year will be $\$100 \cdot 0.034 \cdot 12 = \$40.8$. The annual yield of this investment is 40.8%.

PRACTICE

Find the annual yield for each of the following simple interest rates:

5. 6% per month.

6. 2% per week.

7. 3% semiannually.

8. 4% quarterly (1/4 of a year).

1.5 COMPOUND INTEREST

Simple interest is calculated and paid once at the end of the time interval specified by the rate. Banks usually calculate interest more frequently than that. The interest is then added to the principal to make a new principal. In effect, you earn interest on your interest. Interest that is computed on the principal and any accumulated interest is called **compound interest.**

There is a compound interest formula, but before I tell you about it, I want to make sure that you understand how the compounding of interest works. Suppose that you put \$100 into a bank account that pays 5% interest compounded quarterly. This means that the interest is calculated and added to the principal four times a year. The interest is calculated as a 5% simple interest each time. Here's how the calculations go.

Three months from the date you deposit the \$100, the first interest calculation is made:

$$I = PRT$$

$$= \$100 \cdot 0.05 \cdot 0.25$$

The T value is 0.25 because 3 months is $1/4 = 0.25$ of a year.

$$I = \$1.25$$

The $1.25 in interest is then added to the principal to make a new principal of $101.25. This new principal is used in the next calculation of interest 3 months later:

$$I = \$101.25 \cdot 0.05 \cdot 0.25$$
$$= \$1.27$$

Now this interest is added to the $101.25 to make a new principal of $101.25 + $1.27 = $102.52. In another 3 months, another interest calculation is made. This time the principal used is the new value of $102.52:

$$I = \$102.52 \cdot 0.05 \cdot 0.25$$
$$= \$1.28$$

This interest is added to the principal to make a new principal of $102.52 + $1.28 = $103.80. The final interest calculation in 3 months is made using this new principal:

$$I = \$103.80 \cdot 0.05 \cdot 0.25$$
$$= \$1.30$$

The total amount of money in the account at the end of the year is then $103.80 + $1.30 = $105.10. The account has earned a total of $105.10 − $100 = $5.10 in interest.

What is the annual yield of 5% compounded quarterly?

____

____

____

(Answer: 5.1%)

PRACTICE

Complete the following table of calculations for the interest earned on $100 at a rate of 6% compounded quarterly:

	Principal	Interest	New Principal
1.	$100	$1.50	$101.50
2.	$101.50	______	$103.02
3.	$103.02	______	________
4.	________	______	________
5.	What is the annual yield?		

If we had to go through all these calculations for a rate that was compounded daily, we'd be figuring interest until the cows come home. Fortunately, there is a nice formula that does all the work for us. It is the **compound interest formula:**

$$A = P(1 + R/N)^{NT}$$

Don't panic! This looks like a monster, but actually it is a friendly monster. I'll explain what each letter stands for and then we'll do some examples.

In the compound interest formula,

R = annual interest rate
P = principal
T = time in years
N = number of compounding periods per year
A = total amount accumulated after T years

We can apply the compound interest formula to our first example of \$100 invested for 1 year at 5% compounded quarterly. We have

$R = 0.05$
$P = \$100$
$T = 1$ year
$N = 4$ compounding periods per year

and

$$\begin{aligned} A &= P(1 + R/N)^{NT} \\ &= \$100 \cdot (1 + 0.05/4)^{4 \cdot 1} \\ &= \$100 \cdot (1 + 0.0125)^4 \\ &= \$100 \cdot (1.0125)^4 \\ &= \$100 \cdot 1.051 \\ &= \$105.10 \end{aligned}$$

A calculator is a must for compound interest calculations!

Here's another example. Suppose you invest \$200 at 4% compounded monthly for 2 years. How much money will you have at the end of this time?

Using the compound interest formula, we obtain

$R = 0.04$
$P = \$200$
$T = 2$ year
$N = 12$ compounding periods per year

and

$$\begin{aligned} A &= \$200 \cdot (1 + 0.04/12)^{12 \cdot 2} \\ &= \$200 \cdot (1.0033)^{24} \\ &= \$200 \cdot (1.082) \\ &= \$216.40 \end{aligned}$$

You will have \$216.40 in your account at the end of the 2 years.

What is the annual yield for 4% compounded monthly?

____

____

____

(Answer: 4.03%. Use $P = \$100$ and $T = 1$.)

PRACTICE

Compute the total amount accumulated for each of the following compound interest rate investments:

6. \$3000 at 6% compounded monthly for 3 years.

7. \$2000 at 3% compounded daily ($N = 360$) for 2 years. Banks often use a 360-day year for financial calculations because 360 is divisible by 12 (months), 4 (quarters), and 2 (semiannual). The 360-day year is called the **banker's year.**

8. \$100 at 7% compounded semiannually for 20 years.

9. \$100 at 7% compounded quarterly for 20 years.

10. \$100 at 7% compounded monthly for 20 years.

11. What are the annual yields for the investments in Practice Exercises 8–10?

We can now provide an answer to the question posed in Situation 2. Here are the calculations for each bank on a principal of \$2000 for a time period of 6 months.

First National Bank (5% simple interest)

$$I = PRT$$
$$= 2000 \cdot 0.05 \cdot 0.5$$
$$= \$50$$

Total amount accumulated = \$2000 + \$50 = \$2050

Ravine Bank and Trust (4% compounded daily, $N = 360$)

$$A = \$2000(1 + 0.04/360)^{360 \cdot 0.50} \qquad (T = 1/2 \text{ year!})$$
$$= \$2040.40$$

Apple Tree Savings Bank (4.98% compounded quarterly)

$$A = \$2000(1 + 0.0498/4)^{4 \cdot 0.50}$$
$$= \$2050.11$$

You will earn more money at the Apple Tree Savings Bank than at either of the other two banks.

Situation 3

Understanding Your Bank Card Bill

Your bank card bill came in the mail today. It shows that you made the following purchases:

Date	Store or Business	Amount
04/09	Software, Inc., Laramie, WY	99.67
04/23	Jitterbug, Denver, CO	42.09
05/02	Pizza Yurt, Boody, IL	34.00

Your total charges are \$175.76. The bank card statement says the bank that gave you the card charges a 1.496% monthly rate on purchases. Your statement says that you must pay a minimum of \$20.00 by June 1. If you make no other purchases on your bank card and pay \$20 each month, how long will it take for you to pay all that you owe? How much interest will you pay? How much less interest will you pay if you pay \$40 a month?

1.6 INSTALLMENT BUYING

Situation 3 is an example of installment buying. When you purchase something on installments, then you pay for your purchase over a period of time. Of course, you are usually charged interest on the amount of money that you owe. Situation 3 describes a case of **fixed installment buying** where your periodic payments (in this case, monthly) are all the same amount. Let's see how this works with the credit card bill in Situation 3.

Suppose you pay the \$20 minimum requested by the bank card bill on June 1. This reduces what you owe to $\$175.76 - 20 = \$155.76$. Now the bank will charge you simple interest on this remaining balance for the month of June:

$$\text{Interest for June} = \$155.76 \cdot 0.01496 \cdot 1$$

$$I = \$2.33$$

Now you owe $\$155.76 + \$2.33 = \$158.09$. When you make your July 1 payment of \$20, what you owe is reduced to $\$158.09 - \$20 = \$138.09$. Now the bank will charge you simple interest on this remaining balance for the month of July:

$$I = \$138.09 \cdot 0.01496 \cdot 1$$

$$= \$2.07$$

Now you owe $\$138.09 + \$2.07 = \$140.16$. Your payment of \$20 on August 1 will reduce this to \$120.16. Now the bank will charge you simple interest for the month of August:

$$I = \$120.16 \cdot 0.01496 \cdot 1$$

$$= \$1.80$$

Now you owe $\$120.16 + \$1.80 = \$121.96$. Your payment of \$20 on September 1 will reduce this to \$101.96.

This process continues until the balance owed reaches zero. The previous calculations and the calculations through the entire process are shown in the table below.

PAYING OFF A BANK CARD BILL
(ORIGINAL BALANCE = $175.76; MONTHLY PAYMENT = $20;
MONTHLY INTEREST RATE = 1.496%)

Date of Payment	Previous Balance (PB)	Interest on PB	Payment	New Balance
June 1	$175.76	$0	$20	$155.76
July 1	$155.76	$2.33	$20	$138.09
August 1	$138.09	$2.07	$20	$120.16
September 1	$120.16	$1.80	$20	$101.96
October 1	$101.96	$1.53	$20	$83.49
November 1	$83.49	$1.25	$20	$64.74
December 1	$64.74	$0.97	$20	$45.71
January 1	$45.71	$0.68	$20	$26.39
February 1	$26.39	$0.39	$20	$6.78
March 1	$6.78	$0.10	$6.88*	$0

* The last payment is made in the amount necessary to reduce the balance to zero.

How many months did it take to pay all that you owed?

____

(Answer: 9)

How much interest did you pay? _______________

____

If we add up all the values in the interest column in the table, we find that we paid a total of $11.12 in interest.

PRACTICE

If we pay $40 a month instead of $20, the table is shortened because we pay off the amount owed faster. Fill in the blanks in the table below.

PAYING OFF A BANK CARD BILL
(ORIGINAL BALANCE = $175.76; MONTHLY PAYMENT = $40;
MONTHLY INTEREST RATE = 1.496%)

Date of Payment	Previous Balance (PB)	Interest on PB	Payment	New Balance
June 1	$175.76	$0	$40	$135.76
July 1	$135.76	$2.03	$40	$97.79
August 1	$97.79	(1) ________	$40	$59.25
September 1	(2) ________	(3) ________	$40	(4) ________
October 1	(5) ________	(6) ________	(7) ________	(8) ________

9. How much total interest is paid?

10. How much less interest do you pay when you make $40-a-month payments instead of $20-a-month payments?

1.7 APR

One popular use of fixed installment buying is the automobile loan. Many people do not have enough cash to buy a new car, so they borrow the money from a bank, savings and loan, credit union, or lending institution working with the car dealer.

Suppose that you have agreed to pay $15,000 for a new car. Your credit union will loan you 90% of the purchase price of the car. How much money will the credit union loan you?

Suppose, in addition, that the credit union will charge you 8% simple interest and you must pay back the loan in monthly payments over 5 years. The credit union calculates the interest using the simple interest formula and gets

$$
\begin{aligned}
I &= PRT \\
&= \$13{,}500 \cdot 0.08 \cdot 5 \\
&= \$5400
\end{aligned}
$$

This interest is then added to the amount of the loan to get $13,500 + $5400 = $18,900. There are 60 months in 5 years, so your monthly payment is $18,900/60 = $315. The $5400 in interest is being paid at a rate of 5400/60 = $90 per month.

What is wrong with the preceding calculation?

Right! The credit union calculated the interest on the entire $13,500 over the 5 years. But, just as in the bank card payment calculations, you would have reduced the amount you owed with each payment. Let's look at a few lines of the loan payoff table.

Suppose that you buy the car on June 1 and make your first payment on July 1. Interest will be computed for June.

PAYING OFF A CAR LOAN
(ORIGINAL BALANCE = $13,500; MONTHLY PAYMENT = $315;
ANNUAL INTEREST RATE = 8.00%)

Date of Payment	Previous Balance (PB)	Interest on PB	Payment	New Balance
July 1	$13,500.00	$90.00	$315	$13,275.00
August 1	$13,275.00	$88.50	$315	$13,048.50
September 1	$13,048.50	$86.99	$315	$12,820.49

Notice that the interest is getting smaller each month, but the \$315-per-month payment is based on a monthly interest charge of \$90. So, under the \$315-per-month payment scheme, you would be paying more than 8% simple interest on the amount owed.

In 1969, the U.S. Congress passed the *Truth in Lending Act* that requires all lenders to state the true annual interest rate, called the **annual percentage rate,** or **APR.** The APR can be computed easily from the stated interest rate, also called the **add-on interest** rate, and the number of payments using the **APR formula:**

$$\text{APR} = \frac{2 \cdot N \cdot r}{N + 1}$$

In the APR formula, N is the number of monthly payments and r the add-on interest rate. So, for our car loan, we can find the APR. It is

$$\begin{aligned}\text{APR} &= \frac{2 \cdot 60 \cdot 0.08}{60 + 1} \\ &= 9.6/61 \\ &= 0.157\end{aligned}$$

The APR for the car loan is 15.7%.

When you are comparing rates on loans, make sure that you know the APR for each lender. By comparing APRs, you can easily determine the lowest rate.

We can also compute the APR from the monthly payment. Suppose that you borrow \$12,000 to buy a car and will make 60 monthly payments of \$250. You will make a total of $\$250 \cdot 60 = \$15{,}000$ in payments. Thus, you will pay $\$15{,}000 - \$12{,}000 = \$3000$ in interest in 5 years. The money paid back in excess of the amount borrowed is called the **finance charge.** One year's interest will be $\$3000/5 = \$600$. To find the add-on interest rate, we just need to answer the question, "\$600 is what percent of \$12,000?" The add-on rate is 5%.

The APR can then be found by the APR formula:

$$\begin{aligned}\text{APR} &= \frac{2 \cdot N \cdot r}{N + 1} \\ &= \frac{2 \cdot 60 \cdot 0.05}{60 + 1} \\ &= 6/61 \\ &= 0.098 = 9.8\%\end{aligned}$$

PRACTICE

Fern bought a new car for \$10,000. She borrowed 90% of the \$10,000 from a bank.

1. How much money did Fern borrow?

She will make monthly payments of \$195 for 5 years to pay off the loan.

2. How much money will she pay to the bank in 5 years?

3. What is the finance charge?

4. What is the add-on rate of Fern's loan?

5. What is the APR of Fern's loan?

6. Elmo borrowed $2000 to buy a boat. He will make monthly payments of $67.22 for 3 years. What is the APR of this loan?

7. Myrna borrowed $18,000 to buy aluminum siding for her dairy barn. She will make 48 monthly payments of $405. What is the APR of this loan?

Situation 4

Buying a House

You have found your dream house. Now all you need is to find the $120,000 to pay for it. Your neighborhood Good Neighbor National Bank advertises a conventional 25-year mortgage with an annual rate of 8.5%. The bank requires a 20% down payment and payment of two points at closing. If the bank agrees to give you a loan for your dream house, how much will it cost you now? How much will it cost you per month to pay off the loan? Altogether, how much will it cost to buy your dream house?

A house or condominium is probably the most expensive single purchase you will ever make. Let's take a close look at the terminology and mathematics associated with this financial rite of passage.

1.8 MORTGAGES

A **mortgage** is a long-term installment loan on a house, condominium, or other piece of property. Under the terms of a mortgage, the property serves as security for the loan. The two most popular types of mortgages are **conventional** and **adjustable-rate** mortgages. In a conventional (or **fixed-rate**) mortgage, the annual interest rate is fixed for the entire term of the loan. Conventional rate mortgages are customarily available for 15-, 20-, 25-, and 30-year terms. Adjustable-rate mortgages, on the other hand, may change their annual interest rates every period specified by the loan. Here we will examine only conventional rate mortgages.

Let's look at the mortgage offered by the Good Neighbor Bank in Situation 4. We can identify several distinguishing features of this conventional mortgage.

THE DOWN PAYMENT

Conventional mortgages usually require that the borrower provide a certain percentage of the cost of the property being purchased. This gives the lender some guarantee that the borrower is serious about the purchase. In Situation 4, the down payment is 20% of the purchase price of the house or 20% of $120,000. What is the down payment for Situation 4?

___

___

___

(Answer: $24,000)

POINTS

Many lenders also require the borrower to pay one or more points at the time of **closing** (the final step in the process of buying or selling property). One point is 1% of the amount borrowed. In Situation 4, the Good Neighbor Bank wants two points at closing. Thus, since you will be borrowing $120,000 − $24,000 = $96,000, you will have to pay 2% of $96,000 at closing. What is 2% of $96,000?

(Answer: \$1920)

Thus, in order to obtain the mortgage, you will need to have at least enough money to cover the down payment and the points. You will need at least \$24,000 + \$1920 = \$25,920. (You will probably need even more than this to cover various professional fees like legal fees, appraisals, and so on.)

PRACTICE

Calculate how much money you will need to cover the down payment and points for each of the following conventional mortgages:

1. Price of property: \$30,000
Terms of mortgage: 30 years at 8.5% + 3 points with 10% down

2. Price of property: \$300,000
Terms of mortgage: 25 years at 7.5% + 1 point with 20% down

3. Price of property: \$130,000
Terms of mortgage: 20 years at 8.5% + 2 points with 10% down

4. Price of property: \$130,000
Terms of mortgage: 25 years at 8.0% + 3 points with 20% down

A mortgage is paid off or **amortized** by monthly payments over the term of the loan. The payment is the same each month, so we can produce a table for paying off a mortgage just like we did for the smaller installment loans in Section 1.6 (In fact, those tables are called **amortization tables.**) But with 360 monthly payments in 30 years, we'd get pretty tired. Fortunately, computers can make these tables or any part of the table for us with little effort.

The monthly payment itself can be computed using the **mortgage payment formula:**

$$M = P \cdot \frac{(R/12) \cdot (1 + R/12)^N}{(1 + R/12)^N - 1}$$

In the mortgage payment formula, M is the monthly payment, P the amount borrowed, R the annual simple interest rate, and N the number of monthly payments.

You need to have a calculator that can work with large powers to use this formula. If your calculator isn't able to work with large powers, you can use the table of $(1 + R/12)^N$ values given in Appendix A.

In Situation 4, the annual rate is 8.5%, the term is 25 years or $25 \cdot 12 = 300$ monthly payments, and the amount borrowed is \$96,000. Using the mortgage payment formula, we can find the monthly payment.

Since the term $(1 + R/12)^N$ appears twice in the formula, I'll calculate it first:

$$\begin{aligned}(1 + R/12)^N &= (1 + 0.085/12)^{300} \\ &= (1.0070833)^{300} \\ &= 8.3103\end{aligned}$$

Then the monthly payment is

$$\begin{aligned}M &= P \cdot \frac{(R/12) \cdot (1 + R/12)^N}{(1 + R/12)^N - 1} \\ &= 96{,}000 \cdot \frac{(0.085/12) \cdot 8.3103}{8.3103 - 1} \\ &= \$773.02\end{aligned}$$

Carry as many decimal places as your calculator gives you in this calculation to minimize errors. Your numbers may differ from mine by a few cents.

PRACTICE

5–8. Compute the monthly payments for each of the mortgages in Practice Exercises 1–4.

We can now compute the total cost of our dream house:

Down payment = \$24,000

Points = \$1920

360 monthly payments of \$773.02 = \$278,287.20

Total cost = \$304,207.20

Wow! That is expensive!

Suppose that we keep the down payment and the points the same, but change the rate or term of the mortgage. The following table will give you some numbers for comparison:

TOTAL COST OF A \$120,000 HOUSE WITH 20% DOWN PAYMENT AND TWO POINTS

Term	Rate	Monthly Payment	Total Cost
25 years	8.5%	\$773.02	\$304,207.20
20 years	8.5%	\$833.11	\$225,866.40
15 years	8.5%	\$945.35	\$196,083.00
25 years	8.0%	\$740.94	\$248,202.00
25 years	7.5%	\$709.43	\$238,749.00

In general, what are the effects of lowering the term of a mortgage while keeping the rate constant?

____

____

____

In general, what are the effects of lowering the rate of a mortgage while keeping the term constant?

____

____

____

1.9 WHAT DO YOU KNOW?

If you have worked carefully through this chapter, then you can now work ratio and proportion and percentage problems. You also understand the terms *markup* and *markdown* and can compute simple and compound interest. You also can do the mathematics involved in installment loans and home mortgages. The exercises in this section are designed to test and refine your skills in the mathematics of the marketplace.

1. There are 16 cars parked in a fast-food restaurant's parking lot. There are seven Fords, six Hondas, and three Toyotas.
   (a) Find the ratio of Fords to Hondas.
   (b) Find the ratio of Hondas to Toyotas.
   (c) Find the ratio of Fords to cars.
   (d) Find the ratio of Japanese cars to U.S. cars.
   (e) Find the ratio of U.S. cars to Fords.

2. A car goes 300 miles on 12 gallons of gasoline. How far can the car go on 20 gallons of gasoline?

3. If eggs sell for \$2 a dozen, how many eggs can you buy with \$5?

4. Sam can paint 20 feet of fence in 2 hours. If Sam paints at this pace for 3.5 hours, how many feet of fence will he paint?

5. Five ducks can eat a loaf of bread in 10 minutes. How long will it take three ducks to eat a loaf of bread?
6. An elf can bake 25 cookies in 5 minutes. How many cookies can he bake in 3 hours?
7. A recipe for a casserole that feeds four calls for 12 ounces of chocolate. How many ounces of chocolate are needed if we want to feed six?
8. Three cabbages cost 40 cents. How much will it cost to buy five cabbages?
9. What is 45% of 230?
10. What is 49% of 200?
11. What is 0.8% of 0.9?
12. What is 1.78% of 0.007?
13. 400 is what percent of 678?
14. 50 is what percent of 10?
15. 0.89 is what percent of 0.90?
16. 24 is what percent of 2.4?
17. 60 is 34% of what?
18. 12.3 is 40% of what?
19. 9 is 4% of what?
20. 8 is 0.05% of what?
21. There are three elephants and seven zebras at a water hole. What percent of the animals at the water hole are elephants?
22. Elvin correctly answered 35 of the 45 questions on the exam. What percent of the questions on the exam did Elvin get right?
23. There are six Hondas, four Toyotas, eight Dodges, and 11 Fords in a school parking lot.
    (a) What percentage of the cars in the lot are Hondas?
    (b) What percentage of the cars in the lot are Japanese cars?
    (c) What percentage of the American cars in the lot are Fords?
24. Mrs. Garcia told Ralph that if he got more than 90% of the questions right on his exam, then he would get an "A." The test had 45 questions and Ralph got 40 of them right. Did Ralph get an "A"? Explain your answer.
25. Fern has found that it takes about 10 acres of range land to supply enough feed for three cows. If Fern has a herd of 157 cows, how many acres of range land will she need to feed them?
26. A retailer bought a TV for $200. She then sold the TV for $239.99.
    (a) What was her markup?
    (b) What was her percent markup?
27. Byron bought a new car for $14,000. The auto dealer from whom Byron bought the car paid the car manufacturer $13,200 for the car.
    (a) What was the auto dealer's markup?
    (b) What percent was the markup on the dealer's cost?
28. Sylvia, a computer software retailer, marks up her merchandise 35%. If she bought a software package for $10.00, for what price will she sell it?
29. Just Hats, Inc. has decided to mark down every item in its stores by 40%. Compute the sale price of each of the following items:
    (a) A hat regularly priced at $49.
    (b) A hat regularly priced at $490.
    (c) A hat regularly priced at $4900.
30. A floral print necktie is labeled "only $3.99, 60% off the regular price." What was the regular price of the tie?
31. Billy, the owner of Billy's TV & Appliances, bought an electric range from its manufacturer for $200. Billy marked up the price $36. Unfortunately, the range didn't sell, so Billy marked down the price 20%. The range sold at this price.
    (a) At what price did Billy finally sell the range?
    (b) What was Billy's final percent markup on the range?
32. Fern is studying the advertisements in her Sunday newspaper. She is looking for the best price on a Nycon 3000 XCT camera. Mario's Camera Shop is selling the camera at 10% above dealer's cost. HugeMart says that they are taking 35% off the regular price of $499.99. Fern finds out that Mario paid $300 for the camera.
    (a) What is the lowest available price to Fern for the Nycon 3000 XCT?
    (b) Who is offering the lowest price?
33. A case (24 bottles) of champagne is sold by a French vineyard to an exporter for $288. The exporter marks up the price 20% on his cost and sells the case to an importer. The importer marks up the price 25% on her cost and sells the case to a retailer. The retailer marks up the price 30% on her cost and puts the 24 bottles on the shelf. How much will one bottle of this champagne cost?
34. Flanders puts $200 into a bank account paying 5.25% simple interest.
    (a) If he doesn't take any money out of his account or add to it, how much money will Flanders have in the account in 1 year?
    (b) If he doesn't take any money out of his account or add to it, how much money will Flanders have in the account in 5 years?
    (c) If he doesn't take any money out of his account or add to it, how much money will Flanders have in the account in 10 years?
    (d) If he doesn't take any money out of his account or add to it, how much money will Flanders have in the account in 50 years?
35. How much interest will be earned per month on $1 million dollars if it is put into a bank account paying 6.2% simple interest?
36. If Floyd puts $3000 into an account paying 1.2% simple interest per week, how much money will he have in the account in 3 years?
37. Find the annual yield of each of the following simple interest rates:
    (a) 6% per year.
    (b) 6% per quarter.

(c) 6% per month.

(d) 6% per week.

(e) 6% per day (use a 360-day year).

(f) 6% per hour (use a 360-day year).

38. Find the annual yield of each of the following compound interest rates.

(a) 6% compounded semiannually.

(b) 6% compounded quarterly.

(c) 6% compounded monthly.

(d) 6% compounded weekly.

(e) 6% compounded daily (use a 365-day year).

(f) 6% compounded hourly (use a 365-day year).

39. Lucinda put $1200 into an account paying 4% interest compounded quarterly and $1800 into an account paying 3.7% interest compounded monthly. How much total interest will Lucinda earn in 2 years?

40. Which investment will earn the most interest in 1 year?

(a) $1000 at 12% simple interest.

(b) $1500 at 8% simple interest.

(c) $1000 at 10% compounded monthly.

(d) $1500 at 9% compounded weekly.

41. Georgia bought a sofa for $300. She financed her purchase through her bank at an annual interest rate of 9.5%. If she makes monthly payments of $50 on this loan, how much total interest will she pay? (*Hint:* Construct a payback table.)

42. George bought a TV for $500. He financed his purchase through his bank at an annual interest rate of 8.5%.

(a) If he makes monthly payments of $150 on this loan, how much total interest will he pay?

(b) How much money will he save if he makes monthly payments of $200?

43. Fred and Fern have a bank card bill totalling $1200. They decide that they will pay it off before charging anything else on their card. They find that they can pay $125 a month. The monthly interest rate is 1.5%.

(a) How long will it take Fred and Fern to pay off their bill?

(b) How much total interest will they pay?

(c) What will be the amount of their last payment?

44. Zork is buying a car for $24,000. He makes a $2000 down payment.

(a) Using an add-on rate of 7% for 4 years, compute Zork's monthly payment.

(b) What is the APR of this loan?

45. Zelda is buying a car for $14,000. She makes a $2000 down payment.

(a) Using an add-on rate of 4.9% for 3 years, compute Zelda's monthly payment.

(b) What is the APR of this loan?

46. Zeela is buying a car for $44,000. She makes a $5000 down payment.

(a) Using an add-on rate of 6% for 5 years, compute Zeela's monthly payment.

(b) What is the APR of this loan?

47. Find the APR of each of the following lending schemes. In each case, the amount borrowed is $12,000.

(a) $234 per month for 5 years.

(b) $270 per month for 4 years.

(c) $350 per month for 3 years.

48. Keesha has $20,000 in a bank account earning 6.2% interest compounded monthly. She is interested in buying a new car that costs $13,000. The auto dealer will sell her the car with no down payment and 60 monthly payments of $226 each. Would it be better for Keesha to take the auto dealer's financing offer and leave her savings untouched, or to take $13,000 out of her account and pay cash for the new car?

49. Compute the monthly payments and total cost for each of the following conventional mortgage plans. In each case, assume that you are purchasing a $100,000 house.

(a) 8.5% + 2 points, 30 years, 10% down payment.

(b) 8.5% + 3 points, 25 years, 20% down payment.

(c) 9.5% + 1 point, 20 years, 20% down payment.

(d) 8.5% + 2 points, 30 years, 10% down payment.

(e) 8.5% + 3 points, 25 years, 20% down payment.

(f) 9.5% + 1 point, 20 years, 20% down payment.

(g) 9.0% + 0 points, 30 years, 10% down payment.

50. Salvatore has $15,000 in cash. He wants to buy a $60,000 home with a 25-year fixed mortgage. He is willing to pay $500 per month. Suppose that Salvatore pays no points or fees but is willing to spend all of his $15,000 on a down payment. Approximately, what is the highest interest rate he can afford?

ANSWERS TO PRACTICE EXERCISES

Section 1.1

1. 7/8, 7 to 8, 7 : 8

2. 7/14, 7 to 14, 7 : 14

3. 14/8, 14 to 8, 14 : 8

4. 14/7, 14 to 7, 14 : 7

5. 8/14, 8 to 14, 8 : 14

6. 8/7, 8 to 7, 8 : 7

7. 7 : 8, 1 : 2, 7 : 4, 2 : 1, 4 : 7, 8 : 7

8. 3/40

9. 40 : 3

10. 3/43

11. 43/40

12. $300/12 = x/15$
$25/1 = x/15$
$25 \cdot 15 = 1 \cdot x$
$375 = x$
The car can go 375 miles on 15 gallons of gasoline.

13. $12/2 = 17/x$
Seventeen apples will cost $2.83.

14. $3/5 = x/24$
It will take Myrna 14.4 days to make 24 dresses.

15. $24/1 = x/1461$ (1461 days in 4 years, $4 \cdot 365 + 1$ leap day) There are 35,064 hours in 4 years.

Section 1.2

1. 35/100
2. 23%
3. 0.234
4. 4.5%
5. 0.36
6. 60
7. What
8. 40
9. What, 40, 800
10. 90, what, 300
11. 56, 45, what
12. 9
13. 78
14. 116
15. 48
16. 750
17. 17,500

Section 1.3

1. $100
2. $245
3. $240
4. $50
5. $340
6. $450
7. $140
8. $175
9. $80
10. 28.57%
11. $162.49
12. $865.83
13. $600
14. $15,750
15. 92.31%
16. 49.75%

Section 1.4

1. $60, $560
2. $288, $888
3. $120, $620
4. $24, $624
5. 72%
6. 104%
7. 6%
8. 16%

Section 1.5

2. $1.52
3. $1.55, $104.57
4. $104.57, $1.57, $106.14
5. 6.14%
6. $\$3000 \cdot (1 + 0.06/12)^{36} = \$3590.04$
7. $\$2000 \cdot (1 + 0.03/360)^{720} = \$2123.67$
8. $\$100 \cdot (1 + 0.07/2)^{40} = \$395.93$
9. $\$100 \cdot (1 + 0.07/4)^{80} = \$400.64$
10. $\$100 \cdot (1 + 0.07/12)^{240} = \$403.87$
11. 7.12%, 7.19%, 7.23%

Section 1.6

1. $1.46
2. $59.25
3. $0.89
4. $20.14
5. $20.14
6. $0.30
7. $20.44
8. $0
9. $4.68
10. $6.44

Section 1.7

1. $9000
2. $11,700
3. $2700
4. 2700/5 years = 540 and $540 is 6% of $9000. So, the add-on rate is 6%.
5. $2 \cdot 60 \cdot 0.06/(60 + 1) = 0.118 = 11.8\%$
6. Total payments = $67.22 · 36 = $2419.92
   Total finance charge = $2419.92 − $2000 = $419.92
   Annual interest paid = $419.92/3 = $139.97
   Add-on interest rate = 7%
   APR = $2 \cdot 36 \cdot 0.07/37 = 13.6\%$
7. Total payments = $405 · 48 = $19,400
   Total finance charge = $19,440 − $18,000 = $1440
   Annual interest paid = $1440/4 = $360
   Add-on interest rate = 2%
   APR = $2 \cdot 48 \cdot 0.02/49 = 3.9\%$

Section 1.8

1. Down payment = $3000
   Points charge = $810 (3% of $27,000)
   Total due at closing = $3810
2. Down payment = $60,000
   Points charge = $2400
   Total due at closing = $62,400
3. Down payment = $13,000
   Points charge = $2340
   Total due at closing = $15,340
4. Down payment = $26,000
   Points charge = $3120
   Total due at closing = $29,120
5. Amount borrowed = $27,000
   Rate = 8.5%
   Number of payments = 360
   Monthly payment = $207.61
6. Amount borrowed = $240,000
   Rate = 7.5%
   Number of payments = 300
   Monthly payment = $1773.58
7. Amount borrowed = $117,000
   Rate = 8.5%
   Number of payments = 240
   Monthly payment = $1015.35
8. Amount borrowed = $104,000
   Rate = 8.0%
   Number of payments = 300
   Monthly payment = $802.69

Chapter Two

Mathematics of Description

What's the Point? Sometimes, it seems as if everything is measured by a number. A glance at today's newspaper will no doubt reveal items like these:

PRESIDENT'S POPULARITY UP TO 70%
CUBS' SHORTSTOP HITTING .400
MEDIAN HOME PRICES TUMBLE
AREA SAT SCORES EXCEED NATIONAL AVERAGE
INFANT MORTALITY RATE LOWEST IN TEN YEARS

As an informed citizen, you will want to know what these headlines mean (and don't mean). You'll want to know where the numbers come from and how they can be interpreted. The activities in this chapter will introduce you to some of the basic principles of **statistics,** the mathematics of description.

Situation 5

Happiness Is Being in the 95th Percentile

Suppose your daughter (or sister, or friend) received her college entrance exam scores in the mail today. ''How did you do?,'' you ask innocently. ''Fantastic!'' she replies with obvious joy, ''I had hoped to score at least two standard deviations above the mean, but I actually scored in the 95th percentile!'' What is she talking about?

____

____

____

If you said that she did well on her exam, then you get the basic idea. However, for better understanding we ought to look at the situation a little more closely.

2.1 STATISTICS

A **statistic** is a single numerical value associated with some collection of objects. Examples of statistics associated with the collection of houses sold in Tensleep, Wyoming last year would include

Highest selling price

Lowest selling price

Average selling price

Number of houses that sold for more than $50,000

Percentage of brick houses sold

Ratio of brick houses sold to frame houses sold

Number of green-roofed houses sold

and so on.

Statistics are numerous and can quantify just about anything that you can think up. How about ''the number of green-roofed houses sold in Tensleep, Wyoming last year that sold for less than $30,000, but were still at least $10,000 more expensive than the lowest priced frame house with at least three conifers growing in the front yard?'' How useful would this last statistic be in comparison to the average selling price?

____

____

____

As you can see, some statistics may be more important than other statistics. The mathematical area called **statistics** is devoted to the identification and interpretation of important statistics. Statistics (the area of mathematics) studies statistics (the numbers).

Statistics (the area of mathematics) can be divided into two parts. **Descriptive statistics** is devoted to the calculation and presentation of statistics. Descriptive statistics uses tables and charts and graphs to show the numerical relationships in a collection of objects.

When the collections are very large, like the number of grasshoppers in Indiana in June, we cannot actually determine a statistic for the whole collection. For example, we could not capture and weigh all the grasshoppers in Indiana. Instead, we get statistics for a portion or **sample** of the collection. **Inferential statistics** uses the descriptive statistical information of the sample to make inferences about the whole collection (called the **population**).

In this text, we will look only at the basics of descriptive statistics.

PRACTICE

1. Describe 10 statistics that might be associated with the collection of pet dogs living within 1 mile of the city hall of Arcola, Illinois.

2. Which of the statistics that you described would be the hardest to determine? Which would be the easiest?

3. If you were a salesman for a company making dog chariots (two-wheeled carts that can hold small children and are harnessed to a dog), in what statistics would you be most interested?

2.2 MEAN, MEDIAN, AND MODE

Probably the most familiar statistic is the average. There are grade point averages, batting averages, bowling averages, average temperatures, average heights, and so on. What people commonly refer to as average is the statistic called the arithmetic **mean**. The mean of a group of numbers is the sum of the numbers divided by the number of numbers in the group.

For example, the mean of this collection of six numbers

$$3, 14, 5, 7, 9, 10$$

is

$$\frac{3 + 14 + 5 + 7 + 9 + 10}{6}$$

or $48/6 = 8$.

PRACTICE

1. There are seven ducks swimming in the Fairview Park pond. Their weights are 5, 5.5, 7, 7.8, 8, 8.1, and 8.5 pounds. What is the mean weight of these seven ducks?

2. Elmer got the following scores on his last five history exams: 78, 87, 70, 90, 82. What is the mean of Elmer's scores?

3. In a small business in Parma, Michigan, there are eight employees. Their annual salaries are $10,000, $10,000, $10,000, $10,000, $10,000, $30,000, $30,000, and $350,000. What is the mean annual salary?

In Practice Exercise 3, you found that the mean salary in the factory in Parma is $57,500. Does the average worker in that factory make $57,500 a year?

___

___

___

The mean is a misleading statistic if we are interested in the average salary of a worker in the Parma factory. If we say the average salary is $57,500, that suggests the average worker makes that much money a year. However, in fact, only one person in the factory makes at least $57,500 and the salaries of five of the eight workers are well below that value.

Because the mean is sensitive to values greatly different from most of the values in a collection, two other statistics, the **median** and **mode,** are also used to describe the **average.**

The **mode** is the most frequent value in the collection. In the case of the Parma factory, the mode is $10,000 because the most number of workers, namely, five, have that salary. We say that $10,000 is the *modal salary*. Notice that $10,000 is a better estimate of the average worker's salary than $57,500.

PRACTICE

4. In Binky's backyard there are six pine trees. They are 4, 4, 4, 4, 5, and 6 feet tall.

(a) What is the modal height of Binky's pine trees?

(b) What is the mean height of Binky's pine trees?

5. The McDole's restaurant in North Platte, Nebraska, sold 300 pineapple burgers on Monday, 400 on Tuesday, 300 on Wednesday, 400 on Thursday, 600 on Friday, 800 on Saturday, and 650 on Sunday.

(a) What was the modal number of pineapple burgers sold per day?

(b) What was the mean number of pineapple burgers sold per day?

6. Find the modal test score:

70, 80, 89, 70, 68, 70, 87, 96, 70, 90, 70, 60, 60, 70, 95, 70, 90, 90

The third statistic that can be used to describe an average is the **median.** The median of a group of numbers is the middle number when the numbers are arranged in order of magnitude.

For example, for the collection of test scores

70, 77, 86, 90, 95

the median is 86 because it lies right in the middle of the list. There are the same number of values above and below the median.

It is important that the numbers be arranged in order before the middle is determined! In this collection,

70, 69, 87, 90, 73

the median is 73, not 87. Explain why.

____

____

____

In the case of the Parma salaries, we see that there isn't a middle value because there are an even number of salaries. In this case, we take the median to be the mean of two middle values (marked with *):

10,000, 10,000, 10,000, 10,000*, 10,000*, 30,000, 30,000, 350,000

Thus, the median salary is ($10,000 + 10,000)/2 = $10,000. Since two of our three averages say that the *average* salary of a Parma factory worker is $10,000, we would be inclined to accept $10,000 as the average salary instead of $57,500.

What is the median of this collection?

30, 40, 50, 30, 10, 40, 50, 30

(Answer: 35)

PRACTICE

7. Find the mean, median, and mode of each of the following collections.

(a) 20, 20, 20, 20, 40, 40, 50, 70, 80

(b) 10, 30, 20, 40, 50, 10, 10, 30, 30

(c) 10, 40, 40, 50, 160, 70, 10, 10, 10, 30

(d) 20, 30, 40, 20, 20, 20, 20, 20, 20, 90

8. Which of the statistics that you found for the collections in Practice Exercise 7 is the most appropriate **average** for its collection? Why?

The mean, median, and mode all attempt to identify the average of a collection of numbers. Because they identify a typical value, not too large and not too small, these statistics are called *measures of central tendency.* Unfortunately, central tendency is not always enough to describe a collection. Consider the following three collections:

Collection A: 50, 50, 50, 50, 50

Collection B: 40, 50, 50, 50, 60

Collection C: 0, 50, 50, 50, 100

You should verify that the mean is 50, the median is 50, and the mode is 50 for all three of these collections. Because all three of our *averages* are the same, we are pretty safe in saying that the average of each of the three collections is 50. However, the collections are different from each other.

In what way are the collections different?

____

____

____

We need a statistic that will distinguish collections whose numbers are very similar, like collections A and B, from those whose numbers are not so similar, like collection C. We need a *measure of variability.*

2.3 VARIANCE AND STANDARD DEVIATION

One obvious statistic that gives an indication of how closely the numbers in a collection are grouped is the **range.** The range of a collection of numbers is the difference between the highest value and the lowest value.

Here are the three collections from the end of the last section along with their ranges:

Collection A: 50, 50, 50, 50, 50; Range $= 50 - 50 = 0$

Collection B: 40, 50, 50, 50, 60; Range $= 60 - 40 = 20$

Collection C: 0, 50, 50, 50, 100; Range $= 100 - 0 = 100$

The range is a good statistic in this case because it realistically reflects the differing variability in the three collections. (Statisticians usually refer to collections of numbers as **distributions.** From now on, I will use distribution to refer to a collection of numbers.)

PRACTICE

Find the range of each of the following distributions:

1. 23, 45, 67, 89, 90, 110

2. 10, 10, 10, 30, 30, 40, 50, 60, 320, 20, 10

3. 1.5, 5.6, 7.8, 3.2, 4.5, 1.2

4. −4, −4, −5, 6, 7, 2, −5, 8, 10

Find the mean and range of each of the following distributions:

Distribution D; 2, 5, 6, 6, 6, 6, 6, 7, 10

Distribution E: 2, 3, 4, 5, 6, 7, 8, 9, 10

Though distributions D and E have the same mean (6) and range (8), they are different in their variability. You can see this by looking at a dot chart of each distribution. Each ● above a number represents one instance of that number in the distribution.

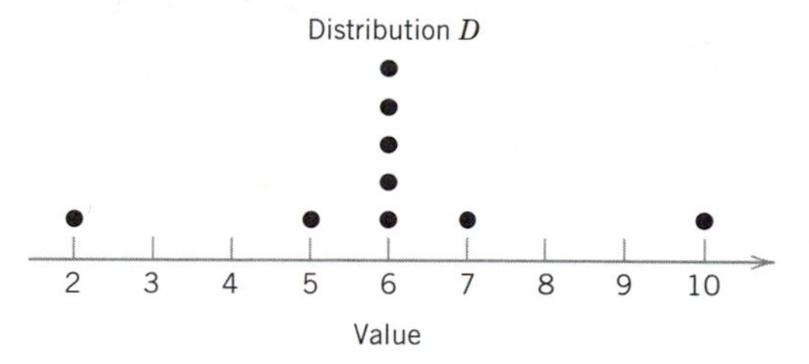

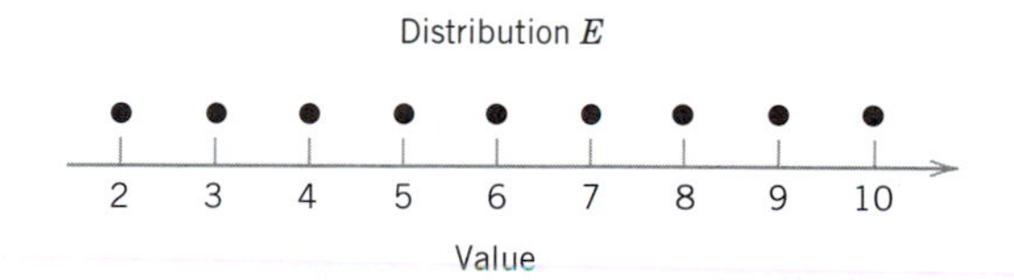

As you can see, most of the ● in distribution D are very close to or at the mean, whereas those in distribution E are scattered. The next two statistics attempt to measure the closeness of the values of a distribution to the mean of that distribution.

The **variance** is the mean of the squares of the differences of the values from the mean. The table below shows how to compute the variance for distribution D.

TABLE FOR COMPUTING VARIANCE OF DISTRIBUTION D
(MEAN = 6; NUMBER OF VALUES = 9)

Value	Difference Between Value and Mean	Square of Difference Between Value and Mean
2	$2 - 6 = -4$	$(-4)^2 = 16$
5	$5 - 6 = -1$	$(-1)^2 = 1$
6	$6 - 6 = 0$	$0^2 = 0$
6	$6 - 6 = 0$	$0^2 = 0$
6	$6 - 6 = 0$	$0^2 = 0$
6	$6 - 6 = 0$	$0^2 = 0$
6	$6 - 6 = 0$	$0^2 = 0$
7	$7 - 6 = 1$	$1^2 = 1$
10	$10 - 6 = 4$	$4^2 = 16$
		34

The sum of the last column is 34. So, the variance, which is the mean of the last column, is $34/9 = 3.78$.

For distribution E, the table to compute the variance is

TABLE FOR COMPUTING VARIANCE OF DISTRIBUTION E (MEAN = 6; NUMBER OF VALUES = 9)

Value	Difference Between Value and Mean	Square of Difference Between Value and Mean
2	$2 - 6 = -4$	$(-4)^2 = 16$
3	$3 - 6 = -3$	$(-3)^2 = 9$
4	$4 - 6 = -2$	$(-2)^2 = 4$
5	$5 - 6 = -1$	$(-1)^2 = 1$
6	$6 - 6 = 0$	$0^2 = 0$
7	$7 - 6 = 1$	$1^2 = 1$
8	$8 - 6 = 2$	$2^2 = 4$
9	$9 - 6 = 3$	$3^2 = 9$
10	$10 - 6 = 4$	$4^2 = 16$
		60

The sum of the last column is 60. So, the variance, which is the mean of the last column, is $60/9 = 6.67$.

Notice that distribution E, which has a greater variability in its values, has the larger variance. Also, notice that the squaring of the differences is necessary because the mean of the second column in both distributions is 0 (as it will be in all distributions), which fails to distinguish one distribution from the other.

As a last example, here is the table to find the variance of the distribution in Practice Exercise 1:

TABLE FOR COMPUTING VARIANCE OF DISTRIBUTION OF PRACTICE EXERCISE 1 (MEAN = 70.67; NUMBER OF VALUES = 6)

Value	Difference Between Value and Mean	Square of Difference Between Value and Mean
23	$23 - 70.67 = -47.67$	$(-47.67)^2 = 2272.43$
45	$45 - 70.67 = -25.67$	$(-25.67)^2 = 658.95$
67	$67 - 70.67 = -3.67$	$(-3.67)^2 = 13.47$
89	$89 - 70.67 = 18.33$	$(18.33)^2 = 335.99$
90	$90 - 70.67 = 19.33$	$(19.33)^2 = 373.65$
110	$110 - 70.67 = 39.33$	$(39.33)^2 = 1546.85$
		5201.34

The variance is $5201.34/6 = 866.89$.

PRACTICE

Find the variance of each of the following distributions:

5. 10, 10, 10, 10, 80

6. 10, 20, 30, 40, 50

7. 20, 30, 40, 40, 40, 40, 30, 10, 10, 80

8. 10, 10, 10, 10, 10, 10

9. 10, -10, 10, -10, 10, -10

The variance certainly reflects variability, but the variance often seems to be a number that is extremely large relative to the values in the distribution (look at the Practice Exercise 1 distribution!). The large relative size of the variance is a result of the squaring of the differences. We need to square the differences to avoid negative differences cancelling out positive differences, but after we find the variance, we can take its square root to whittle down the value to a more reasonable size. The statistic that is the square root of the variance is called the **standard deviation.** The standard deviation represents the average distance of the values in a distribution from the mean of the distribution.

Here are the standard deviations of some of the distributions we've examined so far.

Distribution	Variance	Standard Deviation
D	3.78	$\sqrt{3.78} = 1.94$
E	6.67	$\sqrt{6.67} = 2.58$
Practice Exercise 1	866.89	$\sqrt{866.89} = 29.44$
Practice Exercise 5	784.00	$\sqrt{784.00} = 28.00$
Practice Exercise 6	200.00	$\sqrt{200.00} = 14.14$
Practice Exercise 7	364.00	$\sqrt{364.00} = 19.08$

PRACTICE

Find the standard deviation of each of the following distributions:

10. 10, 20, 20, 20, 10

11. 16, 16, 16, 16, 16

12. 1, 1, 16, 31, 31

13. How does the standard deviation help to distinguish the distributions in Practice Exercises 10, 11, and 12 from one another?

2.4 RELATIVE STANDING

The mean and standard deviation can be used to give you a view of the relationship between an individual value in a distribution and the whole distribution, or **relative standing.** Your daughter, sister, or friend in Situation 5 was alluding to this when she said that she had "hoped to score at least two standard deviations above the mean." We'll look at relative standing in the context of ducks.

Suppose that the mean weight of ducks on the pond in Fairview Park is 6 pounds and the standard deviation is 1 pound. Donna Duck, who lives on the Fairview Park Pond, weighs 10 pounds. Is she a big duck?

___

___

___

Suppose Donna Duck moves to the Lake Sara Wildlife Refuge, where the mean weight of the ducks is 6 pounds and the standard deviation is 3 pounds. Is Donna a big duck?

___

___

___

Donna Duck, weighing in at 10 pounds, is heavier than the average (mean) duck at both Fairview Park and Lake Sara. However, Donna is more outstanding at Fairview Park, where a 10-pound duck is a relatively rare item, than at Lake Sara, where 9-pound ducks are probably quite common.

My justification for the assertions in the last paragraph is a statistical relationship called **Chebyshev's theorem.** Chebyshev's theorem assures us that for any distribution, at least 75% of the values lie within two standard deviations of the mean, at least 89% lie within three standard deviations of the mean, and at least 94% of the values lie within four standard deviations of the mean.

What does this mean?

___

___

___

If μ is the mean and σ the standard deviation of a distribution, then the value that is two standard deviations above the mean is $\mu + 2 \cdot \sigma$. The value that is two standard deviations below the mean is $\mu - 2 \cdot \sigma$. Values that are larger than $\mu - 2 \cdot \sigma$ and smaller than $\mu + 2 \cdot \sigma$ are said to be within two standard deviations of the mean. Similarly, values between $\mu - 3 \cdot \sigma$ and $\mu + 3 \cdot \sigma$ are said to be within three standard deviations of the mean.

For the distribution of duck weights at Fairview Park, Chebyshev's theorem says that at least 75% of the ducks will weigh between 4 and 8 pounds ($6 \pm 2 \cdot 1$ pounds), and at least 89% of the ducks will weigh between 3 and 9 pounds ($6 \pm 3 \cdot 1$ pounds). Thus, Donna will outweigh at least 89% of the other ducks.

At Lake Sara, on the other hand, according to Chebyshev's theorem at least 75% of the ducks will weigh between 0 and 12 pounds ($6 \pm 2 \cdot 3$ pounds). Donna is within this range and, as such, is just an average Lake Sara duck.

PRACTICE

1. Fred, an elephant, weighs 2900 pounds. Fred is one of 1000 elephants in a herd. The mean weight for the herd is 2200 pounds with a standard deviation of 200 pounds. At least how many elephants in the herd weigh less than Fred?

2. Hong scored a 42 on his ceramics exam. The class mean on the exam was 60 with a standard deviation of 5. What percentage of Hong's ceramics classmates scored better than he did on this exam?

We can make the comparison between Donna Duck at Fairview Park and Donna Duck at Lake Sara a little more precise by means of a statistic called the ***z*-score.** The z-score is the number of standard deviations a value is from the mean of a distribution. If the z-score is positive, then the value is above the mean. If the z-score is negative, then the value is below the mean. The z-score is calculated using the ***z*-score formula:**

$$z = \frac{\text{Value} - \text{Mean}}{\text{Standard deviation}}$$

So, Donna's z-score at Fairview Park is

$$z = \frac{10 - 6}{1} = 4$$

and her z-score at Lake Sara is

$$z = \frac{10 - 6}{3} = 4/3 = 1.33$$

Because her z-score is greater at Fairview Park (4) than at Lake Sara (1.33), Donna is a relatively heavier duck at Fairview Park than she is at Lake Sara.

Notice that the z-score of the mean of any distribution is zero. Why?

PRACTICE

3. Compute the z-score for each of the following values in distributions with given means and standard deviations:

(a) Value = 6, Mean = 3, Standard deviation = 2

(b) Value = 6, Mean = 3, Standard deviation = 5

(c) Value = 60, Mean = 40, Standard deviation = 25

(d) Value = 60, Mean = 90, Standard deviation = 25

4. Guenon got an 87 on his last history test. The class mean on this test was 80 with a standard deviation of 3. On today's history test, Guenon received an 89. The class mean was 81 with a standard deviation of 6. Relative to his class, on which test did Guenon have the better score?

5. Consider the following distribution of bowling scores:

102, 104, 140, 130, 120, 109, 200, 235, 157, 180, 160, 150, 120, 280

(a) Compute the mean and standard deviation.

(b) What percentage of the scores is within two standard deviations of the mean? Compare your answer to that predicted by Chebyshev's theorem.

Your daughter (or sister, or friend) in Situation 5 was hoping to score at least two standard deviations above the mean. She was hoping to score better than at least what percentage of the people taking the test?

(Answer: 75%)

2.5 PERCENTILE RANK

Your daughter (or sister, or friend) in Situation 5 was hoping to score at least two standard deviations above the mean, but she actually scored in the 95th percentile. Why was she so happy?

Percentile is another measure of relative standing. It is a very popular way of reporting a person's relative standing on standardized exams. Your percentile represents the approximate percentage of people who scored *below* you on the exam.

If your daughter, sister, or friend scored in the 95th percentile, then about 95% of the people taking the test scored lower than she did. The 95th percentile is a very nice place to be!

Let's look again at the bowling scores from Practice Exercise 5 in the last section:

102, 104, 140, 130, 120, 109, 200,

235, 157, 180, 160, 150, 120, 280

Suppose that you bowled the 180 and are interested in your percentile rank. First, we must arrange the scores in increasing order:

102, 104, 109, 120, 120, 130, 140

150, 157, 160, 180, 200, 235, 280

Now count the number of scores below your score (180). Of the 14 bowling scores, how many are below yours? _____

What percentage of the bowling scores are below yours?

____

____

____

So, your score falls in the 71st percentile because about 71% (actually 71.4%) of the scores are below your score.

PRACTICE

1. In what percentile does the score 45 lie in the following distribution?

40, 45, 56, 78, 47, 89, 34, 35, 67, 90

2. How about 89?

3. How about 47?

Some percentiles have special names. The 25th percentile is also called the **first quartile,** the 50th percentile the **second quartile,** the 75th percentile the **third quartile.**

PRACTICE

Fill in the blanks.

10, 15, 20, 30, 40, 50, 55, 57, 60, 70, 75, 80, 90, 95, 97, 99

For this distribution, the number that lies in

4. The 62nd percentile = ________

5. The first quartile = ________

6. The second quartile = ________

7. The third quartile = ________

8. The 94th percentile = ________

Situation 6

Tornado!

Below is a graph depicting the distribution of tornadoes in Illinois by month for the years 1870–1910.

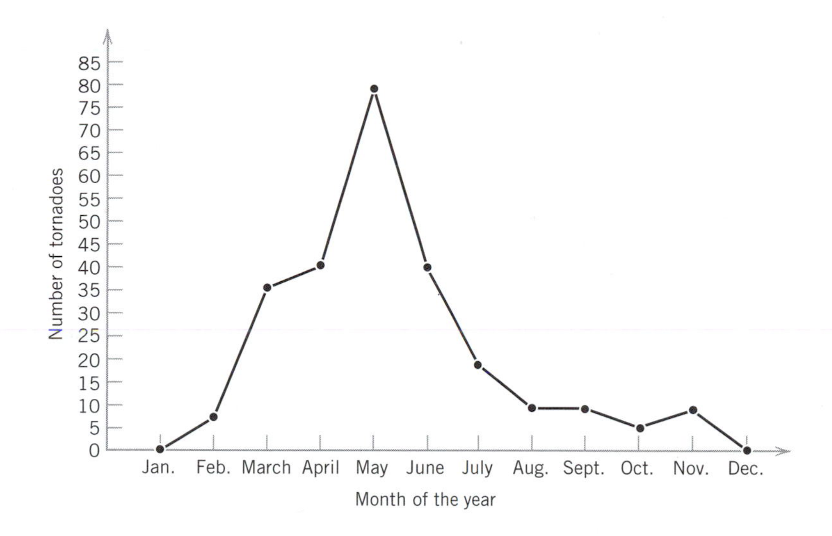

If you were living in Illinois during this period of time, in which month would there be the greatest danger of a tornado? ____

What percentage of the tornadoes during this period occurred in May?

____

____

____

2.6 FREQUENCY DISTRIBUTIONS, HISTOGRAMS, AND FREQUENCY POLYGONS

The graph shown in Situation 6 is one of the common graphs used to display a frequency distribution. A **frequency distribution** gives the number of times (or frequency) a certain value in a distribution occurs. In Situation 6 if we use the number (1 for January, 2 for February, and so on) of the month in which it occurred to denote each tornado, then the distribution of tornadoes might look something like this

$$6, 6, 6, 8, 6, 5, 6, 10, \ldots$$

The portion of the distribution shown here indicates that there were five tornadoes in June, one in May, one in August, and one in October. Thus, for the whole distribution, the number of sixes would indicate the number of June tornadoes. The position of the dot above the month of June in the graph tells us that there were 40 tornadoes in the month of June.

How many tornadoes were there in the month of September? ____

When we connect the dots for all the months with straight-line segments, we obtain a **frequency polygon.** Here is a frequency polygon showing the number of various colored ducks in a pond.

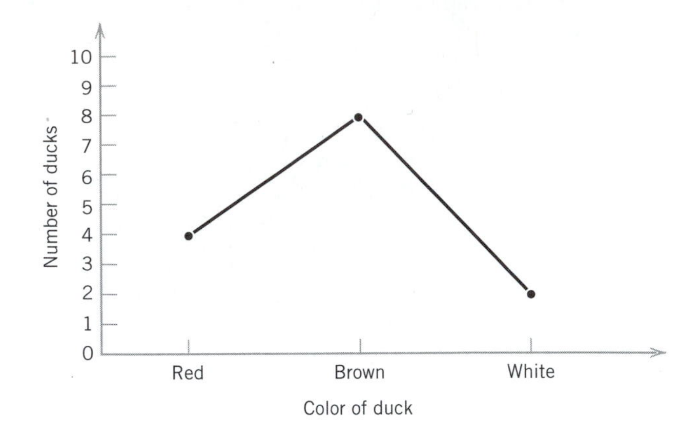

How many red ducks are in the pond? _____

How many white ducks are in the pond? _____

How many brown ducks are in the pond? _____

The frequency polygon gives us some useful statistical information. We can see from the duck graph that the modal duck color is brown. Why?

____

____

____

We can also see that there are 14 ducks in the pond. Why?

____

____

____

Finally, we can compute the percentage of white ducks in the pond because we know that 2 of the 14 ducks are white. Thus, 14.3% of the ducks are white.

You can make a frequency polygon from any distribution. For example, here is the distribution of eye colors of a fifth grade class. In this distribution, b denotes blue eyes, r brown eyes, and h hazel eyes.

$$b, r, r, r, r, b, r, b, r, h, h, b, b, r, r, r, r, r, r, r$$

To make a frequency polygon, we first tally the number of occurrences of each eye color. We make these tallies into a table:

Eye Color	Frequency
b	5
r	13
h	2

This table is also called a **frequency distribution.** Now, we make our graph. The categories or classes of things counted are listed horizontally and the frequencies vertically.

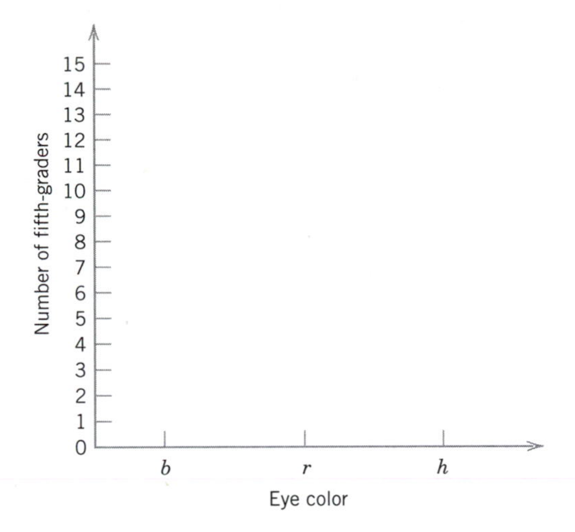

Next, we put a dot (or a star or any other mark you like) above each eye color across from its frequency. To complete the frequency polygon, connect the dots with line segments.

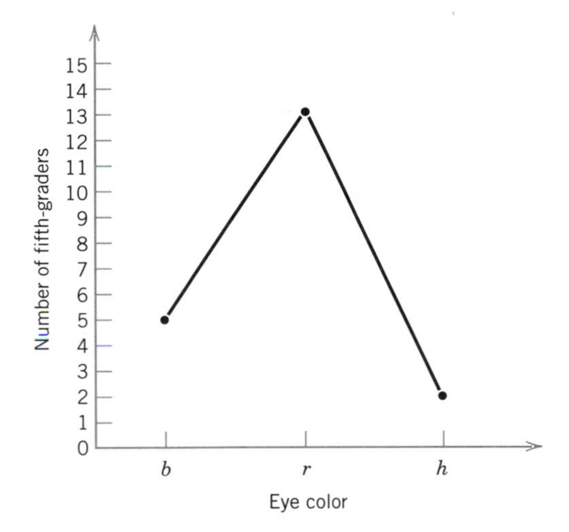

PRACTICE

Make frequency polygons for each of the following distributions:

1. Number of siblings of players on a high school football team.

2, 0, 2, 0, 2, 0, 2, 4, 4, 3, 1, 1, 0, 0, 0, 1, 1, 0, 2, 2, 0, 2, 1, 6, 1, 1, 1, 0, 0

2. Grades on an exam in a freshman history class.

A, B, C, C, C, D, C, D, F, F, A, B, C, D, C, D, B, B

3. What percentage of the football players in Practice Exercise 1 have exactly two siblings?

4. What percentage of the history class in Practice Exercise 2 got a C on the exam?

An alternative to the frequency polygon is the **histogram.** The histogram consists of parallel bars that rise from the categories listed horizontally to a height corresponding to the frequency of that category. Here are histograms for the duck and eye color frequency polygons.

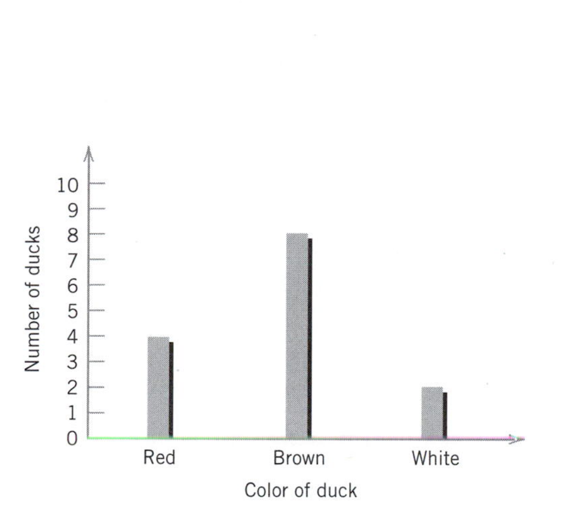

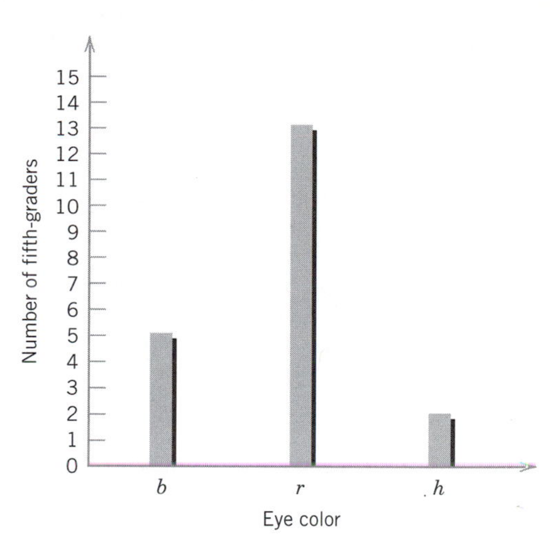

PRACTICE

5. Draw a histogram for the distribution in Practice Exercise 1.

6. Draw a histogram for the distribution in Practice Exercise 2.

Getting back to Situation 6, we can see that the month of May had the highest frequency of tornadoes (80) and these 80 tornadoes represented 31.4% (80 out of 255) of the total.

Frequency polygons and histograms are two of the most commonly used statistical graphs. When we wish to display percentages rather than frequencies, a circle graph is often used.

2.7 CIRCLE GRAPHS

A **circle graph,** also known as a **pie chart,** shows the percentage that each category of a distribution contributes to the whole distribution by means of appropriately sized sectors (or slices) of a circle.

As an example, here is the frequency distribution for the fifth-grader's eye color distribution of the previous section.

Eye Color	Frequency
b	5
r	13
h	2

There are 20 children in the class, so we can easily compute the percentage for each eye color. I'll add another column to the table that shows this percentage.

Eye Color	Frequency	Percentage
b	5	25%
r	13	65%
h	2	10%

We then divide up a circle into regions that encompass 25, 65, and 10% of its area. This can be done most efficiently with a computer spreadsheet software package (see Chapter 12). However, if a circle graph drawing package isn't available, then you can use a compass, a protractor, and the fact that a circle has 360 degrees.

Begin by drawing any radius (a line segment from the center to the outer rim) of the circle.

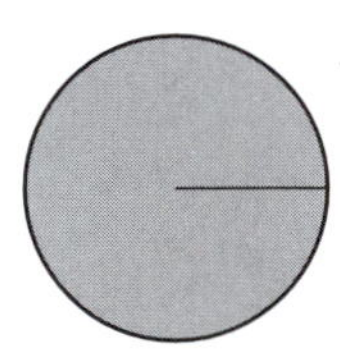

Take the percentage of the first category and compute that percent of 360. In the eye color example, that would be 25% of 360, which is 90. Use the protractor to mark off that angle from the line you first drew. Draw a new radius at this angle and mark the region.

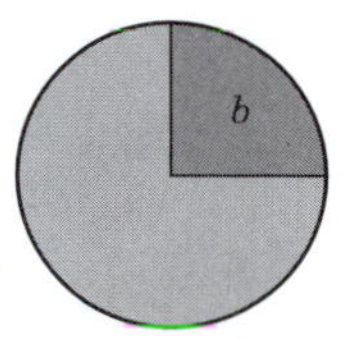

Repeat the procedure using your second category and measuring from the last line you drew. In our example, we now measure an angle of 65% of 360, which is 234 degrees.

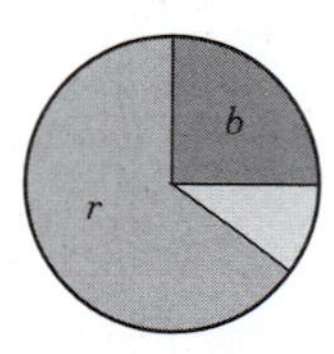

Repeat this procedure until you have used every category. The last category will require no measurement because it takes up what is left. Our eye color circle graph looks like this.

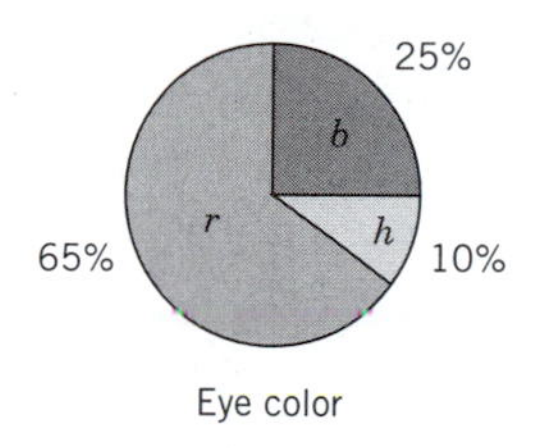

Eye color

PRACTICE

1. Draw a circle graph for the distribution of siblings given in Practice Exercise 1.

2. Draw a circle graph for the distribution of grades given in Practice Exercise 2.

Situation 7

What Did You Get?

Your friend Zork announces that he just correctly answered 40 out of 100 questions on his Solar Energy 101 exam. You tell Zork that you are sorry and you are sure he will do better next time. But Zork is not upset! He tells you that the class mean on the exam was 35 with a standard deviation of 2. Zork's instructor *curved* the exam. As a result, Zork got an A. How does Zork's instructor justify giving an A to Zork when he correctly answered only 40% of the questions on the exam?

When teachers curve tests, they usually are making use of a special frequency distribution called the **normal distribution.** The normal distribution has some very useful mathematical properties and is a commonly occurring distribution. To see how Zork got an A on his Solar Energy 101 exam, we will first look at normal distributions.

2.8 THE NORMAL DISTRIBUTION

A **normal distribution** is characterized by a bell-shaped frequency polygon like this:

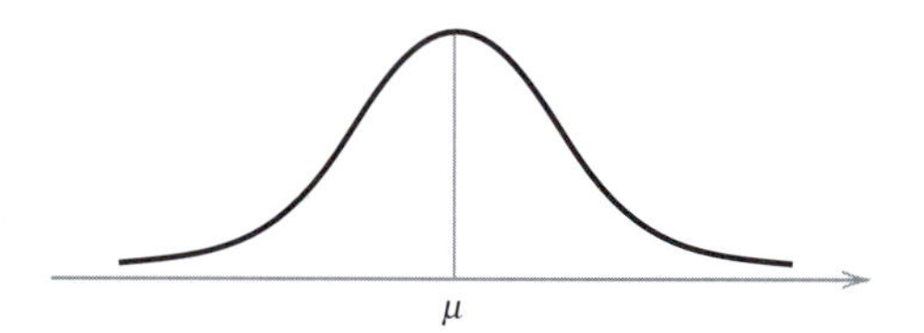

The jagged edges of the frequency polygon have been *smoothed*. The highest point of the bell-shaped curve is reached at the mean of the distribution. Also, the curve is perfectly symmetric on either side of the mean. Here are some more smoothed frequency polygons, some of which represent normal distributions (μ marks the mean in each case).

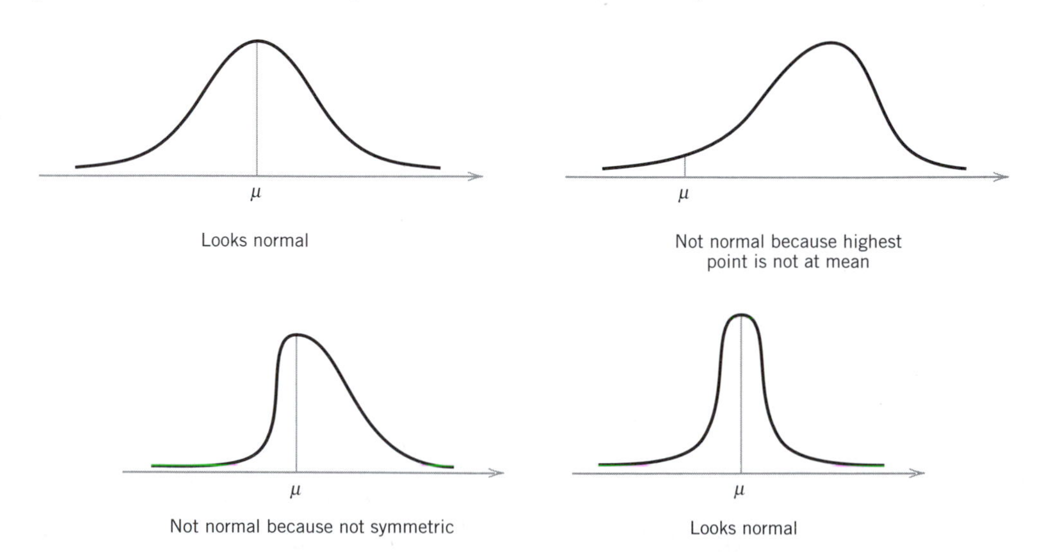

As you can see, the normal curve can be tall or short, narrow or broad, but in all cases, the mean marks the highest point and the curve is symmetric about the mean.

PRACTICE

Which of the following do not represent normal distributions (μ marks the mean in each case)? Explain.

1. 

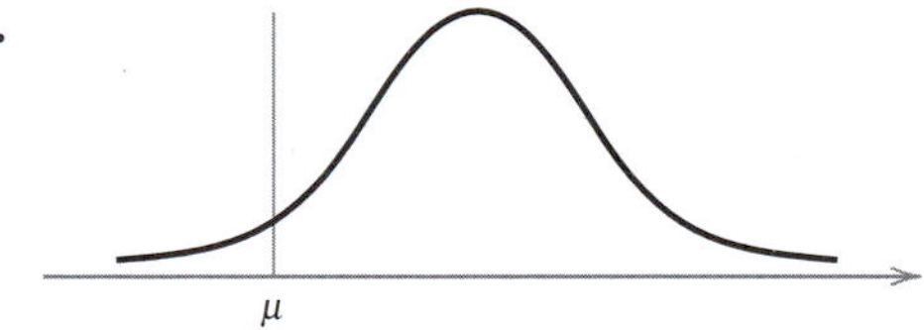

2.

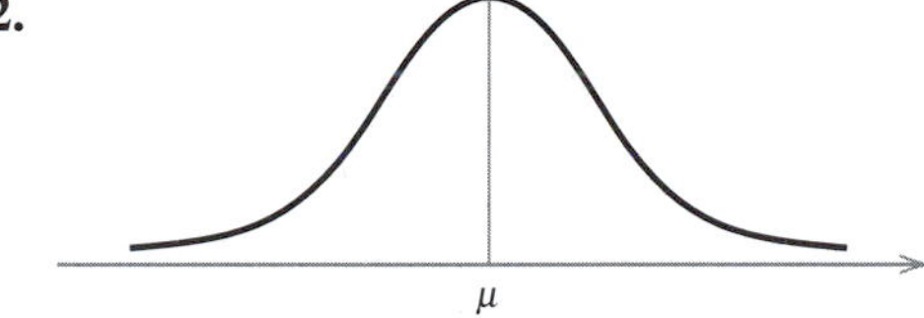

3.

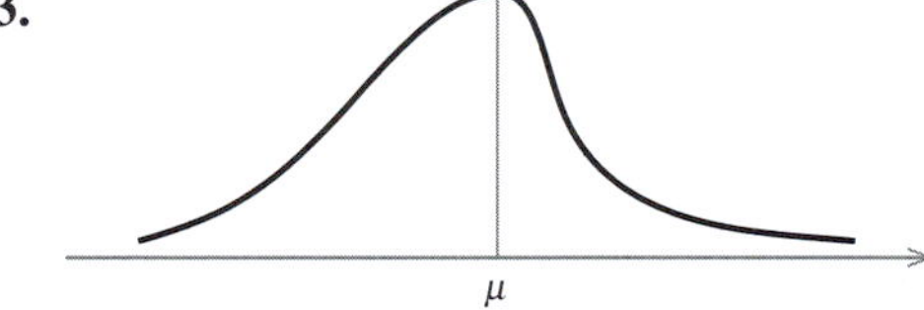

4.

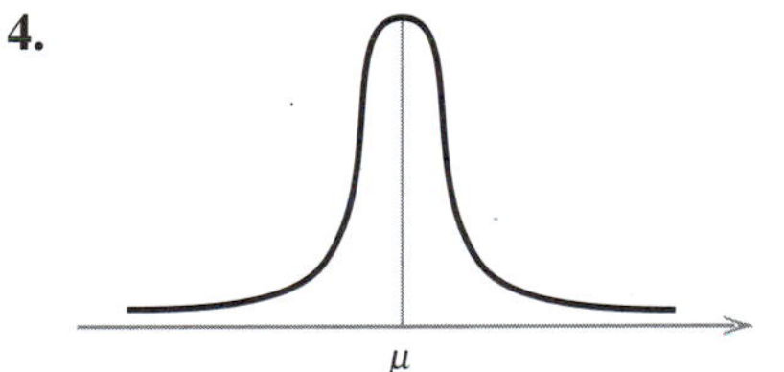

A surprising number of distributions, including students' test scores, are normal or approximately normal. If a distribution is normal, then it has a very peculiar and powerful property. The area under the normal curve between any two points is precisely the percentage of values in the distribution that lie between those two points.

Here's an example. Suppose we look at the area between 20 and 30 in the normal distribution whose curve is shown below.

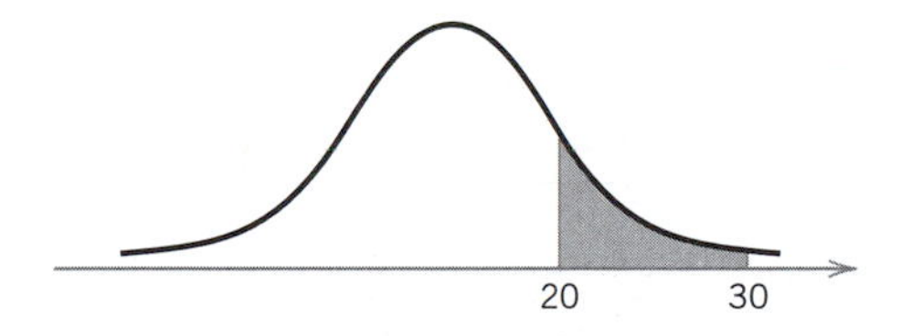

The total area under a normal curve is taken to be 1.00, because 100% of the values in the distribution are represented by the whole curve. Also, because the normal curve is symmetric about the mean, the total area under half of the curve is 0.50.

If the area between 20 and 30 is 0.23, then about 23% of the values in the distribution lie between 20 and 30.

You are probably wondering how we go about finding the areas under parts of the normal curve. Well, fortunately we don't have to work too hard at that because we will always have a table of areas readily available. You should find the table in Appendix B now and refer to it as you read on.

Notice that the table in Appendix B gives the area between the zero z-score and a positive z-score. Let's see how to read the table.

If we want the area between the zero z-score and the z-score 1.34, we first look down the left-most column until we find 1.3. Now we look to the right until we are under the column marked 0.04. The entry in the table, 0.4099, is the area of the curve between 0 and 1.34.

PRACTICE

Find the area between 0 and the indicated z-score using the table in Appendix B.

5. 2.34

6. 0.08

7. 2.01

8. 1.57

9. 3.09

10. −2.34

11. −2.01

In Situation 7, Zork scored a 40 on the exam. The class mean was 35 with a standard deviation of 2. What was Zork's z-score?

(Answer: $(40 - 35)/2 = 2.50$)

Using the table in Appendix B, we find that the area under the normal curve between Zork's z-score (2.50) and the z-score of the mean (0) is 0.4938. This means that 49.38% of the class scores were between the mean (35) and Zork's (40).

Now, 50% of the class scores were below the mean. Why?

____

____

____

This means that Zork's exam score was higher than 50% + 49.38% = 99.38% of those in his class. Zork scored in the 99th percentile on this exam, and that is why the instructor awarded Zork an A.

When a teacher assumes that the grades on an exam are normally distributed and then assigns grades accordingly, we say that he or she *curved* the exam. The curve refers to the normal curve.

Another example using the normal distribution will help you to see the wonderful relationships contained in it.

Suppose we know that the weights of ducks on a pond in Iowa are normally distributed with a mean of 5 pounds and a standard deviation of 1.2 pounds. Also suppose that there are 1000 ducks on the pond. What percentage of the ducks weigh more than 6.5 pounds? About how many ducks weigh more than 6.5 pounds?

Begin by computing the z-scores of the relevant duck weights. The z-score for 5 is 0 because 5 is the mean. The z-score for 6.5 is $(6.5 - 5)/1.2 = 1.25$.

Now draw a normal curve, locate the z-scores on it, and then shade the area of interest.

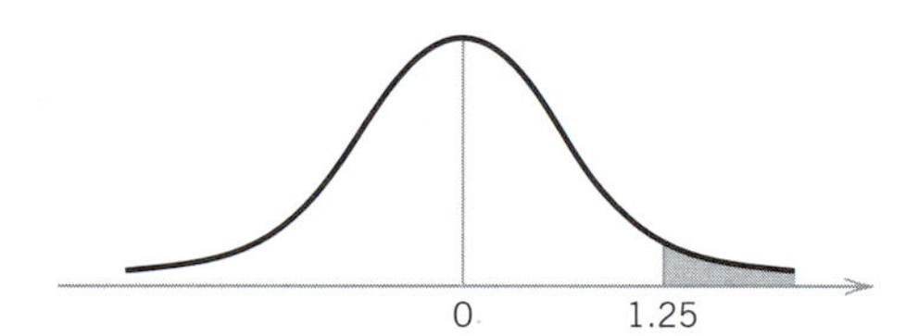

I've shaded the area to the right of 1.25 because we are interested in the percentage of ducks that weigh *more* than 6.5 pounds.

Now the table in Appendix B will give us the area between 0 and 1.25. That area is 0.3944. We know that the total area to the right of 0 is 0.50. Thus, the shaded area is $0.50 - 0.3944 = 0.1056$. About 10.56% of the ducks weigh more than 6.5 pounds.

Finally, since there are 1000 ducks on the pond, we can say that approximately 106 (10.56% of 1000) ducks weigh more than 6.5 pounds.

PRACTICE

All of the following refer to the normal distribution of weights of ducks on a pond in Iowa (1,000 ducks; mean = 5, standard deviation = 1.2).

12. What percentage of the ducks weigh less than 6 pounds?

13. What percentage of the ducks weigh between 4 and 6 pounds?

14. What percentage of the ducks weigh more than 3 pounds?

15. What percentage of the ducks weigh between 6 and 7 pounds?

16. About how many ducks weigh less than 3.5 pounds?

2.9 What Do You Know?

If you have worked carefully through this chapter, then you can compute several basic statistics, including the mean, median, mode, range, variance, and standard deviation. You can construct frequency polygons, histograms, and circle graphs. You also can compute percentiles and z-scores and work with normal distributions. The exercises in this section are designed to test and refine your statistical skills.

1. Describe 10 statistics that might be associated with the collection of pizzas sold by a local pizzeria on a Saturday

night. Tell how you would obtain each statistic and why it might be useful.

2. Describe 10 statistics that might be associated with a major league baseball team. Tell how you would obtain each statistic and why it might be useful.
3. Describe 10 statistics that might be associated with gardening. Tell how you would obtain each statistic and why it might be useful.

Find the mean, median, mode, range, and standard deviation of each of the following distributions:

4. 10, 10, 10, 10, 20, 30, 40
5. 100, 300, 400, 200, 400, 300, 200, 100
6. 15, 20, 15, 15, 15, 15, 15, 15, 15
7. 20, 30, 50, 10, 30, 20, 20, 10
8. 1, 1, 1, 1, 1, 1, 1, 1, 1, 2, 2, 3, 3, 3, 3, 3, 3
9. 5, 5, 5, 5, 5, 5, 5, 5, 5
10. 8, −8, 8, −8, 8, −8, 8, −8
11. In a small developing country, the king owns most of the land and has great personal wealth. The rest of the people in the country exist at a level near poverty. Which of the *averages* (mean, median, or mode) would most likely be the best estimate of the average personal wealth for this country?
12. Fern calculated the mean, median, and mode for the distribution of heights of her bean plants. Fern remarked, "Only one of these 'averages' was actually the height of one of my plants." Which *average* has to be an actual value in the distribution?
13. Ray harvested 30,000 bushels of wheat this year. The mean for his county was 26,124 bushels with a standard deviation of 3000 bushels. Last year Ray harvested 25,000 bushels of wheat. The mean for his county last year was 24,000 with a standard deviation of 1500 bushels. Relative to the other farmers in his county, did Ray have a better harvest this year or last year?
14. The mean price for a new house sold in McHenry County was $120,000 with a standard deviation of $4600. Maria bought a house in McHenry County for $106,000. If housing prices are *not* normally distributed, roughly what percent of the houses sold in McHenry County cost more than Maria's?
15. Morita participated in her company's annual golf outing. The scores posted by the golfers were:

    88, 102, 77, 96, 101, 89, 92, 110, 70, 99, 87, 102

    If Morita scored in the first quartile, what was her golf score?
16. Consider the following distribution of bowling scores:

    102, 120, 140, 130, 120, 120, 200,

    235, 157, 180, 160, 150, 120, 280

    (a) Compute the mean and standard deviation.

    (b) What percentage of the scores are within two standard deviations of the mean? Compare your answer to that predicted by Chebyshev's theorem.
17. Consider the following distribution of golf scores:

    102, 88, 100, 100, 99, 67, 100, 100, 98, 77, 99, 88

    (a) Compute the mean and standard deviation.

    (b) What percentage of the scores are within three standard deviations of the mean? Compare your answer to that predicted by Chebyshev's theorem.

Draw a frequency polygon, histogram, and circle graph for each of the following distributions.

18. A parking lot has 20 Toyotas, 20 Hondas, 15 Fords, 30 Chevrolets, and 15 Dodge Minivans parked in it.
19. Victor's grades on his math quizzes were A, B, C, C, C, C, A, C, B, A, C, D, A.
20. A professional golfer's scores:

    70, 66, 70, 78, 65, 75, 67, 72, 70, 67
21. The salaries of employees in a furniture factory.

Salary Range	Number of Employees
0–$10,000	8
10,001–$20,000	12
20,001–$30,000	5
30,001–$40,000	2
Over $40,000	3

Exercises 22–26 refer to the following frequency polygon, a graph depicting the distribution of tornadoes in Illinois by month for the years 1870–1910.

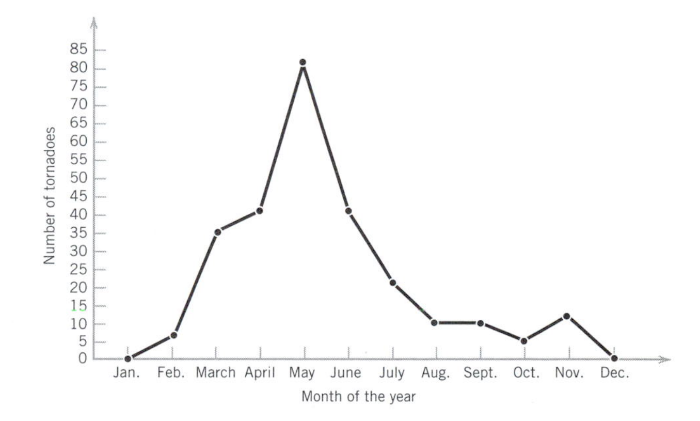

22. How many tornadoes were there in the month of April?
23. How many tornadoes were there in the summer (June, July, and August)?

24. Find the mean number of tornadoes per month.
25. Based on your answer to Exercise 24, which month(s) is(are) an average month for tornadoes?
26. What is the modal month for tornadoes?

Exercises 27–29 refer to the circle graph shown below.

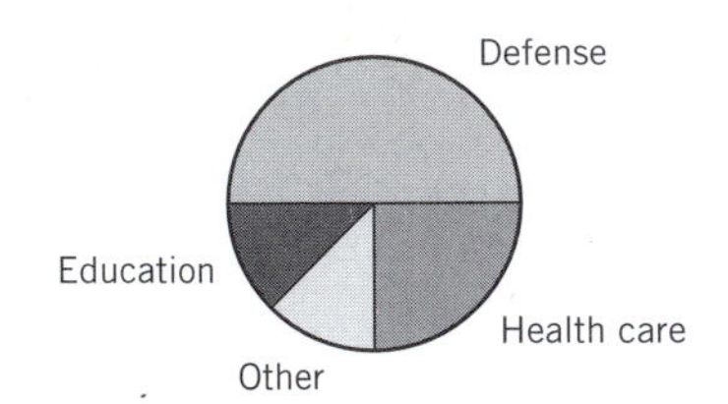

27. What percentage of the money is spent on defense?
28. What percentage of the money is spent on education?
29. What percentage of the money is spent on health care?
30. The heights of emus are normally distributed with a mean of 6.8 feet and a standard deviation of 0.86 feet. What percentage of emus are
    (a) Taller than 6 feet?
    (b) Shorter than 5 feet?
    (c) Between 5 and 6 feet tall?
    (d) Shorter than 6 feet?
    (e) Taller than 5 feet?
    (f) Between 4 and 7 feet tall?
31. The mean price for a new house sold in McHenry County was $120,000 with a standard deviation of $4600. Maria bought a house in McHenry County for $106,000. If housing prices are normally distributed, roughly what percent of the houses sold in McHenry County cost more than Maria's? Compare your answer to the answer that you obtained for Exercise 14.
32. Dr. Pellera uses the following grading scale:

    A student in the 83rd percentile or higher gets an A.
    A student in the 66th percentile or higher gets a B.
    A student in the 34th percentile or higher gets a C.
    A student in the 17th percentile or higher gets a D.
    Students below the 17th percentile get an F.

    On a recent test, the class mean was 62 out of 100 with a standard deviation of 3.5. Dr. Pellera decides to *curve* these test scores. What grade will each of the following students receive?
    (a) Monty, who scored 65 on the test.
    (b) Masria, who scored 62 on the test.
    (c) Abdul, who scored 70 on the test.
    (d) Serena, who scored 68 on the test.
    (e) Alvin, who scored 57 on the test.
33. The amount of time spent studying for a mathematics test (in hours) by students at Local University is normally distributed with a mean of 3 hours and standard deviation of 20 minutes. There are 2000 students taking mathematics at Local University.
    (a) About how many of the students study more than 4 hours for a mathematics test?
    (b) About how many students study less than 2.5 hours for the mathematics test?
    (c) Fern studied 3 hours and 15 minutes for the test. How many students studied for a shorter period of time than Fern?

ANSWERS TO PRACTICE EXERCISES

Section 2.1

1. Here are some possible statistics. The number of brown dogs. The weight of the largest dog. The average weight of the dogs. The number of poodles. The ratio of collies to German shepherds. The percentage of black dogs that have blue eyes. The number of dogs that prefer Spam to Jello. The average daily water intake of the dogs. The average number of minutes spent chasing their own tail. The length of the shortest tail.

Section 2.2

1. $\dfrac{5 + 5.5 + 7 + 7.8 + 8 + 8.1 + 8.5}{7} = \dfrac{49.9}{7} = 7.13$ pounds
2. $\dfrac{78 + 87 + 70 + 90 + 82}{5} = \dfrac{407}{5} = 81.4$
3. $(10{,}000 + 10{,}000 + 10{,}000 + 10{,}000 + 10{,}000 + 30{,}000 + 30{,}000 + 350{,}000)/8 = \$57{,}500$
4. 4 feet, 4.5 feet
5. 300 burgers and 400 burgers are both modes because each appears twice. We say that this collection is **bimodal.** 492.86 burgers.
6. 70
7. (a) Mean = 40, median = 40, mode = 20
   (b) Mean = 25.56, median = 30, mode = 10 and 30
   (c) Mean = 43, median = 35, mode = 10
   (d) Mean = 30, median = 20, mode = 20

Section 2.3

1. 87
2. 310
3. 6.6
4. 15
5. Mean = 24, variance = 784

6. Mean = 30, variance = 200
7. Mean = 34, variance = 364
8. Mean = 10, variance = 0
9. Mean = 0, variance = 100
10. Mean = 16, variance = 24, standard deviation = 4.90
11. Mean = 16, variance = 0, standard deviation = 0
12. Mean = 16, variance = 180, standard deviation = 13.42
13. The means of the three distributions are the same, but the standard deviations help us to distinguish between the distributions. The standard deviation indicates how much variability there is among the scores in the distribution. The distribution in Practice Exercise 11 has no variability, and the distribution in Practice Exercise 12 has greater variability than the distribution in Practice Exercise 10.

Section 2.4

1. 890, since $2900 > 2200 + 3 \cdot 200$
2. At least 89%, $42 < 60 - 3 \cdot 5$
3. (a) 1.5
   (b) 0.6
   (c) 0.8
   (d) −1.2
4. Guenon did better on the earlier test where his z-score was 2.33. On today's test, his z-score was only 1.33.
5. (a) Mean = 156.214, standard deviation = 50.38
   (b) 92.9%

Section 2.5

1. 30th
2. 80th
3. 40th
4. 75
5. 40
6. 60
7. 90
8. 99

Section 2.6

1. Frequency distribution:

Number of Siblings	Number of Players
0	10
1	8
2	7
3	1
4	2
5	0
6	1

Notice that we needed to provide a count for 5!
Frequency polygon:

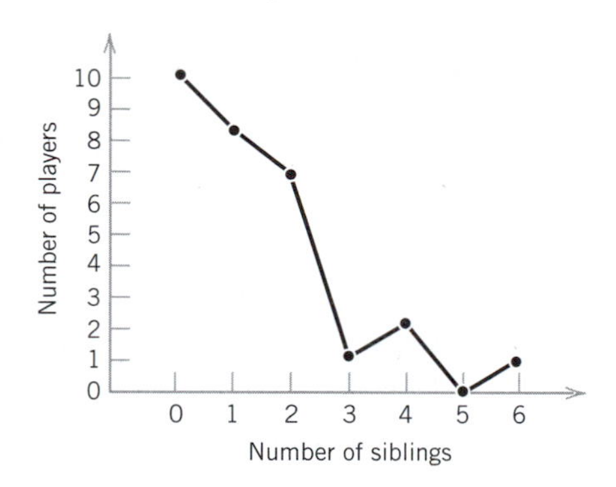

2. Frequency distribution:

Grade on Exams	Number of Freshmen
A	2
B	4
C	6
D	4
F	2

Frequency polygon:

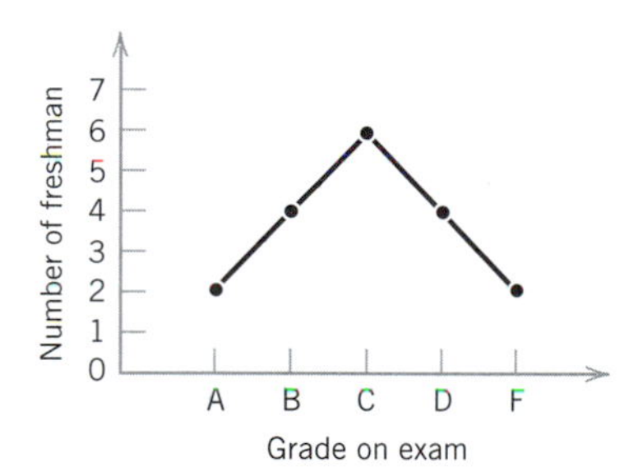

3. 7 out of 29 players have exactly two siblings; 24.14%
4. 6 out of 18 Freshmen got a C; 33.33%
5.

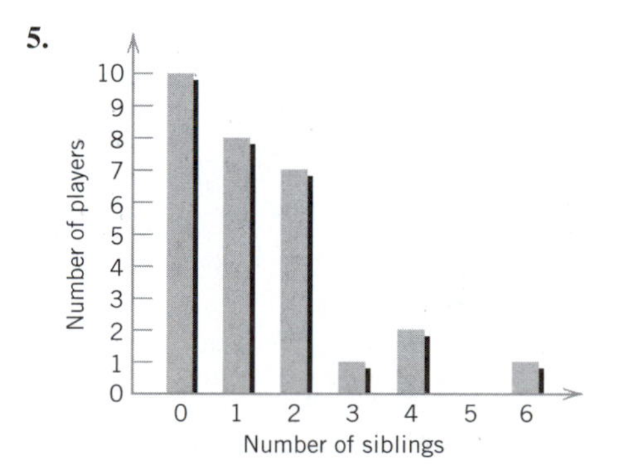

6.

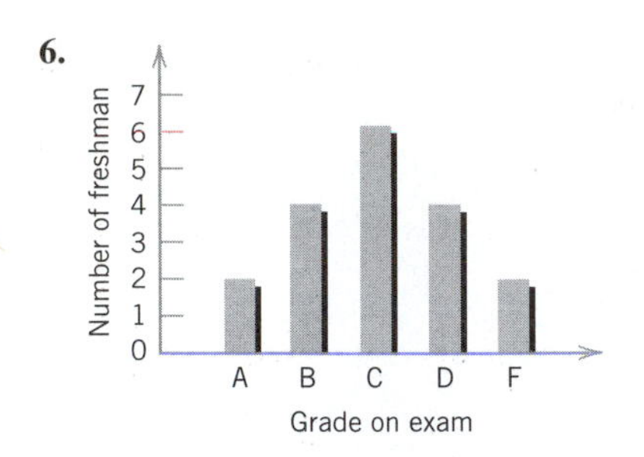

Section 2.7

1.

Number of Siblings	Number of Players	Percentage
0	10	35
1	8	28
2	7	24
3	1	3
4	2	7
5	0	0
6	1	3

Notice that I have rounded the percentages to whole percentages so that I can actually measure the right amount with my protractor.
Circle graph:

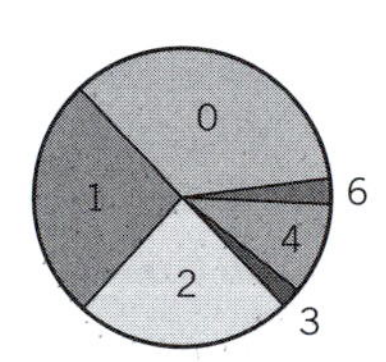

Number of siblings

2.

Grade on Exams	Number of Freshmen	Percentage
A	2	11
B	4	22
C	6	34
D	4	22
F	2	11

I rounded the C percentage up to 34% so that the total percentage would be 100%.
Circle graph:

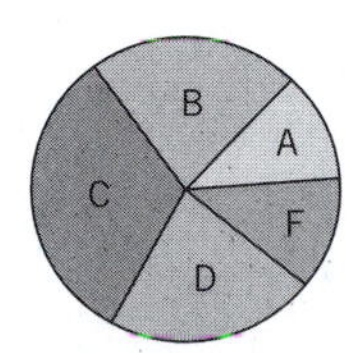

Grade on exam

Section 2.8

1. Not normal: highest point is not the mean
2. Normal
3. Not normal: not symmetric
4. Normal
5. 0.4904
6. 0.0319
7. 0.4778
8. 0.4418
9. 0.4990
10. 0.4904, same as 2.34 because of symmetry
11. 0.4778
12. z-score for 6 = (6 − 5)/1.2 = 0.83

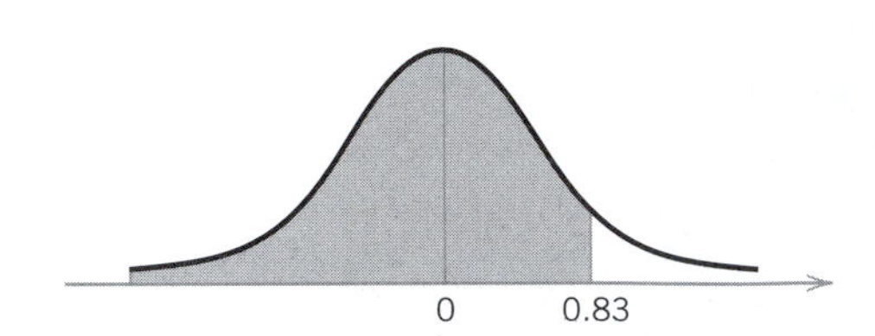

Area between 0 and 0.83 = 0.2967
Shaded area = 0.50 + 0.2967 = 0.7967
79.67% of the ducks weigh more than 6 pounds.

13. z-score for 6 = (6 − 5)/1.2 = 0.83
z-score for 4 = (4 − 5)/1.2 = −0.83

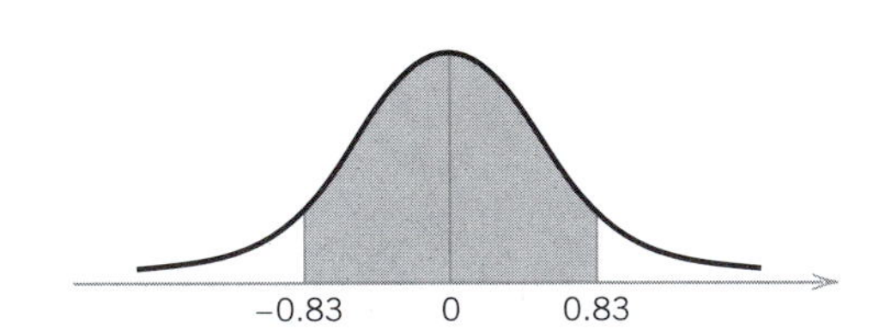

Area between 0 and 0.83 = 0.2967
Area between 0 and −0.83 = 0.2967
Shaded area = 0.2967 + 0.2967 = 0.5934
59.34% of the ducks weigh between 4 and 6 pounds.

14. z-score for 3 = (3 − 5)/1.2 = −1.67

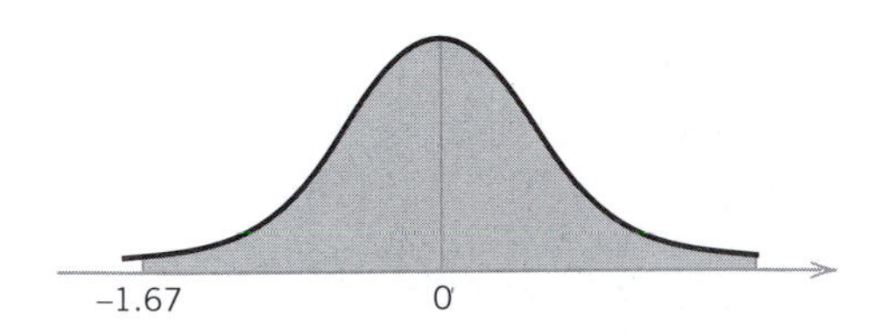

Area between 0 and −1.67 = 0.4525
Shaded area = 0.50 + 0.4525 = 0.9525
95.25% of the ducks weigh more than 3 pounds.

15. z-score for 6 = (6 − 5)/1.2 = 0.83
z-score for 7 = (7 − 5)/1.2 = 1.67

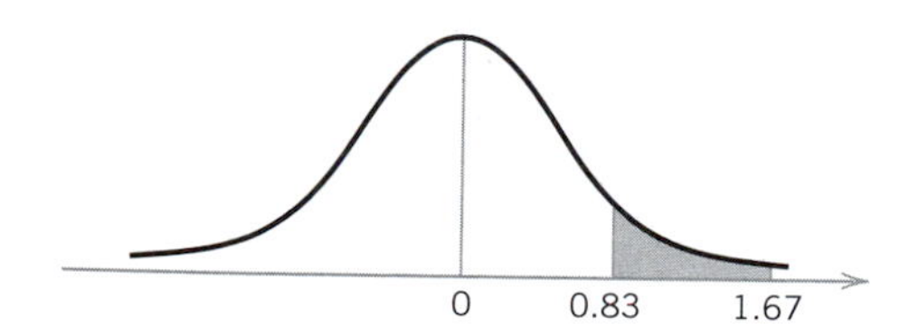

Area between 0 and 0.83 = 0.2967
Area between 0 and 1.67 = 0.4525
Shaded area = 0.4525 − 0.2967 = 0.1558
15.58% of the ducks weigh between 6 and 7 pounds.

16. z-score for 3.5 = (3.5 − 5)/1.2 = −1.25

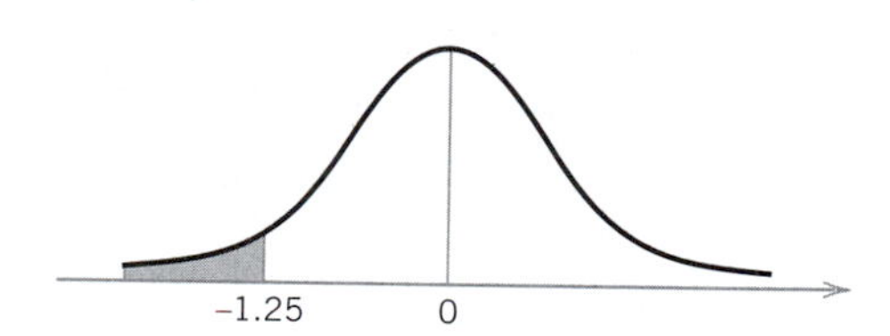

Area between 0 and −1.25 = 0.3944
Shaded area = 0.50 − 0.3944 = 0.1056
10.56% of the ducks weigh less than 3.5 pounds.
About 106 ducks weigh less than 3.5 pounds.

Chapter Three

Mathematics of Collections

What's the Point? People have a tendency to organize their universe into collections or groups. You live in a city, a collection of streets, houses, businesses, and people. You have a family, a collection of related people. You have a course load, a collection of courses. You eat meals, collections of food. Mathematicians recognize the importance of collections and have developed some techniques for talking about and working with collections. This chapter introduces the fundamental concepts of set theory, the mathematics of collections.

Situation 8

How Many People Drink Organic Decaf?

Belson Zork is the proprietor of the Thankyu Haiku coffee shop. Belson offers his customers three types of very strong coffee: organically grown espresso, organically grown dark roast, and organically grown decaffeinated coffee. In order to plan his purchases from his coffee bean distributor, Belson asked his nephew to take a survey of his customers in order to determine their coffee preferences.

This is what Belson's nephew reported:

"I asked 100 customers about their coffee preferences. Sixty customers said that they liked espresso. Twenty-four customers said that they liked dark roast. Thirty-nine customers said that they like decaffeinated coffee. Sixteen customers said that they liked espresso and decaffeinated coffee. Fourteen customers said that they liked espresso and dark roast. Thirteen customers said that they liked dark roast and decaffeinated coffee. And ten customers said that they liked all three coffees."

Belson considered what his nephew told him and then said:

"Wait a minute! You said that you talked to 100 customers, but the numbers you just gave me add up to more than 100. What's going on?"

Belson's nephew said that the numbers he related were correct, and that he was now going to meet some friends at the video arcade.

What is going on here? Could the nephew be correct when he says that his numbers are right? How many people liked only organically grown decaf?

To answer these questions, we must take a closer look at the mathematical notion of a set.

3.1 SETS IN WORDS, SYMBOLS, AND PICTURES

A **set** is a well-defined collection of objects. A collection is well-defined if there is a clear and unambiguous way of deciding whether or not something is in the collection.

For example, the collection of U.S. Senators currently in office is well-defined. To decide whether or not someone is a current U.S. Senator, we need only check the official list in Washington, D.C. The collection of whole numbers between 2 and 45 is also well-defined. We can easily decide that 14 is in this collection, but Abraham Lincoln is not. The collection of tall people is not well defined because being tall is ambiguous and depends on your point of view. A person who is 6 feet tall is tall from the point of view of a 5-year-old, but not that of a 7-footer! On the other hand, the collection of people who are 6 feet tall is well-defined. Why?

____

____

____

The objects that are in a set are called the **elements** or **members** of the set. Jimmy Carter is a member of the set of former U.S. Presidents. He is also an element of the set of men. He is not an element of the set of elephant trainers.

The symbol $\in$ means *is an element of,* and $\notin$ means is not an element of. Thus, we can write

Jimmy Carter $\in$ the set of former U.S. Presidents

and

Jimmy Carter $\notin$ the set of elephant trainers

Sets can be described by listing their members or describing some characteristics that their members share. For example, consider the set of cartoon characters Huey, Dewey, and Louie. We can describe this set by listing its members between two braces, like this

{Huey, Dewey, Louie}

or we can describe this set by saying something about the members, like this

{x: x is a nephew of Donald Duck}

The set-builder notation {x: x is a nephew of Donald Duck} is read, "the set of all x such that x is a nephew of Donald Duck." If we use the set-builder notation, then we must be sure that we describe precisely the members of the set. The set

{x: x is a cartoon duck}

is not the same as {Huey, Dewey, Louie} because Daffy Duck is an element of {x: x is a cartoon duck) but not {Huey, Dewey, Louie}.

Here are some more sets described in two ways. Make sure you see that each pair describes the same set.

{Alaska, Alabama, Arizona, Arkansas}

{x: x is a state beginning with the letter "A"}

{Cubs, White Sox}

{x: x is a major league baseball team in Chicago}

{1, 3, 5, 7, 9, 11, 13, 15, 17, 19, 21, 23, 25, 27, 29, 31, 33, 35, 37, 39, 41, 43}

{x: x is an odd whole number less than 44}

{0, 1, 2, 3, . . . 1499} (The ". . ." means that all the numbers from 4 to 1498 are included.)

{x: x is a whole number less than 1500}

PRACTICE

List the members of each of the following sets:

1. {x: x is the capital of Peru}

2. {x: x is the capital of a state beginning with the letter "I"}

3. $\{x: x$ is a whole number between 3 and 25$\}$

Describe each of the following sets using set-builder notation:

4. $\{2, 4, 6, 8, 10\}$

5. {Iceland, Ireland, Great Britain}

6. {307-555-1212, 303-555-1212, 313-555-1212, 312-555-1212}

Answer true or false.

7. $14 \in \{1, 2, 3, \ldots 20\}$

8. Huey $\in \{x: x$ is a nephew of Donald Duck$\}$

9. $\{4\} \in \{\{1\}, \{2\}, \{3\}, \{4\}\}$

Another way to describe sets is with **set diagrams.** A set diagram consists of one or more plane geometric shapes inside of a square or rectangle. The geometric shapes represent sets. The members of the sets are located inside of the shapes.

Here is a set diagram of the set {Huey, Dewey, Louie}:

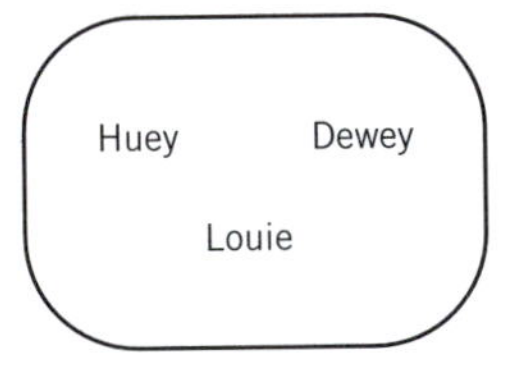

Here is a set diagram for the set $\{a, b, c, d, e, f, g\}$

a b c d e

f g

We can also depict a set using set diagrams even if we do not know what its elements are. For example, here is a set diagram for the set A.

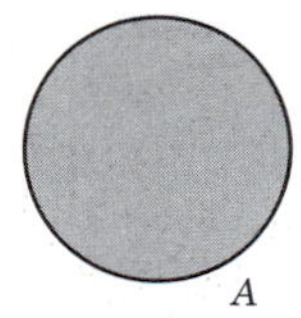

The shading inside of the circle marked with an A indicates that the elements of the set A lie within the shaded area.

Here is a set diagram for the set B.

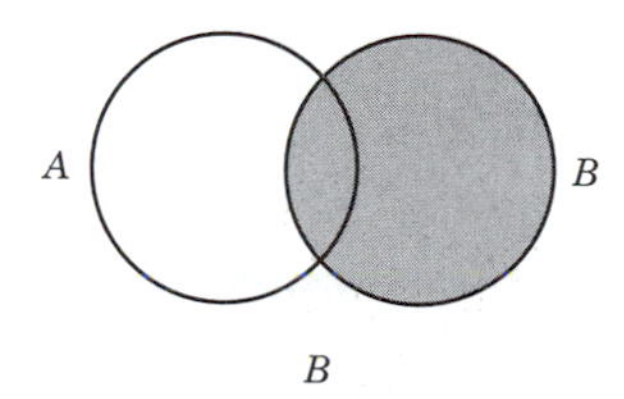

Notice that the entire area within the B circle is shaded even if part of it encroaches into the A circle.

3.2 SUBSETS

A set B is a **subset** of a set C if every element of B is also an element of C. For example, {Huey, Dewey, Louie} is a subset of

{Donald, Huey, Dewey, Daisy, Louie}

because Huey, Dewey, and Louie are all elements of {Donald, Huey, Dewey, Daisy, Louie}.

On the other hand, {Huey, Dewey, Louie} is *not* a subset of

{Donald, Huey, Dewey, Daisy}

because Louie is not an element of {Donald, Huey, Dewey, Daisy}.

The symbol for subset is $\subseteq$. Thus,

{Huey, Dewey, Louie} $\subseteq$ {Donald, Huey, Dewey, Daisy, Louie}

and

{Huey, Dewey, Louie} $\not\subseteq$ {Donald, Huey, Dewey, Daisy}

(The slash through the subset symbol means that the left set is not a subset of the right set.) Notice that {Huey, Dewey, Louie} $\subseteq$ {Huey, Dewey, Louie}. In general, we can say that every set is a subset of itself. We can use this fact to define what it means for two sets to be equal.

Two sets are **equal** if they are subsets of each other. So {Huey, Dewey, Louie} = {x: x is a nephew of Donald Duck} because {Huey, Dewey, Louie} $\subseteq$ {x: x is a nephew of Donald Duck} and {x: x is a nephew of Donald Duck} $\subseteq$ {Huey, Dewey, Louie}.

PRACTICE

Answer true or false.

1. $\{2, 3, 4\} \subseteq \{1, 2, 3, 4\}$

2. $\{a, c, b, f\} \subseteq \{f, a, v, g, b, n\}$

3. $\{x$: x is a bear$\} \subseteq \{x$: x is an animal$\}$

4. $\{x$: x is a bird$\} \subseteq \{x$: x is a duck$\}$

5. $\{x$: x is a television set$\} \subseteq \{x$: x is a television set$\}$

6. $\{4, 3\} \subseteq \{1, 2, 3, 4\}$

7. $\{4, 3\} \in \{1, 2, 3, 4\}$

8. $\{4\} \subseteq \{\{1\}, \{3\}, \{4\}\}$

9. $\{4\} \in \{\{1\}, \{3\}, \{4\}\}$

The subset relationship can be shown using set diagrams. We can draw the shape representing the subset completely inside the shape of the set of which it is a subset. The set diagram below shows $B \subseteq A$.

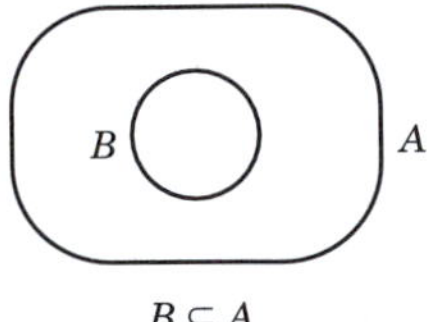

$B \subseteq A$

The **empty set** (or **null set**) is the set with no elements. The symbol for the empty set is $\varnothing$. The empty set can be described in many ways. Here are a few of them:

$\varnothing = \{x$: x is an elephant elected to the U.S. presidency$\}$
$\varnothing = \{x$: x is a number bigger than 3 and smaller than 2$\}$
$\varnothing = \{x$: x is a wombat sitting on your head$\}$

3.3 CARDINAL NUMBERS

The **cardinal number** of a set is the number of elements in that set. We will denote the cardinal number of A by $n(A)$. Thus, if A is the set {1, 2, 3, 4, 5}, then $n(A) = 5$. If B is the set {Huey, Dewey, Louie}, then $n(B) = 3$.

Here are some more sets and their cardinal numbers.

Set *A*	$n(A)$
{1, 3, 5, 7, 9}	5
{x: x is a state in the United States}	50
{x: x is a planet in the solar system}	9
{q, r, s, t, u, v}	6
$\varnothing$	0

Write as many subsets of the set {a, b, c} as you can.

Did you get them all? Here they are: {a}, {b}, {c}, {a, b}, {a, c}, {b, c}, {a, b, c}, and $\varnothing$. Thus, {a, b, c} has eight subsets. It can be shown that for any set A, if $n(A) = k$, then A has 2^k subsets.

PRACTICE

Let A = {1, 2, 3, 4, 5}, B = {a, b, c, d}, and C = {x: x is a state that begins with the letter "A"}. Find

1. $n(A)$

2. $n(B)$

3. $n(C)$

4. All the subsets of B (there are $2^4 = 16$ subsets of B)

3.4 UNION AND INTERSECTION

Given two sets, A and B, we create two new sets from the elements of A and B. These new sets are called the **union** of A with B, symbolized by $A \cup B$, and the **intersection** of A and B, symbolized by $A \cap B$.

Here are the definitions of these two sets in symbols:

Union

$$A \cup B = \{x\colon x \in A \quad \text{or} \quad x \in B \quad \text{or both}\}$$

Intersection

$$A \cap B = \{x\colon x \in A \quad \text{and} \quad x \in B\}$$

Here are the definitions of union and intersection in pictures:

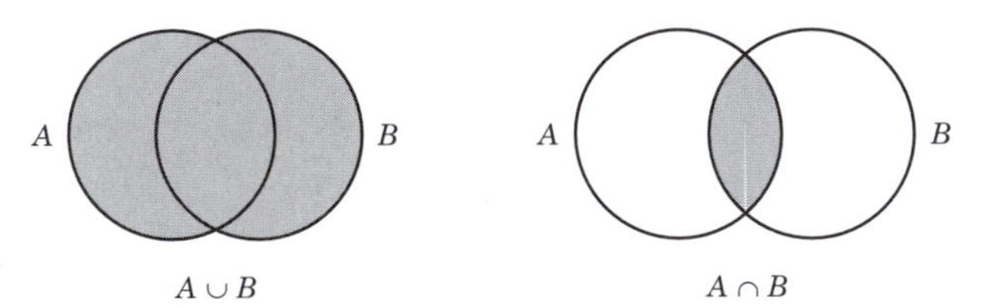

Look carefully at the symbolic definitions and pictures. They should tell you exactly what the union and intersection of two sets mean. Now try to answer the following question.

Suppose A = {○, ●, ♥} and B = {♥, ♦, ♣, ♠, ✾}. List the elements of $A \cup B$ and $A \cap B$.

____

____

____

Did you say that $A \cup B = \{○, ●, ♥, ♦, ♣, ♠, ✱\}$ and $A \cap B = \{♥\}$? Great! You understand that the intersection of two sets is the set of elements that the two sets have in common and the union is the set of elements that belong to one or both of the two sets.

Here are some more examples.

$$\{1, 2, 3\} \cap \{2, 3, 4, 5\} = \{2, 3\}$$

because 2 and 3 are the only elements that belong to *both* sets.

$$\{1, 2, 3\} \cup \{2, 3, 4, 5\} = \{1, 2, 3, 4, 5\}$$

because 1, 2, 3, 4 and 5 belong to *one or both* of the sets.

$$\{1, 2, 3\} \cap \{4, 5, 6\} = \emptyset$$

because the two sets have no elements in common.

$$\{1, 2, 3\} \cup \{4, 5, 6\} = \{1, 2, 3, 4, 5, 6\}$$

because 1, 2, 3, 4, 5, and 6 belong to *one or both* of the sets.

PRACTICE

Find the intersection and union of each of the following pairs of sets:

1. $A = \{a, b, c\}$, $B = \{a, d, f, g\}$

2. $A = \{x\text{: } x \text{ is a bear}\}$, $B = \{x\text{: is a cartoon character}\}$

3. $A = \emptyset$, $B = \{2, 3, 4\}$

4. $A = \{1, 2, 3, 4, 5\}$, $B = \{2, 4, 6, 8\}$

The cardinal numbers of set intersections and unions may be expressed in pictures. Suppose that set A has 10 elements, set B has 12 elements, and the set $A \cap B$ has 4 elements. This situation can be depicted as

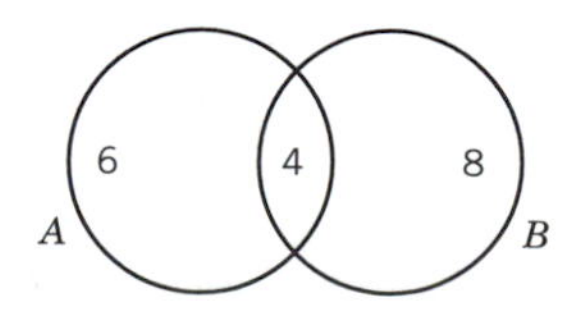

Notice that there must be a total of 10 in the A circle and 12 in the B circle. The 4 occurs in the $A \cap B$ region. Why is the following picture incorrect?

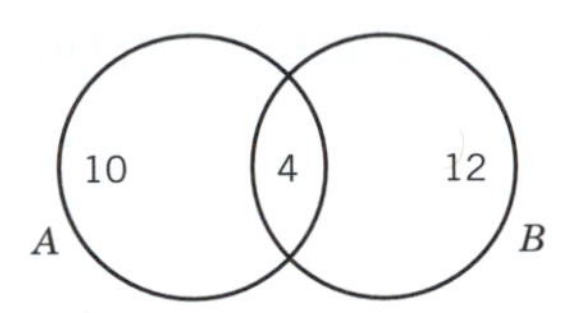

____

____

____

Right! This picture shows a total of 14 inside of A, but we were told that A had only 10 elements.

PRACTICE

Fill in the appropriate numbers for each diagram.

5. $n(A) = 19$, $n(B) = 23$, $n(A \cap B) = 7$

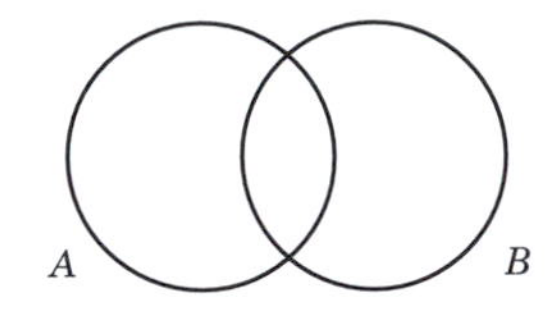

6. $n(A) = 19$, $n(B) = 29$, $n(A \cap B) = 19$

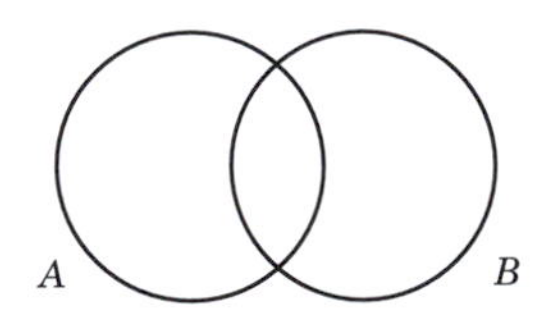

7. $n(A) = 39$, $n(B) = 29$, $n(A \cap B) = 0$

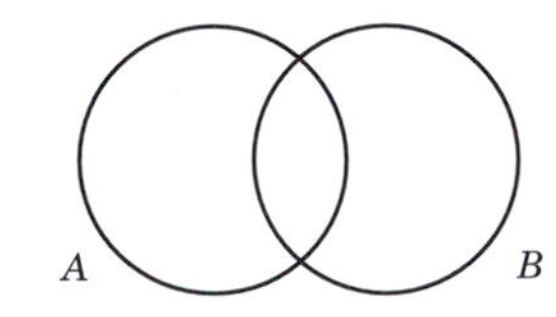

Let's look again at the report on coffee drinkers by Belson Zork's nephew. The nephew said,

> *"I asked 100 customers about their coffee preferences. Sixty customers said that they liked espresso. Twenty-four customers said that they liked dark roast. Thirty-nine customers said that they liked decaffeinated coffee. Sixteen customers said that they liked espresso and decaffeinated coffee. Fourteen*

customers said that they liked espresso and dark roast. Thirteen customers said that they liked dark roast and decaffeinated coffee. And ten customers said that they liked all three coffees.''

Using the concept of intersection and a set diagram, we can begin to make sense of the nephew's report. We'll first draw three intersecting circles to represent the three sets of coffee drinkers:

E = the set of those who said they liked espresso
R = the set of those who said they like dark roast
D = the set of those who said they like decaf

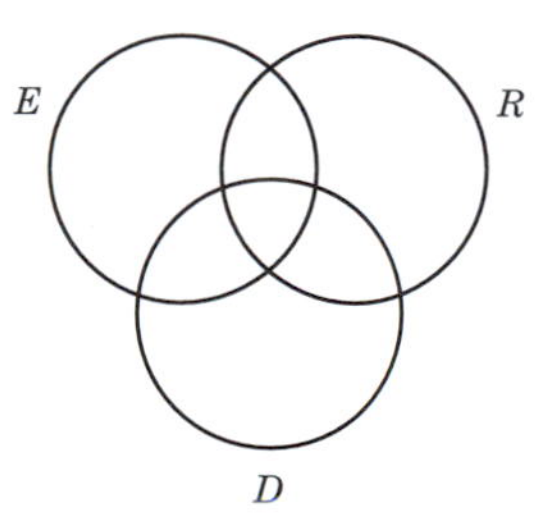

We are told that 10 of the people liked all three coffees, so we will put a 10 in the region common to all three circles:

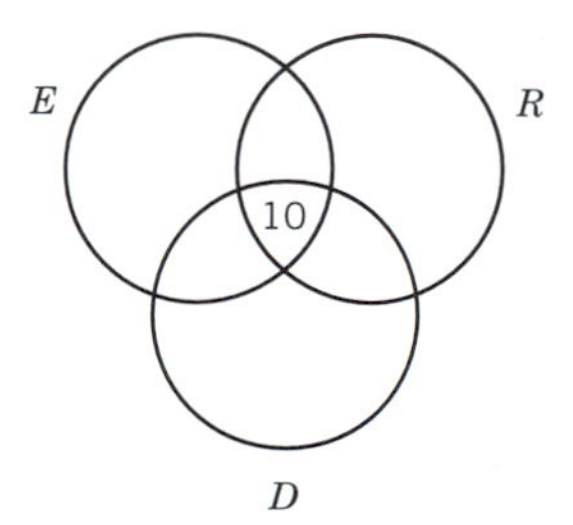

Now, 13 customers said that they liked dark roast and decaf, but 10 of them have already been listed because these 10 like espresso as well. Thus, since the area where the dark roast and decaf circle overlap must contain a total of 13, we put in a 3 as shown below.

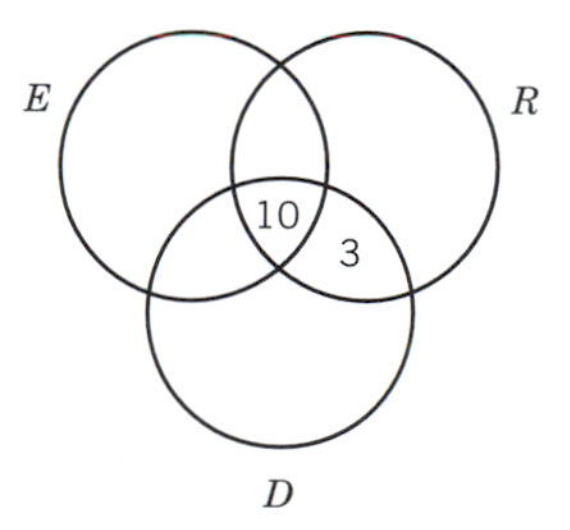

In a similar fashion, we account for the 16 customers who like espresso and decaf, and the 14 who like espresso and dark roast.

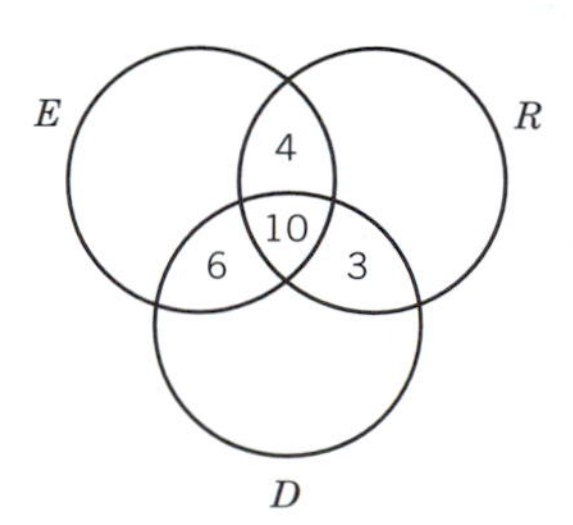

Now, 60 customers said they liked espresso. Our picture already has a total of 20 in the espresso circle. So, we need to add in 40 more.

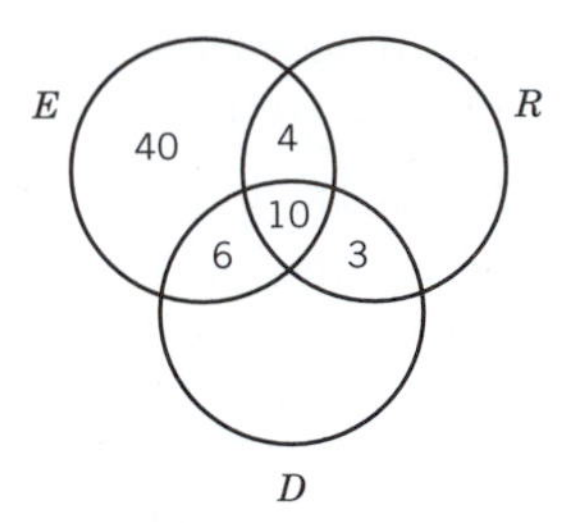

This completes the organization of the espresso circle. Next, we complete the dark roast and decaf circles.

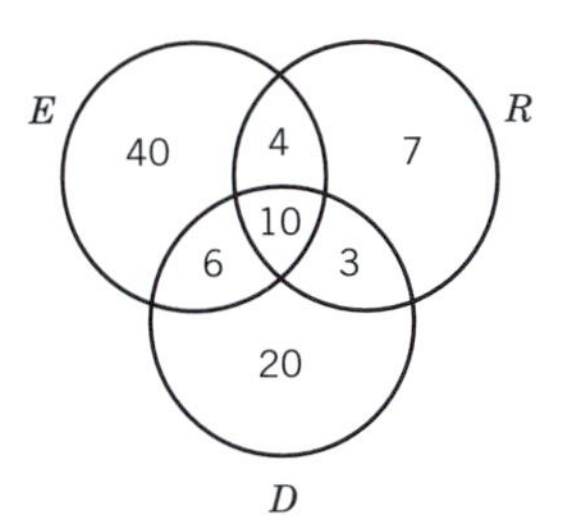

Finally, we know that there were 100 customers surveyed, but the total number of people listed in our picture is only 90. Thus, there must have been 10 customers who said they didn't like any of the coffees (perhaps they came in for the prune Danish?). We put that number outside of all the circles and enclose the whole diagram in a box.

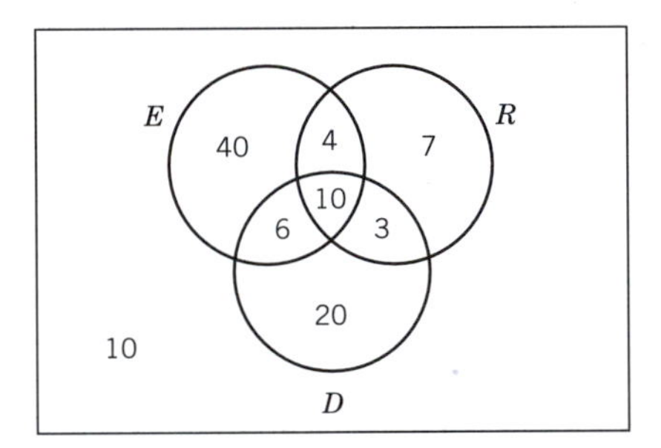

This set diagram has completely organized the results of the coffee drinker survey. From it we can answer a variety of questions about the set of people surveyed. For example, because there is a 40 in the part of the espresso circle that doesn't intersect with either of the other two circles, we know that 40 people liked espresso coffee but not either of the other two.

PRACTICE

Use the coffee drinker survey set diagram to fill in the missing entries in the following table:

Description of Set	Cardinal Number of Set
People who like only espresso.	40
8. People who like only decaf.	______

9. ______________________________ 7

10. People who like dark roast and espresso, but not decaf. _______

11. People who like decaf and espresso, but not dark roast. _______

12. ______________________________ 3

People who didn't like decaf. 61

13. People who didn't like dark roast. _______

Here are the results of another survey. We'll use these results to construct another set diagram.

A survey of 150 college students yielded the following results:

120 were taking an English course.
75 were taking a mathematics course.
90 were taking an anthropology course.
50 were taking English and mathematics.
40 were taking mathematics and anthropology.
70 were taking English and anthropology.
20 were taking all three courses.

Here is the set diagram for these data. Make sure that you see exactly why each of the numbers is correct!

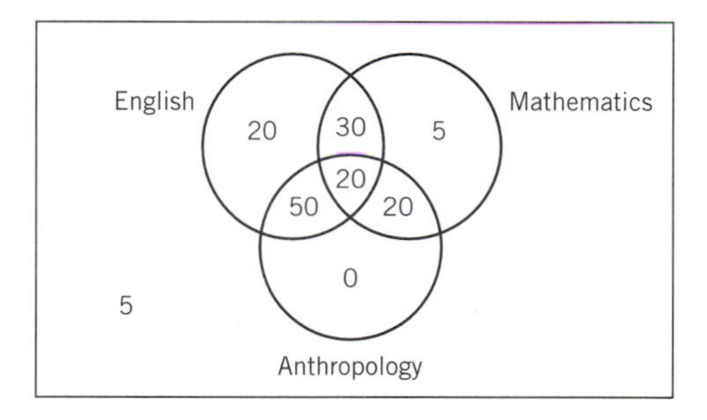

Explain where the 30 in this diagram came from.

____

____

____

Right! We know that 50 students were taking English and mathematics, but of those 50, 20 were also taking anthropology. Therefore, there were 30 who were taking English and mathematics, but not anthropology.

PRACTICE

14. Draw a set diagram for these survey results. A survey of 200 college students yielded the following results:

135 were taking an English course.
100 were taking a mathematics course.
95 were taking an anthropology course.
65 were taking English and mathematics.
45 were taking mathematics and anthropology.
75 were taking English and anthropology.
25 were taking all three courses.

Find the cardinal number of each of the described sets.

15. The set of students who were taking only mathematics.

16. The set of students who were not taking English.

Find each of the following, where
E = the set of students who were taking English
M = the set of students who were taking mathematics
A = the set of students who were taking anthropology

17. $n(E)$

18. $n(M \cap E)$

19. $n(M)$

20. $n(M \cup E)$

21. $n(M \cap (E \cup A))$

3.5 SET COMPLEMENT

In the last section, you found the cardinal numbers of sets described negatively, like "the set of students who weren't taking English," or "the set of people who didn't like decaf." In this section, we will explore the notion of negatively defined sets a little more.

When I asked you to find the number of people who didn't like decaf, you counted up the total number of people outside of the decaf circle in the set diagram. You assumed that I was referring only to the people actually surveyed. But without this assumption, the set of people who didn't like decaf goes beyond the 100 people surveyed by Zork's nephew. There may be a Yanomanö tribesman in the Amazon rain forest who doesn't like decaf. My cousin Bob in Michigan, who wasn't in the survey group, doesn't like decaf either. We didn't want to count these guys in the set of people who didn't like decaf.

It is necessary to specify precisely what we are talking about when we define a set negatively. A way to be specific is to describe a **universe of discourse** (also known as a **universe,** or **universal set**). The universe of discourse (usually denoted by the script letter $\mathcal{U}$) is the set of all (universe) the objects we wish to talk (discourse) about.

In Zork's nephew's survey, the universe (of discourse) was the set of 100 people surveyed. So, for the purposes of that discussion, my cousin Bob and the Yanomanö tribesman were not members of the universe. If we were describing sets of my relatives, then my cousin Bob would be in the universe, but the Yanomanö tribesman would not. Conversely, if we were talking about sets of people who live in the Amazon rain forest, the Yanomanö tribesman would be in the universe, but my cousin Bob would not.

Given a universe of discourse $\mathcal{U}$ and a set of objects, A, from that universe we may describe a set whose members are the objects in the universe that are not in A. This set is called the **complement** of A and denoted A'.

The symbolic definition of A' is

$$A' = \{x: x \notin A \quad \text{and} \quad x \in \mathcal{U}\}$$

where it is understood that A is a subset of $\mathcal{U}$.

The complement of A can be presented in a set diagram like this:

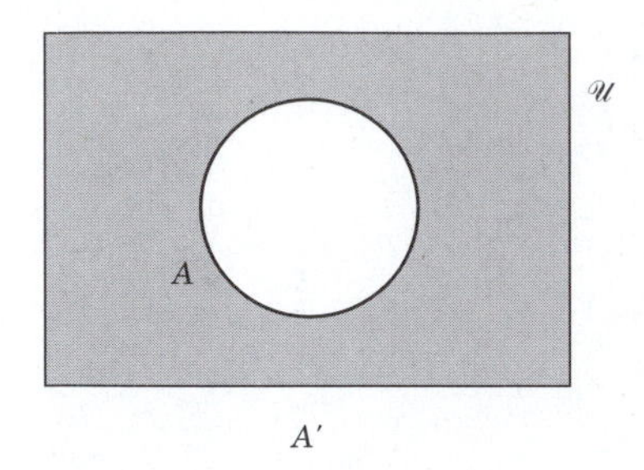

Notice the importance of the universe. Without a universe as a boundary, the picture of A' would spread all over the page, onto your desk, on the floor, out the door, . . .

PRACTICE

Here is another universe of discourse along with some of its subsets and their complements. Make sure that you see why each complement is, in fact, the set of all elements which are not in the set but are in the universe. Then find the other complements as asked. Suppose

$$\mathcal{U} = \{1, 2, 3, 4, 5, 6, 7\}$$
$$A = \{2, 3, 4\}, \qquad A' = \{1, 5, 6, 7\}$$
$$B = \{1, 3, 5, 7\}, \qquad B' = \{2, 4, 6\}$$
$$C = \{6\}, \qquad C' = \{1, 2, 3, 4, 5, 7\}$$
$$D = \varnothing, \qquad D' = \mathcal{U}$$

1. $E = \{2, 3, 4, 5\}, \qquad E' = ?$

2. $F = \mathcal{U}, \qquad F' = ?$

3. $G = \{3, 4, 5, 6, 7\}, \qquad G' = ?$

4. $H = A \cap B, \qquad H' = ?$

5. $K = A \cup C, \qquad K' = ?$

3.6 SET EXPRESSIONS AND FORMULAS

You have now seen all the basic set operations: union, intersection, and complement. In this section, we will use these operations to describe and count the elements of more complicated sets like $(A \cup B')'$.

The most important thing to remember when working with complicated sets is to work inside the innermost group of parentheses first and then gradually work your way out. The examples below should make this process clear to you. They all refer to the following situation:

$$\mathcal{U} = \{a, b, c, d, e, f, g, h, i\}$$
$$Q = \{e, i\}$$
$$S = \{a, b, c, d, e\}$$

First, we will look at the set $(Q \cap S)'$. The complement symbol is outside the right parenthesis, so we will calculate $Q \cap S$ first and then take its complement. Thus,

$$Q \cap S = \{e\}$$
$$(Q \cap S)' = \{e\}'$$
$$= \{a, b, c, d, f, g, h, i\}$$

Now consider $((Q \cap S) \cup Q')'$. Working inside the innermost group of parentheses, we have

$$Q \cap S = \{e\}$$

This set is to be *unioned* with Q'. So, before we can do that, we need to determine Q'.

$$Q' = \{a, b, c, d, f, g, h\}$$

Next, we take the union of $Q \cap S$ with Q'.

$$(Q \cap S) \cup Q' = \{a, b, c, d, e, f, g, h\}$$

Finally, we take the complement of $(Q \cap S) \cup Q'$.

$$((Q \cap S) \cup Q')' = \{i\}$$

As a last example, we will find $(Q \cup (S \cap Q'))'$. Here are the necessary steps.

1. $Q' = \{a, b, c, d, f, g, h\}$
2. $S \cap Q' = \{a, b, c, d\}$
3. $Q \cup (S \cap Q') = \{a, b, c, d, e, i\}$
4. $(Q \cup (S \cap Q'))' = \{f, g, h\}$

PRACTICE

Find the elements of each of the following sets when

$$\mathcal{U} = \{a, b, c, d, e, f, g, h, i\}$$
$$Q = \{e, i\}$$
$$S = \{a, b, c, d, e\}$$

1. $S' \cap Q$

2. $(S' \cap Q)' \cup Q$

3. $(Q \cap (Q' \cap S)')'$

4. $(S' \cap S)'$

When it comes to calculating the cardinal number of a complicated set, two formulas are very useful. They are

1. $n(A') = n(\mathcal{U}) - n(A)$
2. $n(A \cup B) = n(A) + n(B) - n(A \cap B)$

Do these formulas make sense to you?

The first formula, $n(A') = n(\mathcal{U}) - n(A)$, reflects the fact that the complement of a set is the set of everything that isn't in the set, but is still in the universe. Thus, if you count up the number of things in a set A and then subtract that from the total number of things in the universe, the result is the number of things in A'. Check out this formula with the sets

$$\mathcal{U} = \{1, 2, 3, 4, 5, 6\}$$
$$A = \{2, 3\}$$

[You should have gotten $n(A') = n(\mathcal{U}) - n(A) = 6 - 2 = 4$, and $A' = \{1, 4, 5, 6\}$.]

The second formula may not be quite so obvious as the first. Let's take a look at a small example to convince ourselves that the formula is correct. Let

$$\mathcal{U} = \{a, b, c, d, e, f, g, h\}$$
$$A = \{a, b, c\}$$
$$B = \{a, d, e, f, g\}$$

Now,

$$A \cup B = \{a, b, c, d, e, f, g\}$$
$$n(A \cup B) = 7$$

and

$$A \cap B = \{a\}$$
$$n(A \cap B) = 1$$

Furthermore, we have $n(A) = 3$ and $n(B) = 5$. The formula says that

$$n(A \cup B) = n(A) + n(B) - n(A \cap B)$$

or, in this case,

$$7 = 3 + 5 - 1$$

which is true.

Why does the formula require us to subtract $n(A \cap B)$?

Right! Because the element a in $A \cup B$ is counted twice: once as an element of A and once as an element of B. By subtracting $n(A \cap B)$, we compensate for the double counting.

PRACTICE

Use the formulas to find the cardinal numbers as indicated when

$$\mathcal{U} = \{q, r, s, t, u, v, w, x, y, z\}$$
$$A = \{q, r, s, t\}$$
$$B = \{s, t, u, v\}$$
$$C = \{w, x, y\}$$

5. $n(A')$

6. $n(A \cup B)$

7. $n(A \cup C)$

8. $n(C')$

Our formulas will also help us calculate the cardinal numbers of complicated sets like $(A \cup B')'$. First, notice that $(A \cup B')'$ is a complement set. So,

$$n((A \cup B')') = n(\mathcal{U}) - n(A \cup B')$$

But, $A \cup B'$ is a union of two sets, so

$$\begin{aligned} n((A \cup B')') &= n(\mathcal{U}) - n(A \cup B') \\ &= n(\mathcal{U}) - (n(A) + n(B') - n(A \cap B')) \\ &= n(\mathcal{U}) - n(A) - n(B') + n(A \cap B') \end{aligned}$$

And now we can replace $n(B')$ by $n(\mathcal{U}) - n(B)$. Therefore,

$$
\begin{aligned}
n((A \cup B')') &= n(\mathcal{U}) - n(A) - (n(\mathcal{U}) - n(B)) + n(A \cap B') \\
&= n(\mathcal{U}) - n(A) - n(\mathcal{U}) + n(B) + n(A \cap B') \\
&= -n(A) + n(B) + n(A \cap B')
\end{aligned}
$$

Read the last discussion again *very* carefully. Make sure that you understand each step.

PRACTICE

Suppose that $n(\mathcal{U}) = 20$, $n(A) = 12$, $n(B) = 7$, $n(A \cap B) = 2$, and $n(A \cap B') = 10$. Find each of the following.

9. $n(A')$

10. $n(A \cup B)$

11. $n(A \cup B')$

12. $n((A \cap B)')$

There is a nice relationship between the symbols in an expression about sets and the English description of the set. Here it is:

English	Set Symbol
And	$\cap$
Or	$\cup$
Not	$'$

For example, suppose that

$\mathcal{U}$ = the set of all people who work for Gulf Oil
M = the set of men who work for Gulf Oil
C = the set of computer programmers who work for Gulf Oil

Here are some English sentences and their set symbol equivalents. Each is followed by an explanation.

The set of women who work for Gulf Oil = M'

The people who work for Gulf Oil are either women or men. If a person who works for Gulf Oil is not a man, then she is a woman.

The set of male computer programmers who work for Gulf Oil = $M \cap C$

A male computer programmer is a male *and* is also a computer programmer.

People who work for Gulf Oil who are either men or computer programmers or both = $M \cup C$

Set union includes those people who are both men and computer programmers. This is called the **inclusive-or.** Often, in English, we use *or* to mean one or the other, but not both, as in "I'll either sleep in or go to class." This use of or is called the **exclusive-or.**

PRACTICE

Let

$\mathcal{U}$ = the set of students at your college
B = the set of students at your college who are taking a biology course
M = the set of students at your college who are taking a mathematics course

Write each of the following English sentences using set symbols:

13. The set of students at your college who are not taking a biology course.

14. The set of students at your college who are taking a mathematics course and biology course.

15. The set of students at your college who are taking a mathematics course, a biology course, or both.

16. The set of students at your college who are taking a mathematics course but not a biology course.

17. The set of students at your college who are taking neither a mathematics course nor biology course.

18. The set of students at your college who are not taking both a mathematics course and biology course.

19. The set of students at your college who are taking a mathematics course or biology course but not both.

Write each of these in English.

20. M'

21. $M' \cup B$

22. $M \cap B'$

23. $(M \cap B')'$

Situation 9

Dancing on the Ceiling

You have been persuaded by your Aunt Louise, a seventh grade teacher, to serve as a chaperon at the Alan Turing Junior High School Harvest Hop. There are 25 girls and 18 boys attending the Hop. To combat boredom, you begin to reflect on the various possible boy–girl dance pairings that can be made between the boys and girls in attendance.

In your revery, you notice too late that Mr. Mendez, the math teacher, has come over to talk to you.

> *''The function cannot be one-to-one,'' Mr. Mendez says.*
> *''Oh?,'' you reply.*
> *''Indeed,'' Mr. Mendez continues, ''so long as the girls comprise the domain, we can never achieve a one-to-one map. Onto maybe, but not one-to-one.''*

Mr. Mendez looks at you expecting an erudite and mathematically sophisticated response. What do you say?

____

____

____

Functions are special kinds of sets that are important to almost every area of modern mathematics. But before we can talk functions to Mr. Mendez, we need to take a look at a special set operation called the Cartesian product.

3.7 CARTESIAN PRODUCT

The **Cartesian product,** symbolized by $\times$, is a way of making a set of pairs of elements from two sets. Here is a formal definition.

Let A and B be two sets, neither of which is the empty set. Then the Cartesian product (or **cross product**) of A with B is $A \times B = \{(a, b): a \in A \text{ and } b \in B\}$.

The elements of $A \times B$ are *ordered pairs* of elements. To get an idea of what an ordered pair of elements is, consider the sets of boys and girls at the Junior High Hop in Situation 9. Suppose that G is the set of girls at the Hop and B the set of boys at the Hop. Then

$$G \times B = \{(g, b): g \in G \quad \text{and} \quad b \in B\}$$

If Maria and Shanta are girls at the Hop and Floyd is a boy at the Hop, then some of the elements of $G \times B$ are (Maria, Floyd) and (Shanta, Floyd). These two pairs may be viewed as dancing couples. So, $G \times B$ could be a set of possible dancing couples at the Hop.

The pairs in $G \times B$ are ordered in the sense that a girl is always listed first. This means that $B \times G$ is a different set from $G \times B$. The pair (Floyd, Maria) is an element of $B \times G$, but is not an element of $G \times B$, because in $B \times G$ the boy is listed first and in $G \times B$ the boy is listed second.

Does the ordering really make a difference? Well, mathematically it always does, and in life it may. We could use $G \times B$ to represent the set of dancing pairs made when the girl chooses her partner, and then $B \times G$ would represent the set of dancing pairs when the boy does the choosing.

The pairs (Maria, Floyd) and (Shanta, Floyd) are only two of many pairs that can be made from the sets of girls and boys at the Hop. Here are some complete Cartesian products using sets with smaller cardinal numbers.

$$A = \{a, b, c\}, \qquad B = \{1, 2\}$$

$$A \times B = \{(a, 1), (a, 2), (b, 1), (b, 2), (c, 1), (c, 2)\}$$

$$B \times A = \{(1, a), (1, b), (1, c), (2, a), (2, b), (2, c)\}$$

$$C = \{\bigcirc, \bullet\}, \qquad D = \{1, 2, 3, 4, 5\}$$

$$C \times D = \{(\bigcirc, 1), (\bigcirc, 2), (\bigcirc, 3), (\bigcirc, 4), (\bigcirc, 5), (\bullet, 1), (\bullet, 2), (\bullet, 3), (\bullet, 4), (\bullet, 5)\}$$

$$D \times C = \{(1, \bigcirc), (2, \bigcirc), (3, \bigcirc), (4, \bigcirc), (5, \bigcirc), (1, \bullet), (2, \bullet), (3, \bullet), (4, \bullet), (5, \bullet)\}$$

Notice that we make *every* possible pair with elements of the left set in the Cartesian product first and the elements of the right set in the product second.

PRACTICE

Form each of the indicated Cartesian products.

1. $A = \{p, q\}$, $\quad B = \{3, 4\}$. Find $B \times A$.

2. $C = \{0\}$, $\quad D = \{0, 1\}$. Find $D \times C$.

3. $E = \{x$: x is a state that begins with the letter ''I''$\}$
$F = \{$Boston, Chicago$\}$
Find $F \times E$.

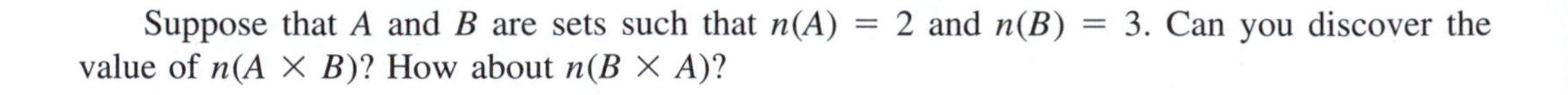

Suppose that A and B are sets such that $n(A) = 2$ and $n(B) = 3$. Can you discover the value of $n(A \times B)$? How about $n(B \times A)$?

Did you find out that $n(A \times B) = 6$ and $n(B \times A) = 6$? If not, take the time now to examine $n(A \times B)$ and $n(B \times A)$ when $A = \{a, b\}$ and $B = \{1, 2, 3\}$. Make sure you can see that $n(A \times B) = 6$ and $n(B \times A) = 6$.

In general, the cardinal number of a Cartesian product is the product of the cardinal numbers of the two sets in the product. In symbols, $n(A \times B) = n(A) \cdot n(B)$. This is true because in order to form $A \times B$, we must match each element in A with each of the $n(B)$ elements of B. There are $n(A)$ elements in A, so there are $n(A) \cdot n(B)$ matches. (See Chapter 4 for more on counting the elements in collections.)

PRACTICE

Suppose that A and B are subsets of a universe $\mathcal{U}$ such that $n(\mathcal{U}) = 50$, $n(A) = 16$, $n(B) = 24$, and $n(A \cap B) = 8$. Find each of the following:

4. $n(A \cup B)$

5. $n(A \times B)$

6. $n(B \times A)$

7. $n(A')$

8. $n(A' \times B)$

9. Let G = the set of girls at the Junior High Hop in Situation 9, and B = the set of boys at the Junior High Hop in Situation 9. Find $n(G \times B)$. Describe this set in English.

3.8 RELATIONS AND FUNCTIONS

A **relation** is any set of ordered pairs. A **relation from the set A to the set B** is any subset of $A \times B$. In this book, relations will be symbolized by an $\mathcal{R}$. The $\mathcal{R}$ may have a subscript to distinguish one relation from another. Here are some relations from the set $A = \{q, r, s\}$ to the set $B = \{4, 5, 6, 7\}$.

$$\mathcal{R}_1 = \{(q, 4), (r, 4), (s, 4), (s, 6), (q, 5)\}$$

$$\mathcal{R}_2 = \{(q, 4), (r, 4), (s, 4)\}$$

$$\mathcal{R}_3 = \{(q, 4), (r, 5), (s, 6)\}$$

$$\mathcal{R}_4 = \{(q, 4), (r, 5), (s, 6), (q, 7)\}$$

$$\mathcal{R}_5 = \{(r, 5), (s, 6)\}$$

A **function F from a set A to a set B** is a relation from A to B that has the following two properties:

1. Every element of A appears as the first element of an ordered pair of F.
2. Each first element of a pair in F can be paired with only one second element. For example, if (a, d) is a pair in some function, then neither (a, f), (a, j), nor (a, anything other than d) can be a pair in the function.

Which of the relations just shown are functions from $\{q, r, s\}$ to $\{4, 5, 6, 7\}$?

Only $\mathcal{R}_2$ and $\mathcal{R}_3$ are functions. $\mathcal{R}_1$ is not a function because s is paired with both 4 and 6, q is paired with both 4 and 5. $\mathcal{R}_4$ is not a function because q is paired with both 4 and 7. $\mathcal{R}_5$ is not a function because the element q does not appear as the first element of any ordered pair in $\mathcal{R}_5$.

Notice that it is OK for a *second* element to be paired with different first elements. Thus, $\mathcal{R}_2$ is a function even though all the second elements are the same.

If F is a function from A to B, then A is called the **domain** of F and B the **codomain** of F. The **range** of F is the set of elements from B that actually appear as second elements in the ordered pairs of F.

For the function $\mathcal{R}_2$, the domain is $\{q, r, s\}$, the codomain is $\{4, 5, 6, 7\}$, and the range is $\{4\}$. For the function $\mathcal{R}_3$, the domain is $\{q, r, s\}$, the codomain is $\{4, 5, 6, 7\}$, and the range is $\{4, 5, 6\}$.

PRACTICE

For each of the following relations from the set A to the set B, decide whether or not $\mathcal{R}$ is a function. If $\mathcal{R}$ is a function, find its domain, codomain, and range.

1. $A = \{1, 2, 3\}, \quad B = \{4, 5, 6\}$
   $\mathcal{R} = \{(1, 4), (1, 5), (2, 4), (3, 6)\}$

2. $A = \{1, 2, 3\}, \quad B = \{4, 5, 6\}$
   $\mathcal{R} = \{(1, 4), (2, 4), (3, 6)\}$

3. $A = \{1, 2, 3, 4, 5, 6, 7\}, \quad B = \{a, b, c\}$
   $\mathcal{R} = \{(1, a), (2, a), (3, a), (4, a), (5, a), (6, a), (7, a)\}$

4. $A = \{1, 2, 3, 4, 5, 6, 7\}, \quad B = \{a, b, c\}$
   $\mathcal{R} = \{(1, a), (2, a), (3, a), (3, b), (4, a), (5, a), (6, a), (7, a)\}$

5. $A = \{\text{Maria}\}, \quad B = \{x: x \text{ is a boy at the Hop}\}$
   $\mathcal{R} = A \times B$

A function F is **onto** if the codomain of F and range of F are the same sets. A function F is **one-to-one** if no two ordered pairs in F with different first elements have the same second element. Here are some example functions from A to B to illustrate the concepts of one-to-one and onto.

$$A = \{a, b, c\}, \qquad B = \{1, 2\}$$
$$F_1 = \{(a, 1), (b, 1), (c, 2)\}.$$

F_1 is onto because its codomain and range are the same. F_1 is not one-to-one because both $(a, 1)$ and $(b, 1)$ belong to F_1.

$$A = \{a, b, c\}, \qquad B = \{1, 2, 3, 4\}$$
$$F_2 = \{(a, 1), (b, 2), (c, 3)\}.$$

F_2 is not onto because its codomain, $\{1, 2, 3, 4\}$, and range, $\{1, 2, 3\}$, are not the same. F_2 is one-to-one because no pairs with different first elements and the same second elements belong to F_2.

$$A = \{a, b, c\}, \qquad B = \{1, 2, 3\}$$
$$F_3 = \{(a, 1), (b, 2), (c, 3)\}.$$

F_3 is onto because its codomain and range are the same. F_3 is one-to-one because no pairs with different first elements and the same second elements belong to F_3.

$$A = \{a, b, c\}, \qquad B, = \{1, 2, 3\}$$
$$F_4 = \{(a, 1), (b, 2), (c, 2)\}.$$

F_4 is not onto because its codomain, {1, 2, 3}, and range, {1, 2}, are not the same. F_4 is not one-to-one because both b and c are matched with 2 in F_4.

PRACTICE

For each of the following functions, decide whether the function from A to B is onto, one-to-one, both onto and one-to-one, or neither onto nor one-to-one.

6. $A = \{q, r, s\}, \quad B = \{3, 4, 5, 6\}$
$F = \{(q, 3), (r, 6), (s, 4)\}$

7. $A = \{q, r, s\}, \quad B = \{3, 4, 5\}$
$F = \{(q, 3), (r, 5), (s, 4)\}$

8. $A = \{q, r, s\}, \quad B = \{3, 4\}$
$F = \{(q, 3), (r, 3), (s, 4)\}$

9. $A = \{q, r, s\}, \quad B = \{3, 4, 5\}$
$F = \{(q, 3), (r, 5), (s, 5)\}$

10. $A = \{q, r, s\}, \quad B = \{3, 4, 5, 6, 7, 8\}$
$F = \{(q, 5), (r, 6), (s, 8)\}$

We can now respond to Mr. Mendez's remarks at the Hop. Recall that he said, ''Indeed, so long as the girls comprise the domain we can never achieve a one-to-one map. Onto maybe, but not one-to-one.''

Mr. Mendez saw the dancing partners as ordered pairs in a function with the set of girls comprising the domain (indicating that the girls chose their partners). Since there were more girls than boys, some girls would have to dance with the same boy as another girl. Thus, the set of dancing pairs would include pairs with different first elements but the same second elements. Hence, the dancing pair function would not be one-to-one. However, because there were at least 18 girls, every boy could be chosen. Thus, the dancing pair function could be onto.

Explain under what circumstances the dancing pair function conceived by Mr. Mendez would not be onto.

____

____

____

3.9 WHAT DO YOU KNOW?

If you have worked carefully through this chapter, then you can perform basic set operations, find the cardinal number of a set, determine whether or not a relation is a function, and determine if a function is onto or one-to-one. The exercises in this section are designed to test and refine your skills in set theory, the mathematics of collections.

Which of the following collections are well defined?

1. The collection of Rome Beauty apples on your desk right now
2. The collection of states that begin with the letter "I"
3. The collection of happy people in Toledo, Ohio on March 4, 1994
4. The collection of fractions that are equivalent to 1/2
5. The collection of students who have passed a course in finance
6. The collection of students who are overweight
7. The collection of mallard ducks who were in Louisiana yesterday
8. The collection of houses that sold for more than $100,000 in 1993

List the members of each of the following sets:

9. $\{x: x$ is the capital of Chile$\}$
10. $\{x: x$ is a dog owned by you or any member of your family$\}$
11. $\{x: x$ is a whole number between 4 and 12$\}$
12. $\{x: x$ is a President of the United States whose last name begins with the letter "C"$\}$
13. $\{x: x$ is a name of a team in the National Football League$\}$
14. $\{x: x$ is the capital of Thailand$\}$

Describe each of the following sets using set-builder notation:

15. {Rabbit, Eeyore, Piglet, Tigger, Winnie-the-Pooh}
16. {Canada, United States, Mexico}
17. {San Diego, San Francisco, Los Angeles, Sacramento, Mill Valley, . . .}
18. $\{2, 4, 6, 8, 10\}$
19. {Australia}
20. {Coffee, Coca-Cola, Pepsi-Cola, Mountain Dew, . . .}

True or false?

21. $4 \in \{x: x$ is a number greater than 3$\}$
22. $\{4\} \subseteq \{x: x$ is a number greater than 3$\}$
23. $4 \subseteq \{x: x$ is a number greater than 3$\}$
24. $\{4\} \in \{x: x$ is a number greater than 3$\}$
25. $3 \in \{1, 2, 4, 5, 6\}$
26. $4 \in \{1, 2, 4, 5, 6\}$
27. $\{4\} \subseteq \{1, 2, 3, 4, 5, 6\}$
28. $\{3\} \subseteq \{1, 2, 4, 5, 6\}$

Suppose that $\mathcal{U} = \{1, 2, 3, 4, 5, 6, 7\}$, $A = \{1, 3, 5, 7\}$, $B = \{2, 4, 6\}$, and $C = \{3, 5, 6\}$. Answer true or false.

29. $2 \in A \cup B$
30. $C \subseteq A$
31. $5 \in A \cap C$
32. $\{2, 5\} \subseteq A \cup C'$

Suppose that $\mathcal{U} = \{a, b, c, d, e, f, g\}$, $A = \{a, e\}$, $B = \{a, b, d\}$, and $C = \{b, c, f, g\}$. Answer true or false.

33. $e \in A \cup B$
34. $C \subseteq A$
35. $B \cap C \subseteq A \cap C$
36. $\{a, g\} \subseteq A \cup C'$

Draw a set diagram for each of the following:

37. $W \cap K$
38. $W \cup K$
39. W'
40. $W' \cap K$
41. $W \cap K'$
42. $W' \cup K$
43. $W \cup K'$
44. $(W' \cap K)'$
45. (a) Under what conditions will a set H not be a subset of a set G?

    (b) Use your answer to part (a) to explain why the empty set is a subset of every set.

46. List all the subsets of the set $\{q, r, s\}$.

47. List all the subsets of the set $\{s, a, d, i\}$.

48. Let A be the set $\{a, b, \{a, c\}, c\}$. Find a subset of A that is also an element of A.

49. (a) Draw a set diagram for $(A \cap B)'$.
(b) Draw a set diagram for $A' \cup B'$.
(c) Compare the diagrams you drew in parts (a) and (b).

50. (a) Draw a set diagram for $(A \cup B)'$.
(b) Draw a set diagram for $A' \cap B'$.
(c) Compare the diagrams you drew in parts (a) and (b).

If $\mathscr{U} = \{a, b, c, d, e, f, g, h, i\}$, $A = \{a, e, i\}$, and $B = \{h, i, g, f\}$, find each of the following cardinal numbers:

51. $n(A)$

52. $n(B')$

53. $n(B \cap A)$

54. $n(B' \cup A)$

55. $n(A \cap (B \cup A'))$

If $\mathscr{U} = \{a, b, c, d, e, f, g, h, i\}$, $A = \{d, e, f\}$, and $B = \{h, e, i, g, f\}$, find each of the following cardinal numbers:

56. $n(A)$

57. $n(B')$

58. $n(B \cap A')$

59. $n(B' \cup A)$

60. $n(A \cap (B \cup A'))$

61. Fill in the appropriate numbers in each region of the set diagram if $n(\mathscr{U}) = 50$, $n(A) = 25$, $n(B) = 19$, $n(A \cap B) = 8$.

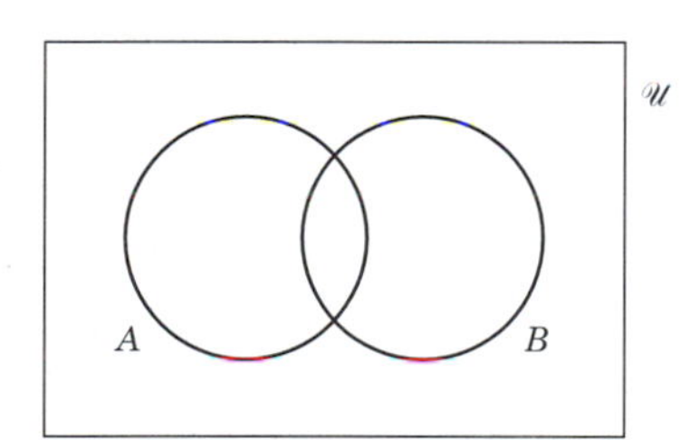

62. Fill in the appropriate numbers in each region of the set diagram if $n(\mathscr{U}) = 50$, $n(A) = 25$, $n(B') = 27$, $n(A \cap B) = 8$.

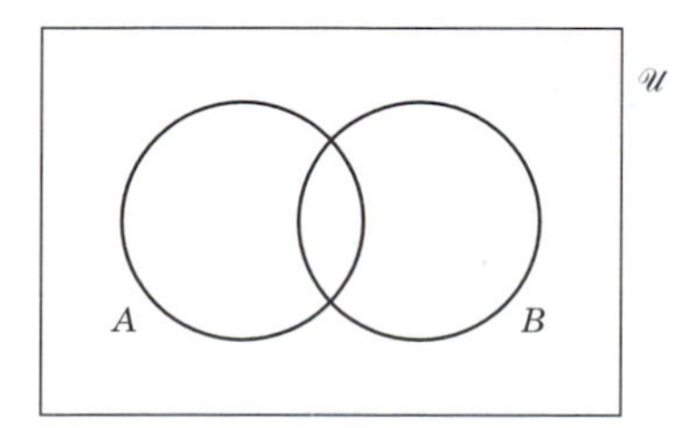

63. Fill in the appropriate numbers in each region of the set diagram if $n(\mathscr{U}) = 50$, $n(A) = 25$, $n(B) = 19$, $n(A \cup B) = 30$.

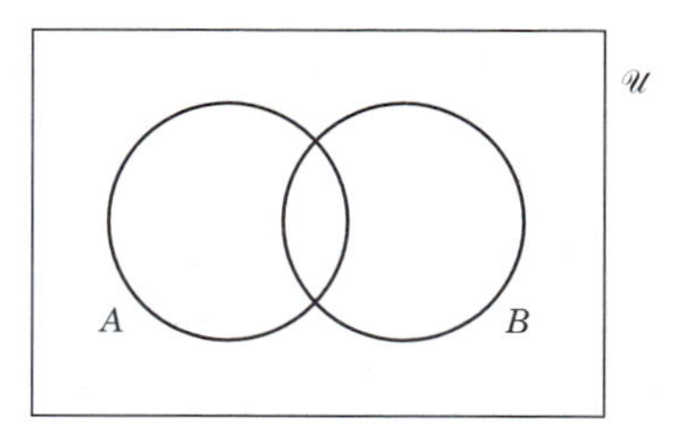

64. Fill in the appropriate numbers in each region of the set diagram if $n(\mathscr{U}) = 50$, $n(A') = 25$, $n(B) = 27$, $n(A \cup B) = 42$.

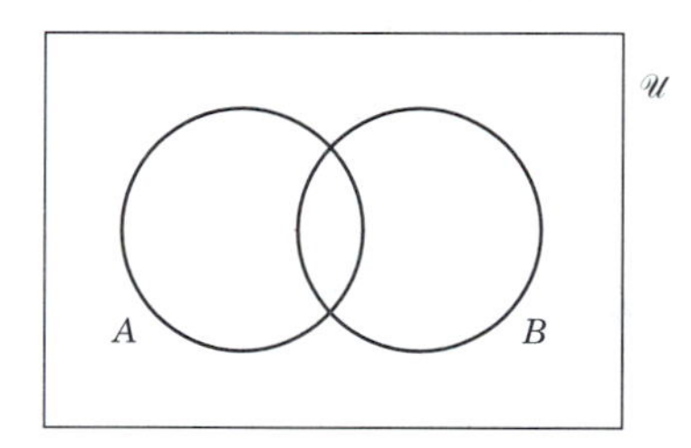

In a survey of 300 college students, it was found that

100 liked to sleep late on Saturday.
40 liked anchovy pizza.
80 liked spring break.
30 liked spring break and to sleep late on Saturday.
20 liked anchovy pizza and spring break.
10 liked anchovy pizza and to sleep late on Saturday.
2 liked Spring Break, to sleep late on Saturday, and anchovy pizza.

65. How many of the students surveyed liked anchovy pizza but not spring break?

66. How many of the students surveyed liked to sleep late on Saturday and anchovy pizza, but didn't like spring break?

67. How many of the students surveyed didn't like to sleep late on Saturday, anchovy pizza, or spring break?

A marketing specialist asks shoppers at a mall about three different names for a new candy bar. She discovers that

60 like the name "Chocolate Chocolate."
70 like the name "Gourmet Gonzo."
30 like the name "LiteFree."
10 like both "Chocolate Chocolate" and "Gourmet Gonzo."
8 like both "LiteFree" and "Gourmet Gonzo."

4 like both ''Chocolate Chocolate'' and ''LiteFree.''

1 person liked all three names.

90 people didn't like any of the names.

68. How many people were asked about the names?

69. How many people liked only one name?

70. How many people didn't like the name ''LiteFree''?

71. How many people liked only the name ''Gourmet Gonzo''?

72. How many people liked exactly two of the names?

A mathematics professor has been keeping track of the excuses students give for missing her class. She discovers that

65 said that their dog had the flu.

70 said that their car wouldn't start.

30 said that they needed to bake cookies.

12 said that their dog had the flu and that they needed to bake cookies.

8 said that their car wouldn't start and that they needed to bake cookies.

5 said that their dog had the flu and that their car wouldn't start.

1 person used all three excuses.

95 people didn't use any of the excuses.

73. How many people were in the professor's records?

74. How many people had only one excuse?

75. How many people didn't say that their car wouldn't start

76. How many people said only that they had to bake cookies?

77. How many people had exactly two excuses?

In a survey of 300 college students, it was found that

100 liked to sleep late on Saturday.

40 liked anchovy pizza.

80 liked spring break.

30 liked to sleep late on Saturday and spring break.

20 liked anchovy pizza and spring break.

10 liked anchovy pizza and to sleep late on Saturday.

2 liked to sleep late on Saturday, spring break, and anchovy pizza.

Let $\mathcal{U}$ be the set of students surveyed, $S = \{x: x$ liked to sleep late on Saturday$\}$, $A = \{x: x$ liked anchovy pizza$\}$, and $B = \{x: x$ liked spring break$\}$. Find

78. $n(A)$

79. $n(A \cup S)$

80. $n(S \cap B)$

81. $n(S')$

82. $n(S \cup B')$

A marketing specialist asks shoppers at a mall about three different names for a new candy bar. She discovers that

65 like the name ''Chocolate Chocolate.''

70 like the name ''Gourmet Gonzo.''

30 like the name ''LiteFree.''

12 like both ''Chocolate Chocolate'' and ''Gourmet Gonzo.''

8 like both ''LiteFree'' and ''Gourmet Gonzo.''

5 like both ''Chocolate Chocolate'' and ''LiteFree.''

1 person liked all three names.

95 people didn't like any of the names.

Let $\mathcal{U}$ be the set of people asked about candy bar names, $C = \{x: x$ liked the name ''Chocolate Chocolate''$\}$, $G = \{x: x$ liked the name ''Gourmet Gonzo''$\}$, and $L = \{x: x$ liked the name ''LiteFree''$\}$. Find

83. $n(C)$

84. $n(G \cap L)$

85. $n(L')$

86. $n(G' \cup L)$

87. $n(L' \cap C')$

Suppose $\mathcal{U} = \{x: x$ is a whole number between 0 and 25$\}$, $A = \{x: x$ is an even number between 0 and 25$\}$, $B = \{3, 6, 9\}$, and $C = \{1, 23\}$. Find each of the following:

88. $A \cap B$

89. A'

90. $C \times B$

91. $n(A \cup C)$

92. $n(A \times C)$

93. $n(A \cap B')$

Suppose $\mathcal{U} = \{x: x$ is a whole number between 0 and 25$\}$, $A = \{x: x$ is an odd number between 0 and 25$\}$, $B = \{3, 6, 9\}$, and $C = \{1, 23\}$. Find each of the following:

94. $A \cap B$

95. A'

96. $C \times B$

97. $n(A \cup C)$

98. $n(A \times C)$

99. $n(A \cap B')$

For each of the following relations from the set A to the set B, decide whether or not $\mathcal{R}$ is a function. If $\mathcal{R}$ is a function, find its domain, codomain, and range.

100. $A = \{1, 2, 3\}$, $B = \{4, 5, 6\}$
$\mathcal{R} = \{(1, 4), (3, 5), (2, 4), (3, 6)\}$

101. $A = \{1, 2, 3\}$, $B = \{4, 5, 6\}$
$\mathcal{R} = \{(1, 4), (2, 5), (3, 6)\}$

102. $A = \{1, 2, 3, 4, 5, 6, 7\}$, $B = \{a, b, c\}$
$\mathcal{R} = \{(1, a), (2, a), (3, b), (4, a), (5, a), (6, a), (7, a)\}$

103. $A = \{1, 2, 3, 4, 5, 6, 7\}$, $B = \{a, b, c\}$
$\mathcal{R} = \{(1, a), (2, a), (3, a), (3, b), (4, a), (5, a), (6, a), (7, a)\}$

104. $A = \{x: x \text{ is an odd positive number}\}$, $B = \{x: x \text{ is an even positive number}\}$

$$\mathcal{R} = A \times \{2\}$$

105. $A = \{x: x \text{ is an odd positive number}\}$, $B = \{x: x \text{ is an even positive number}\}$

$$\mathcal{R} = \{(1, 2), (3, 4), (5, 6), (7, 8), \ldots\}$$

For each of the following functions F from A to B, decide if F is onto, one-to-one, both, or neither:

106. $A = \{q, r, s\}$, $B = \{s, t, u\}$
$F = \{(q, s), (r, s), (s, t)\}$

107. $A = \{2, 3, 4\}$, $B = \{2, 3, 4, 5, 6\}$
$F = \{(2, 3), (3, 5), (4, 6)\}$

108. $A = \{(a, b, c\}$, $B = \{x, y\}$
$F = \{(a, x), (b, y), (c, x)\}$

109. $A = \{a, b, c\}$, $B = \{x, y, z\}$
$F = \{(a, x), (b, z), (c, y)\}$

110. $A = \{x: x \text{ is a positive odd number}\}$, $B = \{x: x \text{ is a positive even number}\}$

$$F = \{(a, 2): a \in A\}$$

111. $A = \{x: x \text{ is a positive odd number}\}$, $B = \{x: x \text{ is a positive even number}\}$

$$F = \{(a, a + 1): a \in A\}$$

Decide whether or not each relation is a function. If the relation is a function, find its domain and range and determine if it is onto or one-to-one. The universe is the set of people in the world.

112. $F = \{(x, y): x \text{ is the father of } y\}$

113. $F = \{(x, y): y \text{ is the father of } x\}$

114. $F = \{(x, y): x \text{ is the wife of } y\}$

115. $F = \{(x, y): y \text{ is the wife of } x\}$

116. $F = \{(x, y): y \text{ is the dentist of } x\}$

ANSWERS TO PRACTICE EXERCISES

Section 3.1

1. {Lima}
2. {Boise, Springfield, Des Moines, Indianapolis}
3. {4, 5, 6, 7, 8, 9, 10, 11, 12, 13, 14, 15, 16, 17, 18, 19, 20, 21, 22, 23, 24}
4. $\{x: x \text{ is an even whole number between 1 and 11}\}$
5. $\{x: x \text{ is an island nation in the North Atlantic}\}$
6. $\{x: x$ is the directory assistance number for Colorado, Wyoming, Chicago, or Detroit$\}$
7. True
8. True
9. True

Section 3.2

1. True
2. False. c is not a member of $\{f, a, v, g, b, n\}$.
3. True
4. False. A robin is not a duck.
5. True
6. True
7. False. The elements of {1, 2, 3, 4} are 1, 2, 3, and 4.
8. False. 4 is not an element of {{1}, {3}, {4}}.
9. True

Section 3.3

1. 5
2. 4
3. 4
4. $\{a\}, \{b\}, \{c\}, \{d\}, \{a, b\}, \{a, c\}, \{a, d\}, \{b, c\}, \{b, d\}, \{c, d\}, \{a, b, c\}, \{a, b, d\}, \{a, c, d\}, \{b, c, d\}, \{a, b, c, d\}, \varnothing$

Section 3.4

1. $A \cap B = \{a\}$, $A \cup B = \{a, b, c, d, f, g\}$
2. $A \cap B = \{x: x \text{ is a bear and a cartoon character}\}$ = {Yogi Bear, Winnie the Pooh, . . .}
$A \cup B = \{x: x$ is a bear or x is a cartoon character or both$\}$
= {Gentle Ben, Smokey Bear, Yogi Bear, Winnie the Pooh, Donald Duck, . . .}
3. $A \cap B = \varnothing$, $A \cup B = \{2, 3, 4\}$
4. $A \cap B = \{2, 4\}$, $A \cup B = \{1, 2, 3, 4, 5, 6, 8\}$
5. $n(A) = 19$, $n(B) = 23$, $n(A \cap B) = 7$

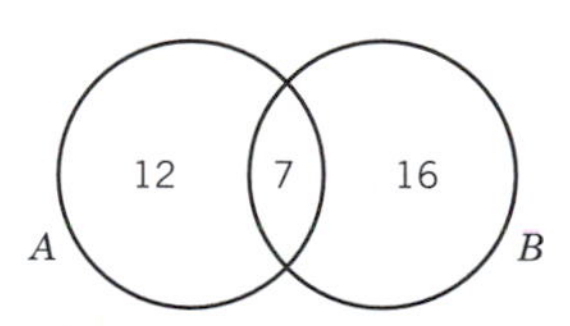

6. $n(A) = 19$, $(B) = 29$ $n(A \cap B) = 19$

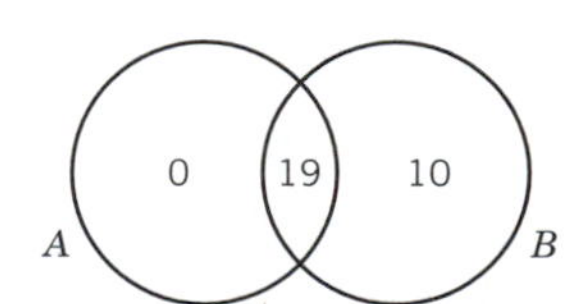

7. $n(A) = 39, \quad n(B) = 29, \quad n(A \cap B) = 0$

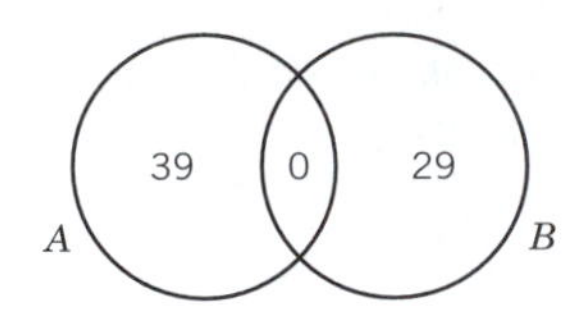

8. 20
9. People who like only dark roast.
10. 4
11. 6
12. People who like dark roast and decaf, but not espresso.
13. 76
14.

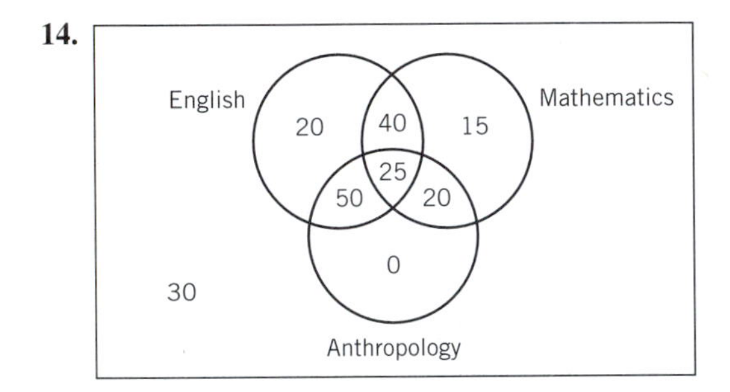

15. 15
16. 65
17. 135
18. 65
19. 100
20. 170 (Count those students who are in M, E, or both.)
21. 85

Section 3.5

1. $\{1, 6, 7\}$
2. $\varnothing$
3. $\{1, 2\}$
4. $A \cap B = \{3\}$, so $(A \cap B)' = \{1, 2, 4, 5, 6, 7\}$
5. $A \cup C = \{2, 3, 4, 6\}$, so $(A \cup C)' = \{1, 5, 7\}$

Section 3.6

1. $S' = \{f, g, h, i\}$
   $S' \cap Q = \{i\}$
2. $S' = \{f, g, h, i\}$
   $S' \cap Q = \{i\}$
   $(S' \cap Q)' = \{a, b, c, d, e, f, g, h\}$
   $(S' \cap Q)' \cup Q = \mathscr{U}$
3. $Q' = \{a, b, c, d, f, g, h\}$
   $Q' \cap S = \{a, b, c, d\}$
   $(Q' \cap S)' = \{e, f, g, h, i\}$
   $Q \cap (Q' \cap S)' = \{e, i\}$
   $(Q \cap (Q' \cap S)')' = \{a, b, c, d, f, g, h\}$
4. $S' = \{f, g, h, i\}$
   $S' \cap S = \varnothing$
   $(S' \cap S)' = \mathscr{U}$
5. $n(A') = n(\mathscr{U}) - n(A) = 10 - 4 = 6$
6. $n(A \cup B) = n(A) + n(B) - n(A \cap B) = 4 + 4 - 2 = 6$
7. $n(A \cup C) = n(A) + n(C) - n(A \cap C) = 4 + 3 - 0 = 7$
8. $n(C') = n(\mathscr{U}) - n(C) = 10 - 3 = 7$
9. $n(A') = n(\mathscr{U}) - n(A) = 20 - 12 = 8$
10. $n(A \cup B) = n(A) + n(B) - n(A \cap B) = 12 + 7 - 2 = 17$
11. $n(A \cup B') = n(A) + n(B') - n(A \cap B')$
    $= n(A) + n(\mathscr{U}) - n(B) - n(A \cap B')$
    $= 12 + 20 - 7 - 10$
    $= 15$
12. $n((A \cap B)') = n(\mathscr{U}) - n(A \cap B) = 20 - 2 = 18$
13. B'
14. $M \cap B$
15. $M \cup B$
16. $M \cap B'$
17. $M' \cap B'$
18. $(M \cap B)'$
19. $(M \cup B) \cap (M \cap B)'$
20. The set of students at your college who are not taking a mathematics course.
21. The set of students at your college who are not taking a mathematics course or are taking a biology course.
22. The set of students at your college who are taking a mathematics course and are not taking a biology course.
23. The set of students at your college who are not taking a mathematics course or are taking a biology course.

Section 3.7

1. $\{(3, p), (3, q), (4, p), (4, q)\}$
2. $\{(0, 0), (1, 0)\}$
3. {(Boston, Indiana), (Boston, Illinois), (Boston, Iowa), (Boston, Idaho), (Chicago, Indiana), (Chicago, Illinois), (Chicago, Iowa), (Chicago, Idaho)}
4. $16 + 24 - 8 = 32$
5. $16 \cdot 24 = 384$
6. $24 \cdot 16 = 384$
7. $50 - 16 = 34$
8. $34 \cdot 24 = 816$
9. $n(G \times B) = 25 \cdot 18 = 450$
   $G \times B$ is the number of different possible opposite sex dancing pairs at the Hop.

Section 3.8

1. $\mathscr{R}$ is not a function.
2. $\mathscr{R}$ is a function. Domain $= \{1, 2, 3\}$, codomain $= \{4, 5, 6\}$, range $= \{4, 6\}$.

3. $\mathcal{R}$ is a function. Domain = {1, 2, 3, 4, 5, 6, 7}, codomain = {a, b, c}, range = {a}.
4. $\mathcal{R}$ is not a function.
5. $\mathcal{R}$ is not a function.
6. F is one-to-one, but not onto.
7. F is one-to-one and onto.
8. F is onto, but not one-to-one.
9. F is neither one-to-one nor onto.
10. F is one-to-one, but not onto.

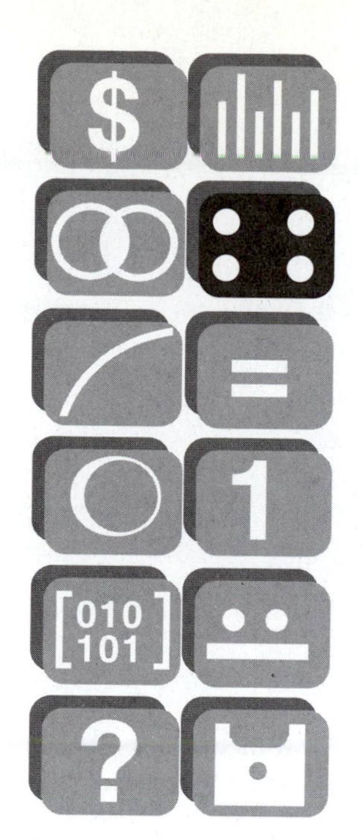

Chapter Four

Mathematics of Prediction

What's the Point? A variety of activities in life involve chance and prediction. When you buy stocks, purchase life insurance, buy a lottery ticket, or bet on a horse race, you are taking a chance. Probability is the mathematical study of chance. The activities in this chapter will help you to learn how to use mathematics to assess possible outcomes of future events.

Situation 10 Silly Kids' Games

You are playing a child's game with a 4-year-old. The game involves moving a token along a multicolored pathway toward a finish line. The first person to reach the end wins. Your move is determined by a spinner like the one shown below.

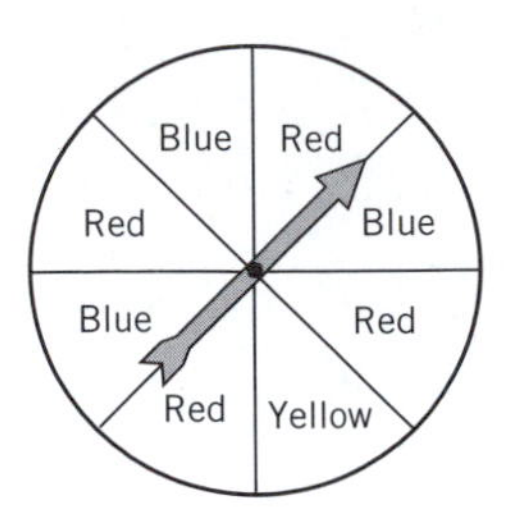

The child notices that she doesn't seem to get yellow very often and asks you why she doesn't. What do you tell her?

If you wrote that there is only one yellow sector whereas the blue and red are more plentiful, then you have an intuitive grasp of probability. Let's look at the situation from a mathematical point of view. You will see that we will be able to predict how often the spinner will land on yellow.

4.1 EXPERIMENTS, SAMPLE SPACES, AND EVENTS

Most of us have performed science experiments. We mixed or measured and then recorded the results. For mathematicians, experiments include a much wider variety of activities. An **experiment** is any activity that has distinguishable outcomes. Some experiments have a finite number of distinguishable outcomes and others have an infinite number of distinguishable outcomes. In this chapter, we look only at those experiments that have a finite number of distinguishable outcomes.

For example, throwing a die is an experiment. The possible outcomes are that the die could come up 1, 2, 3, 4, 5, or 6. Now, it is also possible that my dog Fred could catch the die in midair and then swallow it. Should we call that an outcome? What about the possibility that the die could shatter into a zillion pieces when it hits the floor?

In order to prevent an unlimited supply of imaginative outcomes for every experiment, when we define an experiment, the precise nature of the outcomes will be given. Thus, I'll say that the experiment is to throw a die, assume that it lands with one face up, and then record the number of dots on that face. Now there are only six possible outcomes to the experiment and I don't need to worry about Fred or brittle dice.

It can get pretty tiresome to completely specify every experiment. So, in this book we'll not state everything when the intent of the experiment is clear.

Here are some more experiments along with their outcomes.

1. *Deal-a-card experiment:* Shuffle a deck of standard bridge cards. A standard bridge deck consists of 52 playing cards. The 52 cards are divided into 4 suits of 13 cards each. The 4 suits are hearts, spades, clubs, and diamonds. Within each suit the cards are divided into 13 ranks. The ranks are 2, 3, 4, 5, 6, 7, 8, 9, 10, jack, queen, king, and ace. An individual card is referred to by its rank and suit. Hence, we talk about the "queen of hearts," the "three of spades," the "seven of diamonds," and so on. Deal out one card. Look at the card that was dealt. What is it?
   Outcomes: Ace of hearts. 2 of hearts. 3 of hearts, and so on. (52 total possible outcomes.)
2. *Have-a-baby experiment:* Give birth to a baby. Record the sex of the baby.
   Outcomes: Male. Female. (Two possible outcomes.)
3. *Try-to-win-a-raffle experiment:* You draw a numbered slip of paper from a big barrel of 1000 slips numbered 000 to 999. The number you draw is the winning ticket.
   Outcomes: 000 is the winning ticket. 001 is the winning ticket. 002 is the winning ticket, and so on. (1000 possible outcomes)

PRACTICE

List the outcomes to each of the following experiments:

1. *Visit-the-zoo experiment:* Xeno's town has a small zoo. On exhibit are lions, tigers, bears, and manatees. Xeno decides to see one exhibit during his lunch hour. Which exhibit will Xeno visit?

   The possible outcomes are

   There are ____ possible outcomes.

2. *Ninety-nine-bottles-of-beer-on-the-wall experiment:* Melvin's beer cellar has 99 numbered bottles of beer carefully stacked on shelves on the wall. Melvin takes a bottle from the shelves to give to his friend Rhonda. Which bottle of beer did Rhonda receive?

   The possible outcomes are

   There are ____ possible outcomes.

The set of possible outcomes of an experiment is called the **sample space** of that experiment. A sample space is usually denoted by a capital letter "S." Here are the sample spaces for some of the experiments we've already looked at.

Throw-a-die

$$S = \{1, 2, 3, 4, 5, 6\}$$

Comment: Note that the numbers represent the number of dots showing on the up face of the tossed die.

Deal-a-card

$$S = \{\text{ace of hearts, ace of clubs, ace of spades, ace of diamonds, king of hearts, king of clubs, } \ldots, \text{2 of spades, 2 of diamonds}\}$$

Comment: The ellipses (the three dots) indicate that the same pattern that has been started continues. The list ends with the 2 of diamonds.

Have-a-baby

$$S = \{\text{male, female}\}$$

Try-to-win-a-raffle

$$S = \{000, 001, 002, \ldots, 998, 999\}$$

PRACTICE

3. Write the sample spaces for the visit-the-zoo and 99-bottles-of-beer-on-the-wall experiments.

4. Write the sample space for the experiment of spinning the spinner described in Situation 10. How many total outcomes are there?

Suppose that we are performing the deal-a-card experiment. Further suppose that the queen of hearts is dealt to you. Now answer these yes or no questions. Remember that you have the queen of hearts.

Do you have a queen? ____

Do you have a heart? ____

Do you have a card in a red suit? _____

Do you have the queen of hearts? _____

Your answers should be yes to all these! Thus, you can see that your queen of hearts belongs to several different collections of cards—namely, queens, hearts, face cards, red cards, and its own special collection of queens of hearts. Here are each of these collections written with braces and given a letter name.

(Queens) Q = {queen of hearts, queen of diamonds, queen of clubs, queen of spades}

(Hearts) H = {ace of hearts, king of hearts, queen of hearts, . . . 2 of hearts}

(Face cards) F = {king of hearts, queen of hearts, jack of hearts, king of clubs, . . . king of diamonds, . . . , queen of spades, jack of spades}

(Red cards) R = {ace of hearts, king of hearts, . . . 2 of hearts, ace of diamonds, king of diamonds, . . . , 2 of diamonds}

(Queen of hearts) QH = {queen of hearts}

Each of these subcollections (or subsets) of the sample space is called an **event.** In general, any subcollection or subset of a sample space is an event. The outcomes listed in an event must be taken from the list of outcomes in the sample space.

Here is another example using the throw-a-die experiment. For each event, a description of the event is given. Fill in the blanks with the correct event or description.

	Event	Description of Event
(a)	{2, 4, 6}	An even number comes up.
(b)	{1, 3, 5}	____________________
(c)	{5, 6}	A number bigger than 4 comes up.
(d)	_______	A number less than 4 comes up.
(e)	{3, 6}	A multiple of 3 comes up.

(Answers: (b) An odd number comes up.
(d) {1, 2, 3})

Explain why the event described by *an even number comes up* is not {2, 4, 6, 8, 10, . . .}.

____

____

____

PRACTICE

Fill in the blanks with the correct event or description.

5. For the have-a-baby experiment:

Event: {female}   Description:

6. For the visit-the-zoo experiment:

Event: Description: Xeno visited a big cat exhibit.

7. For the 99-bottles-of-beer-on-the-wall experiment:

Event: {11, 22, 33, 44, 55, 66, 77, 88, 99} Description:

4.2 ONE, TWO, THREE, PROBABILITY

We are about to really get into the mathematics of chance, but before we do, one very important aspect of experiments needs to be examined. Take a look again at the sample spaces for the spinner game and throw-a-die experiment.

(Spinner) $S = \{\text{blue, red, yellow}\}$

(Die) $S = \{1, 2, 3, 4, 5, 6\}$

Suppose that the die is an ordinary one. Would you expect any number to be more likely to come up than any other?

____

____

____

Right! Each side has just the same chance as any other to come up. We say that the outcomes in the throw-a-die experiment are **equally likely.**

Now look at the spinner game. There are eight regions on the spinner and four of them are red. We would guess (and be right!) that red has a better chance of being the region where the needle rests than either blue or yellow. Thus, the outcomes in the spinner game are not equally likely.

Much of what we will do from now on requires equally likely outcomes in a sample space. So, I'll use a gimmick to make the spinner game outcomes equally likely. Relabel the spinner like this:

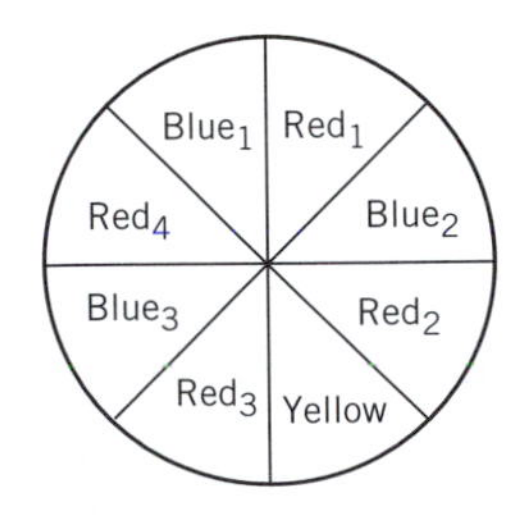

When we say that the spinner landed on red, we are talking about the event red $= \{\text{red}_1, \text{red}_2, \text{red}_3, \text{red}_4\}$. However, no single red region has an advantage over any other region. Under this gimmick, the sample space $\{\text{red}_1, \text{red}_2, \text{red}_3, \text{red}_4, \text{blue}_1, \text{blue}_2, \text{blue}_3, \text{yellow}\}$ for the spinner game has eight equally likely outcomes.

Now, it is time for the first probability formula (which also supplies a definition of probability) you'll want to know. Don't try to memorize it! If you understand the principle here, you'll never forget the formula.

FIRST PROBABILITY FORMULA

The probability of an event of an experiment with equally likely outcomes is the ratio of the number of outcomes in that event to the total number of outcomes in the sample space for the experiment.

Here's an example. Consider the throw-a-die experiment. Suppose we are interested in how likely it is that a 3 will come up. We could reason this way:

1. There are six possible outcomes.
2. Only one of those outcomes is a 3.
3. Thus, the 3 has one chance in six of coming up.

The probability of a 3 coming up is simply a ratio that expresses 3's chance of coming up. The probability is expressed as one-sixth (1/6).

Using the first probability formula, you would say this:

1. There are six outcomes in the sample space {1, 2, 3, 4, 5, 6}.
2. There is one outcome in the event {3}.
3. Thus, the probability of the event is 1/6.

Of course, mathematicians have symbols for all this. $P(E)$ means "the probability of the event E," $n(E)$ "the number of outcomes in the event E," and $n(S)$ "the number of outcomes in the sample space S." So, if E stands for the event of a 3 coming up, we have

1. $n(S) = 6$
2. $n(E) = 1$
3. $P(E) = n(E)/n(S) = 1/6$

(Note that $P(E) = n(E)/n(S)$ is a symbolic version of the first probability formula.)

Let's find the probability of another event in the throw-a-die experiment. Suppose that we let F stand for the event {1, 3, 5}; an odd number comes up. Then we have

1. $n(S) = 6$ (same sample space)
2. $n(F) = 3$ (three outcomes in F)
3. $P(F) = 3/6 = 1/2$

The first probability formula applies to all the experiments we've looked at so far. Fill in the blank steps for each of the following calculations.

Have-a-baby experiment. What is the probability of having a female baby? Recall that the sample space had two outcomes in it. Call the event F = {Female}.

1. $n(S) = 2$
2. $n(F) = 1$
3. $P(F) =$ _____

The answer should have been 1/2, because $P(F) = n(F)/n(S) = 1/2$.

Deal-a-card experiment. What is the probability of dealing a queen? Let the event {queen of hearts, queen of diamonds, queen of clubs, queen of spades} be called Q.

1. $n(S) = 52$

2. $n(Q) =$ _____

3. $P(Q) = 4/52$

I hope that your answer was 4.

A couple of things can be concluded from the first probability formula. First, because an event can never have more outcomes than its sample space, the probability of an event cannot be greater than 1. Second, none of the numbers involved in calculating a probability are negative, so a probability can never be less than 0.

Look out for weird answers! If you get a probability that is less than 0 or bigger than 1, then you made a mistake.

If event E has probability 1/4 and event F has probability 1/78, which event would you least expect to happen?

____

____

____

If you said F, then you understand! E has 1 chance in 4 of happening, but F has 1 chance in 78. F is less likely to happen than E.

In general, the closer the probability of an event is to 1, the greater the likelihood of that event occurring. Probability is a measure of astonishment. The higher the probability of an event, the less astonished you will be if the event actually happens.

PRACTICE

For the throw-a-die experiment, compute the probability of each of the following events:

1. {2}

2. {4, 5}

3. An even number comes up.

For the visit-the-zoo experiment, compute the probability of each of the following events:

4. {lions, bears}

5. {manatees}

6. Xeno visits a land mammal exhibit.

Finally, we can apply our first probability formula to the spinner game. The probability of getting red is calculated as follows:

1. $n(S) = 8$ (remember we are using equally likely outcomes)

2. $n(R) = 4$ (the event is $\{red_1, red_2, red_3, red_4\}$)

3. $P(R) = 4/8 = 1/2$

The probability of getting blue is 3/8. (Work this out. Don't just believe me!) The probability of getting yellow is 1/8. Thus, yellow has the least chance of happening. That is why it doesn't seem to turn up very often.

Now that we have the first probability formula, we can go to the racetrack and reap the benefits of our newfound knowledge. But first, there is one more thing to notice about the probability of event *E*.

The first probability formula tells us that the probability of an event *E* is the ratio of the number of outcomes in *E* to the total number of outcomes in the sample space. However, the sample space is made up of outcomes that are in *E* and outcomes that are not in *E*. There are no other options for an outcome. It is either in *E*, or it is not in *E*.

For example, consider the event $E = \{3, 4\}$ of the throw-a-die experiment. Every outcome in the sample space is either in *E* (3 and 4) or it is not in *E* (1, 2, 5, and 6). If we consider the sample space to be a universe of discourse, then every outcome is either in the event $E = \{3, 4\}$ or its complement, the event $E' = \{1, 2, 5, 6\}$.

The probabilities of an event and its complement always add up to 1. Verify the last sentence for the events $E = \{3, 4\}$ and $E' = \{1, 2, 5, 6\}$ in the throw-a-die experiment.

We can summarize the relationship between the probability of an event and that of its complement in a nice little formula:

$$P(E) + P(E') = 1$$

Also, because the event E' is the event containing the outcomes not in *E*, we often refer to E' as *not E*. Thus, the formula $P(E) + P(E') = 1$ can be used to calculate the probability that an event *doesn't* happen.

For example, the probability that we don't toss a 2 in the throw-a-die experiment can be calculated like this:

Let E = the event that a 2 is tossed = {2}. We know that $P(E) = 1/6$. Now the event that a 2 is not tossed is $E' = \{1, 3, 4, 5, 6\}$. Using our formula, we have

$$\begin{aligned} P(E) + P(E') &= 1 \\ 1/6 + P(E') &= 1 \\ P(E') &= 1 - 1/6 \\ &= 5/6 \end{aligned}$$

Thus, the probability of not getting a 2 is 5/6. Verify that this answer is correct using the first probability formula and the number of outcomes in E'.

PRACTICE

Use the formula $P(E) + P(E') = 1$ to calculate each of the following:

7. The probability of not having a male in the have-a-baby experiment.

8. The probability of not getting a queen in the deal-a-card experiment.

9. The probability of not winning with 123 in the try-to-win-a-raffle experiment.

Situation 11 Blue Flash by a Nose

You are at the Universal Downs Race Course for the Tea Cup, an annual race between the fastest horses in Clapboard County. As you take your seat, you glance at the TV monitors and see the following display:

Tea Cup	
Blue Flash	3 to 2
Red Eye	9 to 1
Green Streak	19 to 1
Yellow Blur	7 to 3
Purple Haze	17 to 3

Ed, a veteran of many Tea Cups, says to you that Blue Flash is the favorite today. What does Ed mean? How does he know?

____

____

____

4.3 ODDS

The horse race described in Situation 11 is also an experiment. The sample space is

$$S = \{\text{Blue Flash wins, Red Eye wins, Green Streak wins, Yellow Blur wins, Purple Haze wins}\}$$

If we are interested only in which horse wins the race, then we can concentrate on the probability of the events {Blue Flash wins}, {Red Eye wins}, {Green Streak wins}, and so on. The probability of each of these events is given after the horse's name on the TV monitor. The probabilities are expressed in a new form called **odds.**

The **odds against an event *E* happening** is the ratio of the probability that E will not happen to the probability that E will happen. For example, consider the throw-a-die experiment. The probability of throwing a 2 is 1/6. The probability of throwing something other than a 2 is 5/6.

Explain why that last probability is 5/6.

____

____

____

The odds against throwing a 2 are thus

$$\frac{5/6}{1/6} = \frac{5}{1}$$

or *5 to 1.*

The definition of odds can also be expressed in a mathematical formula like this:

$$\begin{aligned}\text{The odds against an event } E &= (\text{the probability of not } E)/(\text{the probability of } E)\\ &= P(\text{not } E)/P(E)\\ &= P(E')/P(E)\end{aligned}$$

In the deal-a-card experiment, the odds against dealing a queen are

$$\frac{P(\text{not queen})}{P(\text{queen})} = \frac{48/52}{4/52} = \frac{48}{4} = \frac{12}{1}$$

or *12 to 1*. In the visit-the-zoo experiment, the odds against Xeno visiting the lions on his lunch hour are

$$\frac{P(\text{not lions})}{P(\text{lions})} = \frac{3/4}{1/4} = \frac{3}{1}$$

or *3 to 1*.

PRACTICE

1. In the deal-a-card experiment, find the odds against dealing the ace of clubs.

2. In the try-to-win-a-raffle experiment, find the odds against picking the number of the winning ticket.

Now suppose that the probability of a horse, call her Beetlebird, winning a race is 2/5. What are the odds against her winning the race?

Well, we have the probability of Beetlebird winning, but we don't have the probability that she will not win. Fortunately, the probability of not winning can be calculated from the probability of winning using the formula $P(E) + P(E') = 1$.

The probability of Beetlebird winning is 2/5. Therefore,

$$\begin{aligned}P(\text{Beetlebird winning}) + P(\text{Beetlebird not winning}) &= 1\\ 2/5 + P(\text{Beetlebird not winning}) &= 1\\ P(\text{Beetlebird not winning}) &= 1 - 2/5 = 3/5\end{aligned}$$

The odds against Beetlebird winning are thus

$$\frac{P(\text{not Beetlebird wins})}{P(\text{Beetlebird wins})} = \frac{3/5}{2/5} = \frac{3}{2}$$

or *3 to 2*.

Here is another example. Suppose $P(G) = 13/77$. Then $P(\text{not } G) = 1 - 13/77 = 77/77 - 13/77 = 64/77$. Thus, the odds against G are

$$\frac{P(\text{not } G)}{P(G)} = \frac{64/77}{13/77} = \frac{64}{13}$$

or *64 to 13*.

PRACTICE

Find the odds against each of the given events.

3. H, when $P(H) = 1/12$.

4. K, when $P(K) = 1/45$.

5. L, when $P(L) = 13/15$.

6. M, when $P(M) = 2/3$.

Now that you can express a probability as odds, let's reverse the process.

The odds against each horse winning the Tea Cup were given on the TV monitor. How can we compute the probability of each horse winning the race from these odds?

Let's look at Red Eye first. The odds against Red Eye winning the race are given as 9 to 1. Remember where these two numbers came from. They represent the ratio of the probability that Red Eye won't win to the probability that Red Eye will win. Both those probabilities have the same denominator (the size of the sample space) that divided itself out when the ratio was reduced.

If the last paragraph isn't clear, look again at the calculation of the odds against drawing a queen in the deal-a-card experiment. Notice how the two denominators of 52 divided each other out when the ratio was reduced from

$$\frac{48/52}{4/52} \text{ to } \frac{48}{4} \qquad \left(\frac{48/52}{4/52} = \frac{48}{52} \cdot \frac{52}{4} = \frac{48}{4}\right)$$

Getting back to Red Eye, we can conclude that the probability that Red Eye will not win is $9/x$ and the probability that Red Eye will win is $1/x$, where x represents the unknown size of the sample space.

Finally, since Red Eye either wins or doesn't win, we have $9/x + 1/x = 1$ or $10/x = 1$. The correct value for x is 10 and the probability that Red Eye will win is 1/10.

In general, if the odds against an event are *A to B,* then the probability of that event is $B/(A + B)$.

Complete the table for the rest of the Tea Cup contestants.

Horse	**Odds Against Winning**	**Probability of Winning**
Red Eye	9 to 1	1/10
Blue Flash	3 to 2	2/5
Green Streak	19 to 1	_____
Yellow Blur	7 to 3	_____
Purple Haze	17 to 3	3/20

(Answers: $P(\text{Green Streak wins}) = 1/(19 + 1) = 1/20$
$P(\text{Yellow Blur wins}) = 3/(7 + 3) = 3/10$)

We can compare these probabilities by looking at the decimal equivalent of each ratio:

Horse	Probability of Winning
Red Eye	1/10 = 0.10
Blue Flash	2/5 = 0.40
Green Streak	1/20 = 0.05
Yellow Blur	3/10 = 0.30
Purple Haze	3/20 = 0.15

We can now see that Blue Flash has the greatest probability of winning the race. Blue Flash is called the *favorite.*

PRACTICE

Find the probability of each event when the odds against that event are as shown.

7. 14 to 1

8. 100 to 1

9. 23 to 2

10. 45 to 7

Although most gambling odds are given as odds against an event, you may also encounter the **odds in favor of an event.** Fortunately, the relationship between odds against and odds in favor is very friendly.

The odds in favor of an event are the reverse of the odds against an event.

Thus, the odds in favor of Red Eye winning the Tea Cup are *2 to 3* because the odds against Red Eye winning are *3 to 2.* The odds in favor of drawing a queen in the deal-a-card experiments are *1 to 12,* and so on.

4.4 EMPIRICAL PROBABILITY

Why were the probabilities for the horses in the Tea Cup different?

____

____

____

Did you say that some horses are faster than others? That some jockeys have better skills? That some horses run better in the mud than on hard dirt?

Good! Because no two horses are identical in strength and speed and no two jockeys are identical in skill, the outcomes of the Tea Cup race are not equally likely. Furthermore, it is difficult to assign probabilities to the events because the outcome of the race depends on too many uncontrollable factors like weather, health of the horses and jockeys, and so forth.

However, experts on horse racing do assign probabilities and predict winners. Their judgments are based on experience. They have watched many horse races. They have studied the past races of the horses involved. The probabilities the experts come up with are called **empirical probabilities.** These probabilities are based on the experience of repeated similar experiments. They differ from the **theoretical probabilities** that are based on counting outcomes.

Another kind of empirical probability is a baseball player's batting average. The batting average is the ratio of the number of hits the player has gotten to the number of times she has been to bat. A player who has gotten 4 hits in 20 times at bat has a batting average of $4/20 = 2/10 = .200$. This batting average also serves as the probability that the player will get a hit the next time she bats.

The odds given for horses in a horse race are empirically found in two different ways. When the odds are given in the newspaper before a race, they reflect the opinion of the newspaper's staff of horse-racing experts. The odds that appear on the TV monitor (or toteboard) at the racetrack at the time of the race reflect the opinions of those people who have wagered on the race. The more money that is wagered on a horse, the higher the empirical probability of winning assigned to that horse.

Empirical probability is used extensively in sports gambling and investing in stocks and bonds.

Situation 12

Lotteries

On a whim you decide to buy a $1 state lottery ticket. You must pick six numbers between 1 and 54, inclusive. If the numbers you pick are selected, in any order, by the lottery officials at the official drawing, then you will win 50 million dollars. You select the numbers 4, 15, 12, 21, 28, and 46. Your spouse says that you have no chance of winning. Is that true? What is your chance of winning?

If you said that the odds against you winning are 25,827,164 to 1, then you are exactly right and should skip the rest of this discussion. If not, then we need to start with a simpler kind of lottery and work our way up to the one described in Situation 12.

4.5 FUNDAMENTAL COUNTING PRINCIPLE

The first probability formula makes it necessary to count the number of outcomes in an event in order to calculate its probability. If the sample space of an experiment is large, then the counting can become difficult. It is wise, therefore, to be on the lookout for efficient counting techniques. One counting technique that will prove to be quite useful is the **fundamental counting principle.**

To illustrate the fundamental counting principle, consider the ordering-a-meal experiment.

Ordering-a-meal experiment. You are ordering a meal from the restaurant menu shown below

Soups	**Salads**
Cream of broccoli	Caesar salad
Chicken noodle	Fruit salad
	Spinach salad
	Tossed salad
Entrées	**Desserts**
Roast beef	Chocolate pie
Grilled swordfish	Angel food cake
Ratatouille	Lime frozen yogurt

A *meal* consists of one soup, one salad, one entrée, and one dessert. Select a meal.

The sample space for this experiment is {(cream of broccoli, Caesar salad, roast beef, chocolate pie), (chicken noodle, Caesar salad, roast beef, chocolate pie), (cream of broccoli, fruit salad, roast beef, chocolate pie), . . .}. How many outcomes are in the sample space? One way of beginning to count all these outcomes is to observe that either of the two soup choices can be included with the same combination of salad, entrée, and dessert. For example, you could have cream of broccoli soup or chicken noodle soup in a meal that included tossed salad, ratatouille, and angel food cake. So, we can say that $n(S) = 2 \cdot$ (the number of different collections of salad, entrée, and dessert).

However, if we try to count the soupless collections, we notice that any of the four salads can be included with each of the entrée and dessert choices. So,

The number of different collections of

salad, entrée, and dessert = 4 · (The number of different collections of entrée and dessert)

Finally, each of the three entrées can go with each of the three desserts. So,

The number of different collections of

entrée and dessert = 3 · 3

Putting it all together, we see that the number of different meals is $2 \cdot 4 \cdot 3 \cdot 3 = 72$. Or, to state it in more specific terms,

$$\begin{aligned} n(S) &= \text{(The number of soup choices)} \cdot \text{(the number of salad choices)} \\ &\quad \cdot \text{(the number of entrée choices)} \cdot \text{(the number of dessert choices)} \\ &= 2 \cdot 4 \cdot 3 \cdot 3 \\ &= 72 \end{aligned}$$

What would the number of possible meals be if we added apple pie to the dessert list?

(Answer: $2 \cdot 4 \cdot 3 \cdot 4 = 96$)

The fundamental counting principle says that if an experiment can be viewed as a sequence of independent smaller experiments, then the number of outcomes of that experiment is the product of the number of outcomes of the smaller experiments. Two experiments are **independent** if neither one's outcome depends on the other.

In the ordering-a-meal experiment, we can view the ordering of a meal as a sequence of the experiments ordering-a-soup, ordering-a-salad, ordering-an-entrée, and ordering-a-dessert.

Although we may not like to have chicken noodle soup with spinach salad, for the purposes of counting the meals, the salad choice is independent of the soup choice.

Here is another example of how to use the fundamental counting principle to count sample spaces. Suppose that the model of a new car you are looking at comes in two body styles with your choice of eight exterior colors and four interior colors. How many different cars are possible?

Using the fundamental counting principle, we count each individual aspect of the car and then form their product.

$$\begin{aligned} \text{Total number of cars} &= \text{(number of body styles)} \cdot \text{(number of exterior colors)} \\ &\quad \cdot \text{(number of interior colors)} \\ &= 2 \cdot 8 \cdot 4 \\ &= 64 \text{ cars} \end{aligned}$$

PRACTICE

Use the fundamental counting principle to answer each of the following:

1. Zork is getting dressed for work. An *outfit* for Zork consists of a necktie, shirt, pair of trousers, pair of socks, and pair of shoes. Zork has 12 ties, 5 shirts, 6 pairs of trousers, 15 pairs of socks, and 1 pair of shoes. How many different possible outfits does Zork have?

2. How many different four-letter automatic teller machines (ATM) codes are possible?

3. How many different license plate numbers are possible if each plate consists of two letters followed by three digits (e.g., XW 999, ZZ 145, or AB 098)?

As one more example of the use of the fundamental counting principle, recall the try-to-win-a-raffle experiment. In this experiment, we hoped that our three-digit number would be picked. There were 1000 possible outcomes to this experiment.

This result could have been obtained using the fundamental counting principle. We have 10 choices for our first digit, 10 for our second, and 10 for our third. Thus, there are $10 \cdot 10 \cdot 10 = 1000$ choices.

Sometimes, we need to be just a little bit sneaky in counting a large collection. For example, suppose in the try-to-win-a-raffle experiment none of the slips had three identical digits. That is, suppose 000, 111, 222, 333, 444, 555, 666, 777, 888, and 999 were not possible selections. Now how many possible outcomes are there?

If you said 990, you were right! Here's why.

There are 1000 outcomes in the regular raffle experiment. But, 10 of these outcomes need to be eliminated. Thus, we have $1000 - 10 = 990$ outcomes in the new raffle.

As you can see, the fundamental counting principle is pretty handy for counting when the numbers are large. Would you really want to try to write down all 676,000 different license plates described in Practice Exercise 3?

Let's look at a slightly more complicated application of the fundamental counting principle.

On a college campus, all the telephone numbers have the same prefix (455). The numbers 455-9999 and 455-0000 are reserved for the public safety officer and college president, respectively. Phone numbers whose last four digits begin with a 1 or 0 are reserved for administration and staff. Phone numbers whose last four digits begin with 8 or 9 are reserved for the faculty. All the other possible numbers are reserved for student use. How many different telephone numbers are available for student use?

Using the fundamental counting principle, we can say that we have only one choice for each of the first three digits of our phone number (455). The fourth digit must be a 2, 3, 4, 5, 6, or 7. The fifth, sixth, and seventh digits can be anything from 0 to 9. Thus,

$$\text{Number of student phones} = 1 \cdot 1 \cdot 1 \cdot 6 \cdot 10 \cdot 10 \cdot 10 = 6000$$

PRACTICE

4. How many different four-letter codes are possible that begin and end with the letter "z"?

5. How many different four-letter codes do not have four identical letters?

6. How many different four-letter codes begin with the letter "a" and end with the letter "z"?

4.6 PERMUTATIONS

Here is a slightly different application of the fundamental counting principle.

How many different four-letter codes contain no letter more than once? That is, *abcd* is OK, but *abad* is not because the letter "a" is repeated.

In this case, the fundamental counting principle applies, but we need to be careful. There are 26 choices for the first letter in the code. However, once that first letter has been picked, we cannot use it for the second letter. Thus, there are only 25 choices for the second letter. The third letter must be different from *both* the first and second letter. So, there are 24 choices for the third letter. Finally, the fourth letter must differ from all of the first three letters of the code. Hence, there are 23 choices for the fourth letter. Therefore,

$$\begin{aligned}\text{Number of four-letter codes with no repeated letter} &= 26 \cdot 25 \cdot 24 \cdot 23 \\ &= 358{,}800\end{aligned}$$

Let's look at another related problem. Suppose that you have four framed photographs you wish to display on your shelf. Suppose you have pictures of Aunt Agnes, Cousin Cal, Uncle Ulysses, and Grandma Gladys. For simplicity, denote these pictures by the letters *A, C, U,* and *G,* respectively. Here are three different arrangements of the pictures: *ACUG, GUCA,* and *ACGU.* Write as many other different arrangements as you can.

How do you know if you got them all? The fundamental counting principle tells us that there are 24 different arrangements.

$$\text{Number of different picture arrangements} = 4 \cdot 3 \cdot 2 \cdot 1 = 24$$

The picture-arranging problem and four-letter code problem are two instances of the use of permutations. A **permutation** of a collection of objects is an ordered rearrangement of those objects. When I asked for the number of ways the pictures could be arranged on a shelf, I was asking for the number of permutations of the four pictures.

When arranging objects from a collection, we can use all the objects (like the pictures) or just some of the objects (like the letters in our codes). So, permutations can be rearrangements of all or some members of a collection.

Here are two more examples.

Problem 1 How many different permutations of the letters in the word *duckling* are there?

Solution

There are eight different letters from which to choose. Using the fundamental counting principle, we determine that the number of different arrangements $= 8 \cdot 7 \cdot 6 \cdot 5 \cdot 4 \cdot 3 \cdot 2 \cdot 1 =$ 40,320. ■

Problem 2 Sam, Mario, Felecia, Agnes, Cameron, and Lucinda are the finalists in a contest. The winner will get a free trip to Paris. The runnerup will receive a new car. In how many different ways can the prizes be awarded?

Solution

There are six choices for the trip winner. Once the winner of the trip is selected, there will be five choices for the winner of the car. Thus, the total number of different ways to award the prizes $= 6 \cdot 5 = 30$. ■

Problem 3 How many different four-letter codes have at least one repeated letter?

Solution

Did you say 98,176? If not, here's how. Notice that all four-letter codes either have repeated letters or they don't. Thus, if we take away all the codes with no repeated letters from the total number of codes, we'll have only codes with at least one repeated letter left. That is,

Number of four-letter codes with at least one repeated letter

= Total number of four-letter codes − number of four-letter codes with no repeated letters

$= 26 \cdot 26 \cdot 26 \cdot 26 - 26 \cdot 25 \cdot 24 \cdot 23$

$= 456{,}976 - 358{,}800$

$= 98{,}176$ ■

PRACTICE

1. In how many different ways can we arrange the letters of the word *history*?

2. In how many different ways can the offices of president, vice president, and treasurer be filled for a club with seven members?

3. Five people are playing *musical chairs* with four chairs. How many different seating arrangements are possible when the music stops?

How many different arrangements of the letters in the alphabet are there?

Wow! That *is* a big number! Let's see if we can express it a little more compactly using **factorial notation.**

4.7 FACTORIAL

The factorial notation was invented by Christian Stamp in 1808 to make the representation of the large numbers arising from permutation calculations a little easier to look at. The notation $n!$, read *n-factorial,* means the product of all the positive integers less than or equal to n. Thus,

$$1! = 1$$

$$2! = 2 \cdot 1 = 2$$

$$3! = 3 \cdot 2 \cdot 1 = 6$$

$$4! = 4 \cdot 3 \cdot 2 \cdot 1 = 24$$

$$5! = 5 \cdot 4 \cdot 3 \cdot 2 \cdot 1 = 120$$

Calculate 6!

(Answer: 720)

(Your calculator probably has a factorial button on it. To find 6!, just press 6 and then your factorial button.)

We will need to do some calculations with factorials in our discussion of Situation 12. So, here is a useful observation.

Suppose we needed to calculate 89!/87!. Try it. If you tried to do it by first calculating 89! and next dividing by 87!, then you are very tired by now. Even your calculator will give up on this one! Fortunately, it can be done much more easily.

Let's look at what 89!/87! means.

$$\frac{89!}{87!} = \frac{89 \cdot 88 \cdot 87 \cdot 86 \cdot 85 \cdot 84 \cdot 83 \cdot 82 \cdots 5 \cdot 4 \cdot 3 \cdot 2 \cdot 1}{87 \cdot 86 \cdot 85 \cdot 84 \cdot 83 \cdot 82 \cdots 5 \cdot 4 \cdot 3 \cdot 2 \cdot 1}$$

Can you see there are a lot of common factors that can be divided out in this fraction? In fact, everything from 87 on down to 1 divides out. We are left with $89 \cdot 88 = 7832$.

PRACTICE

Calculate each of the following:

1. $9!$

2. $14!/13!$

3. $77!/74!$

With factorial notation, we can now generalize the permutation calçulations of the last section.

PERMUTATION FORMULAS

1. To find the number of permutations of all the objects in an n-object set, calculate $n!$.
2. To find the number of permutations of k objects from an n-object set, calculate $n!/(n - k)!$.

The number of four-letter code with no repeated letters can then be calculated using the second permutation formula:

$$26!/(26 - 4)! = 26!/22! = 26 \cdot 25 \cdot 24 \cdot 23 = 358{,}800$$

The picture-arranging problem can be solved using the first permutation formula:

$$4! = 4 \cdot 3 \cdot 2 \cdot 1 = 24$$

Now let $P(n, k)$ denote the number of permutations of k objects taken from an n-object collection. Then the second permutation formulas can be written:

$$P(n, k) = n!/(n - k)!$$

It would be nice if the first permutation formula could also be written this way. (Then we would have a single formula for all permutation problems!) Now the first permutation formula tells us that $P(n, n) = n!$, and the second permutation formula, if applied to the case where $n = k$, would give $P(n, n) = n!/(n - n)!$. If these two expressions are to give the same answer, we need to have

$$n! = n!/(n - n)!$$

or

$$n! = n!/0!$$

This can only happen if $0! = 1$. We therefore define $0!$ to be 1. We now have a single permutation formula:

$$P(n, k) = n!/(n - k)!$$

We can use this formula to do all the practice exercises from the last section.

1. $P(7, 7) = 7!(7 - 7)! = 7!/0! = 7!/1 = 7! = 5040$

2. $P(7, 3) = 7!/(7 - 3)! =$ __________________

3. $P(5, 4) =$ ____________________________

(Answers: 2. $7!/4! = 7 \cdot 6 \cdot 5 = 210$
3. $5!/(5 - 4)! = 5!/1! = 5 \cdot 4 \cdot 3 \cdot 2 = 120$)

Let's get back to the lottery described in Situation 12.

4.8 COMBINATIONS

The lottery described in Situation 12 does not require you to pick the six numbers in any particular order. That is, if you pick 4, 12, 15, 21, 28, and 46, but the lottery officials draw numbers 15, 12, 46, 4, 28, and 21, then you will still win. How does the disregard for order alter the calculations? To get an answer to that question, consider a smaller problem.

In a tiny lottery, a player must pick three numbers from the numbers 1 through 5. If the numbers picked by the player match those picked by the lottery officials, then the player wins a case of grapefruit. How many different three-number outcomes are possible? List them.

____

____

____

Here are the 10 outcomes you should have listed: $\{1, 2, 3\}$, $\{1, 2, 4\}$, $\{1, 2, 5\}$, $\{1, 3, 4\}$, $\{1, 3, 5\}$, $\{1, 4, 5\}$, $\{2, 3, 4\}$, $\{2, 3, 5\}$, $\{2, 4, 5\}$, $\{3, 4, 5\}$. Keep in mind that the order of the numbers does not make a difference. $\{1, 3, 5\}$ is the same outcome as $\{3, 1, 5\}$.

How can you know how many different outcomes there are going to be? To answer that question, we need to go back to counting permutations.

Let's pretend that the winning numbers for this little grapefruit lottery are selected by the lottery officials in the following manner.

There are five ping pong balls numbered 1 through 5 in a large barrel. The lottery official reaches in and takes out one ball at a time and reads aloud its number until three balls are drawn. Focus your attention on the number of different possible *readings*. For example, the official could say "4, 2, 3," or "2, 4, 3," and so on.

We know that there are $P(5, 3) = 5!/(5 - 3)! = 5!/2! = 60$ different sequences of numbers that could be read in this way.

However, the lottery official could have used another technique. He could use a scoop that holds three balls to scoop three balls out of the barrel. Then he could read the numbers of each of these three balls. The scoop method has two steps:

1. Scoop the balls.

2. Read the numbers of the scooped balls.

The fundamental counting principle tells us that the total number of possible outcomes from the scoop method will be

(The number of different scoops)

· (the number of ways the numbers on the three scooped balls can be read)

Now we know that the number of different ways to read the numbers of three balls is $P(3, 3) = 3!$. So, the total number of possible outcomes using the scoop method is

(The number of different scoops) · 3!

But, this product should also be $P(5, 3)$ because we are still reading the numbers of three balls. So,

$$\text{(The number of different scoops)} \cdot 3! = P(5, 3)$$
$$= 5!/(5 - 3)!$$
$$\text{(The number of different scoops)} = \frac{5!}{3!(5 - 3)!} = 10$$

Now a *scoop* is really an outcome to the lottery experiment. So, we have calculated the total number of outcomes to be 10. This matches the answer we got by listing.

The reasoning in the last few paragraphs was subtle. Let's look at a related experiment and go through a similar analysis.

The Beach Prize Experiment

Suppose that I wanted to award three different cash prizes ($10, $100, and $1000) to three different people selected from a group of five people standing next to me on a beach. From the previous section, we know I could do this in $P(5, 3) = 5!/(5 - 3)! = 60$ different ways. But, let us think about how we could actually award the prizes right there on the beach. There are at least two distinct ways of doing this.

One way to award the prizes is to pick one person and give him or her a prize and then pick another person and give him or her a prize and continue like this until all the prizes have been awarded. I might say, "Sally wins $10, Fred wins $100, and Maria wins $1000." The permutation formula says we can do this in $P(5, 3) = 5!/(5 - 3)!$ ways.

Another way to award the prizes is to pick three people as prize winners and *then* distribute the prizes to them. In this case, I might say, "Sally, Maria, and Fred have won prizes. Sally wins $10, Fred wins $100, and Maria wins $1000." This method requires two steps. First, pick the group of prize winners. Second, arrange the prize winners according to the prize they win. The fundamental counting principle tells us that this method can be done in this many ways:

(Number of ways to pick the group of three people from the group of five people)

· (number of ways to arrange the group of three)

Notice that the two methods are truly different. Consider the situation of Harvey. With the first method, when Harvey hears that Sally will receive the $10 prize, he can still hope to be picked for one of the other prizes. With the second method, Harvey knows that he is out of the running as soon as the prize winning group is named. Now look at it from Maria's point of view. With the first method, if she hears that Sally is awarded the $10, Maria can hope to win another prize but she doesn't know if she will. But, if she is named as a prize winner, then she must only wait to hear the value of her prize.

The methods are different, but they achieve the same outcomes. Thus, the number of arrangements calculated using either method should be the same. We have then

$5!/(5-3)! =$ (Number of ways to pick the group of three people from the group of five people) · (number of ways to arrange the group of three)

And, since a group of three can be arranged in 3! different ways:

$5!/(5-3)! =$ (Number of ways to pick the group of three people from the group of five people) · 3!

Dividing both sides of the last equation by 3! yields

$$\text{Number of ways to pick the group of three people from the group of five people} = \frac{5!}{3!(5-3)!}$$

This result generalizes to any size collection, so that

$$\text{The number of ways to pick a group of } k \text{ things from a group of } n \text{ things} = \frac{n!}{k!(n-k)!}$$

For simplicity, we will denote the number of ways to pick k people from a group of n things by $C(n, k)$. The symbols $C(n, k)$ can be read, "n choose k." Thus, we have the **combination formula** for the number of ways an unordered group of k things can be selected from a group of n things:

$$C(n, k) = \frac{n!}{k!(n-k)!}$$

PRACTICE

Calculate each of the following:

1. $C(6, 2)$

2. $C(5, 2)$

3. $C(8, 4)$

4. $P(8, 2)$

5. $C(8, 6)$

6. $C(5, 5)$

We will use the formula for $C(n, k)$ whenever we need to calculate the number of ways we can select unordered groups of things. In the case of the tiny lottery, there are $C(5, 3) = 10$ outcomes. We can now calculate the number of outcomes in the lottery described in Situation 12.

In Situation 12, we were to pick six numbers from the numbers 1 through 54. The order of the numbers is unimportant, so the number of different possible groups of six numbers that can be selected from the group of 54 numbers is

$$C(54, 6) = \frac{54!}{6!(54-6)!} = \frac{54!}{6!48!} = \frac{54 \cdot 53 \cdot 52 \cdot 51 \cdot 50 \cdot 49}{6!}$$
$$= 25{,}827{,}165$$

The probability of winning this lottery with one selection is then 1/25,827,165 or approximately .0000000387. If you look at this from the point of view of odds, your odds against winning this lottery are worse than 25 million to 1. (What are your exact odds against winning?) Your spouse is wrong in saying that you have no chance of winning, but the probability of your winning is quite low.

PRACTICE

7. You are dealt three cards from a standard deck. Disregarding the order in which the cards are dealt to you, what is the probability that you will be dealt only hearts?
Hint: The total number of possible three-card hands is $C(52, 3)$. The number of three-heart hands is $C(13, 3)$ because you must select from the collection of 13 hearts.

8. Myrna, Becky, and Amy are triplets in a second grade class of 20 students. Six of the students are going to be chosen at random to take a trip to Washington, D.C. What is the probability that all three of the triplets will be chosen?

Situation 13 Estimating Your Winnings

You are attending a summer carnival. The carnival includes a midway with a variety of games of chance. One of the games is the classic three-cups-and-a-pea game. The game operator places a pea under one of three identical cups. She then quickly scrambles the cups and asks you to pick the cup that hides the pea. If you pick the correct cup, then you win \$6. If you pick the wrong cup, then you lose your \$1 entry fee. If you decide to play 10 games, win or lose, it will cost you \$10 to play. How much money can you expect to have gained or lost after those 10 games?

Do you think you will have more than \$10, less than \$10, or exactly \$10? Why?

____

____

____

4.9 EXPECTATION

Situation 13 asks us to determine what we can expect to win or lose. We are trying to determine the **expected value** of the game. The expected value of the game is the average amount you can expect to win or lose over the long run. Let's look at the game more closely.

The probability that you will pick the cup with the pea is 1/3. Explain why.

____

____

____

Similarly, the probability that you will pick the wrong cup is 2/3. Thus, for each game the probability that you will win \$5 (\$6 minus the \$1 you paid to play the game) is 1/3 and the probability that you will lose \$1 is 2/3. Let's look at the various outcomes for a particular game and the amount of money you will win or lose.

Name the cups A, B, and C, and suppose that the pea is actually under cup A. (The analysis will be similar if the pea is under another cup.) Complete the following table:

THE PEA IS UNDER CUP *A*

The Cup You Pick	Your Winnings/Losses
A	Win \$5
B	Lose \$1
C	Lose \$1

Viewing winnings as positive numbers and losses as negative numbers, we can calculate the average winnings over the various possible outcomes:

$$\text{Average winnings} = \frac{5 + 2 \cdot (-1)}{3} = \$1.00$$

We can rewrite the fraction above as

$$\frac{5}{3} + \frac{2 \cdot (-1)}{3}$$

Another way of looking at this is to say that 1/3 of the time you will win \$5 and 2/3 of the time you will lose \$1. Or,

$$(1/3) \cdot (\$5) + (2/3) \cdot (-\$1) = \$1.00$$

Thus, our expected winnings or expected value per game is \$1.00. Over a 10-game span, we can expect to win $10 \cdot (\$1) = \$10.00$.

It is important not to interpret the word *expect* in this context too literally. The calculated expected value does not constitute a promise or guarantee of future earnings. It is merely an estimate. Although not an oracle, expected value can be of assistance in comparing games of chance.

For example, suppose that there were four cups in the carnival game, but the payoffs were the same as those for the three-cup game. The expected value for the four-cup game would then be

$$(1/4) \cdot (\$5) + (3/4) \cdot (-\$1) = \$0.50$$

We should be more inclined to play the three-cup game because our expected winnings are twice those of the four-cup game.

PRACTICE

1. What is the expected value of a five-cup game if all the payoffs are the same as those for the three-cup game?

2. What is the expected value of a six-cup game if all the payoffs are the same as those for the three-cup game?

3. What is the expected value of a seven-cup game if all the payoffs are the same as those for the three-cup game?

4. What is the expected value of a two-cup game if all the payoffs are the same as those for the three-cup game?

Let's look a little closer at the changing expected values as the number of cups in the game is increased or decreased. As before, if you win, you get $5, and if you lose, you lose $1.

EXPECTED VALUE OF CARNIVAL GAME

Number of Cups	Expected Value
2	$2.00
3	$1.00
4	$0.50
5	$0.20
6	$0.00
7	−$0.14

Notice that for two, three, four, and five cups, your expected winnings are positive, but for seven cups, your winnings are negative. A negative expected value indicates that you will probably lose money no matter how long you play the game. In the six-cup game, your expected winnings are $0. This means that, on average, you can expect to neither win nor lose money. Games that have an expected value of 0 are called **fair games.**

The two-, three-, four-, and five-cup games are not fair because you have a monetary advantage. The seven-cup game is not fair because you have a monetary disadvantage.

Is an eight-cup game with the same payoffs a fair game? Explain.

We can generalize the work we did on the cup game to create a formula for the expected value of any game. Suppose that an experiment has several events $E_1, E_2, E_3, \ldots E_k$, each of which has an associated payoff $M_1, M_2, M_3, \ldots M_k$. Also suppose that the events are **mutually exclusive.** A collection of events is mutually exclusive when any two events, when viewed as sets, have no common elements. In the context of probability, this means that no two of the events can happen simultaneously.

The expected value of this game is given by the **expected value formula:**

$$E = P(E_1) \cdot M_1 + P(E_2) \cdot M_2 + P(E_3) \cdot M_3 + \cdots + P(E_k) \cdot M_k$$

Let's apply the formula to the three-cup carnival game.

The two events with monetary value are (1) picking the cup with the pea (call it E_1), and (2) picking a cup without the pea (call this event E_2). We can make a little table of these events, their probabilities, and their payoffs.

Event	Probability of Event	Payoff for Event
E_1	1/3	$5
E_2	2/3	−$1

Using the formula for expected value, we get

$$E = (1/3) \cdot (\$5) + (2/3) \cdot (-\$1) = \$1.00$$

Exactly the same answer as before!

Here's the table for another game. Compute the expected value of the game using the formula.

Event	Probability of Event	Payoff for Event
E_1	1/4	\$5
E_2	1/2	−\$1
E_3	1/4	\$2

(Answer: $E = (1/4) \cdot (\$5) + (1/2) \cdot (-\$1) + (1/4) \cdot (\$2) = \$1.25$)

Here is another example. Consider the following two-queens game. Two cards are dealt to you from a standard deck. If the first card dealt to you is a queen and the second isn't, then you win \$50. If the first card dealt isn't a queen and the second is, then you win \$100. If both cards dealt to you are queens, then you must pay \$200.

To compute the expected value of this game, we need to make a table of events and payoffs. But before we make the table, it is important to be sure that the events involved are mutually exclusive. Explain why the three events in this game are mutually exclusive.

Here is the table. Event E_1 is the event in which only the first card is a queen, E_2 is the event in which only the second card is a queen, and E_3 is the event in which both cards are queens. The sample space for this game has $52 \cdot 51 = 2652$ outcomes.

TWO-QUEENS GAME

Event	Probability of Event	Payoff for Event
E_1	192/2652	\$50
E_2	192/2652	\$100
E_3	12/2652	−\$200

The expected value for the two-queens game is

$$E = (192/2652) \cdot (\$50) + (192/2652) \cdot (\$100) + (12/2652) \cdot (-\$200)$$
$$= \$9.95$$

Thus, on average, you can expect to win \$9.95 per game if you play this game. The game is not fair. The game is to your advantage. How would you feel about playing this game?

PRACTICE

5. You roll two fair dice. If you get *doubles,* then you win \$72. Otherwise, you must pay \$6. What is the expected value of this game? Is it a fair game?

6. What is the expected value of the lottery described in Situation 12?

7. You draw one card from a standard deck. If it is a heart, then you win \$52. If it is a club, then you win \$26. If it is a spade, then you must pay \$78. What is the expected value of this game?

Situation 14

Do You Have *Gilliganitis*?

One of your relatives has a rare genetic disease called *gilliganitis* that manifests itself as an uncontrollable desire to watch ''Gilligan's Island'' reruns. You are concerned that you too may have *gilliganitis,* so you go to see your physician. She explains to you that 85% of the population actually has *gilliganitis.* There is a simple blood test for the disease that correctly identifies the presence of the disease 98% of the time, but will falsely announce the presence of *gilliganitis* in a person who really doesn't have the disease 1% of the time. You agree to take the blood test. The test results say that you have the disease. What is the probability that you actually have *gilliganitis*?

If you said 0.998 (not 0.98!), then you can go on to Situation 15. If you replied something else, then you had better keep reading. The probability calculation is a little tricky because of the imperfect nature of the blood test. Before we can arrive at the solution to this situation, we will need to lay some groundwork.

4.10 PROBABILITY TREES

Consider the following experiment.

Change-the-balls experiment. A bucket contains three red balls and four yellow balls. A single ball is selected at random from the bucket.

If the ball selected is red, then *it and three more red balls* are put into the bucket. If the ball selected is yellow, then *it and two more yellow balls* are placed in the bucket. In either case, a second ball is then selected from the bucket. The color of the second ball is noted.

The change-the-balls experiment looks like one to which the fundamental counting principle could be applied. However, our calculations are complicated by the changing contents of the bucket. The table below shows the four possible outcomes of this experiment.

CHANGE-THE-BALLS EXPERIMENT

Contents of Bucket	First Ball	New Contents	Second Ball
Three red, four yellow	Red	Six red, four yellow	Red
Three red, four yellow	Red	Six red, four yellow	Yellow
Three red, four yellow	Yellow	Three red, six yellow	Red
Three red, four yellow	Yellow	Three red, six yellow	Yellow

What is the probability of getting a red ball on the second draw?

There are two ways to get a red on the second draw. You could get a red ball on the first draw and then a red on the second. On the other hand, you could get a yellow on the first draw and then a red on the second. Thus,

$$P(\text{red ball on second draw}) = P(\text{red ball first and red ball second}) + P(\text{yellow ball first and red ball second})$$

We can use the fundamental counting principle to calculate P(red ball first and red ball second). There are seven choices for the first ball and 10 for the second. Thus, $n(S) = 7 \cdot 10 = 70$. There are three red balls to choose from on the first draw and six to choose from on the second. Thus, the n(red ball first and red ball second) $= 3 \cdot 6$. Thus,

$$P(\text{red ball first and red ball second}) = \frac{3 \cdot 6}{7 \cdot 10} = \frac{18}{70} = \frac{9}{35}$$

Similarly, we can calculate

$$P(\text{yellow ball first and red ball second}) = \frac{4 \cdot 3}{7 \cdot 9} = \frac{12}{63} = \frac{4}{21}$$

Thus,

$$\begin{aligned} P(\text{red ball on second draw}) &= P(\text{red ball first and red ball second}) + \\ &\quad P(\text{yellow ball first and red ball second}) \\ &= 9/35 + 4/21 = 329/735 \approx .447 \end{aligned}$$

PRACTICE

1. Calculate P(yellow ball on second draw) for the change-the-balls experiment.

In the preceding calculations, you should have noticed that the probability of getting a red ball on your second draw depends on what happened on the first draw.

PRACTICE

2. What is the probability of getting a red ball second if the first ball drawn is red?

3. What is the probability of getting a red ball second if the first ball drawn is yellow?

4. What is the probability of getting a red ball second if you don't know what the color of the first ball was?

Probabilities that vary depending on what other information you have are called **conditional probabilities.** We use the symbol "|" to stand for the phrase *given the event.* Thus, we can write the probability that the second ball is red if the first ball drawn is yellow as $P(\text{red second}|\text{yellow first})$. The symbolic versions of the probabilities in the last practice exercises are

$$P(\text{red second}|\text{red first}) = 6/10 = .60$$

$$P(\text{red second}|\text{yellow first}) = 3/9 = 1/3 = .33$$

$$P(\text{red second}) \approx .447$$

We can show all these events and their probabilities on a **probability tree.** To make the tree more readable, let

R_1 = the event that a red ball is selected first
Y_1 = the event that a yellow ball is selected first
R_2 = the event that a red ball is selected second
Y_2 = the event that a yellow ball is selected second

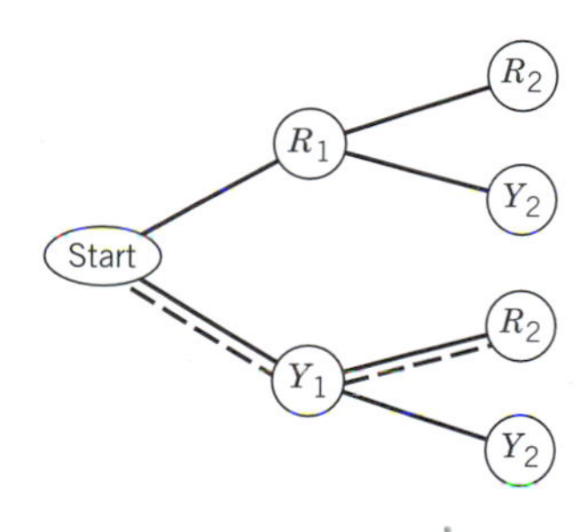

The tree displays all the outcomes to the experiment. You can create an outcome by starting at the start circle and then moving along the lines from left to right until you reach the end of the tree. For example (follow the dotted line on the tree), if you begin at the start circle and then *walk* down to the Y_1 circle and along the line going up to the R_2 circle, you will have traced out the outcome in which a yellow ball is drawn first and a red ball second.

Next we put probabilities on the *branches* like this:

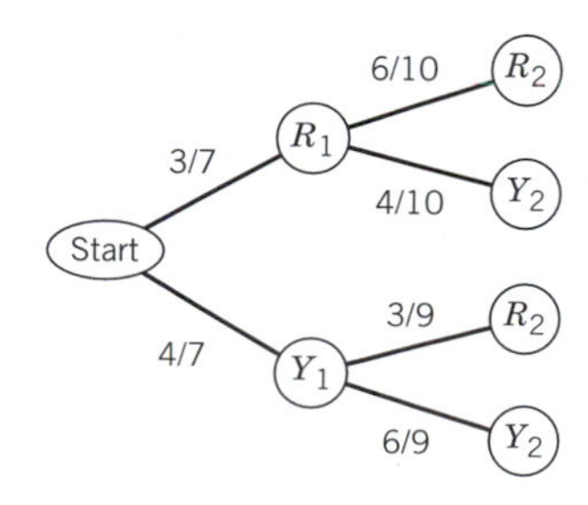

The value on the branch connecting the start circle with the R_1 circle tells us that the probability of getting a red ball on the first draw is 3/7. The value on the branch connecting R_1 to Y_2 informs us that the probability of getting a yellow ball on the second draw *given* that we have gotten a red ball first is 4/10. Thus,

$$P(R_1) = 3/7 \qquad \text{and} \qquad P(Y_2|R_1) = 4/10$$

PRACTICE

Fill in the blanks using the probability tree.

5. $P(Y_1) =$

6. $P(Y_2|\quad\quad) = 6/9$

7. $P(R_2|R_1) =$

8. $P(\quad\quad|Y_1) = 3/9$

To finish the tree, we use the fundamental counting principle to put the probabilities of each outcome next to the last circle in that outcome on the tree. To get these values, we form the product of all the probabilities on the branches connecting the start circle to the final outcome.

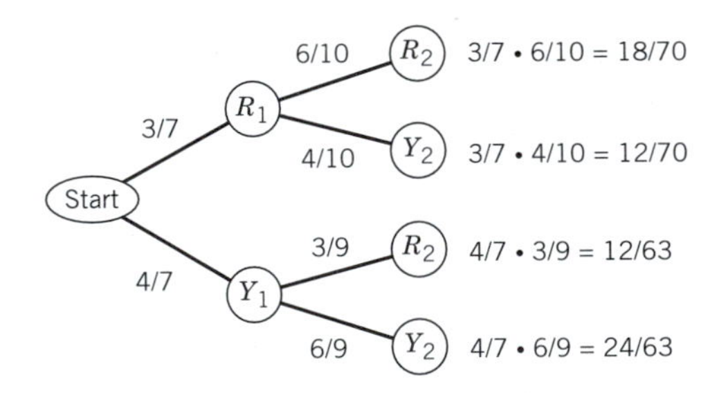

The tree values tell us that

P(red ball first and red ball second) = 18/70

P(red ball first and yellow ball second) = 12/70

P(yellow ball first and red ball second) = 12/63

P(yellow ball first and yellow ball second) = 24/63

Notice that the sum of these probabilities is 1 because these are the only four possible outcomes. The sum of the probabilities along all the branches of a probability tree is always 1.

PRACTICE

A bucket has three blue balls and two white balls in it. A ball is selected at random from the bucket. If the ball selected is blue, then it and three more blue balls are placed in the bucket.

If the ball selected is white, then it is put back into the bucket and two blue balls are removed. In either case, a second ball is selected at random from the bucket.

9. Draw a probability tree for this experiment.

10. What is the probability that a white ball will be selected second if the first ball is blue?

11. What is the probability that the first ball will be blue and the second white?

12. What is the probability that the second ball will be white?

If we replace each probability in the top outcome of our probability tree by its symbolic name, we can see a nice relationship.

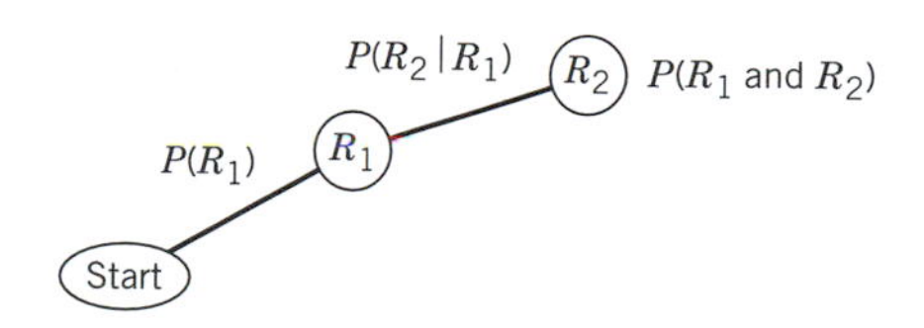

Notice that $P(R_1 \text{ and } R_2) = P(R_1) \cdot P(R_2|R_1)$. If we divide both sides of this equation by $P(R_1)$, we get

$$P(R_2|R_1) = P(R_1 \text{ and } R_2)/P(R_1)$$

Using general events A and B, we have the **conditional probability formula:**

$$P(A|B) = P(A \text{ and } B)/P(B)$$

This formula has a wide variety of applications in situations where conditional probabilities arise. This formula and the probability tree can be used together to solve the problem posed in Situation 14.

4.11 BACKWARDS CONDITIONAL PROBABILITY

Our solution to the *gilliganitis* problem begins with a probability tree for Situation 14. Before we can draw the tree, however, we need to decide the best way to construct it. This comes down to what is first and what is second. In this case, if you have the disease, you had it before you heard about your relative. Thus, the blood test logically follows the presence or absence of the disease. Let

G = the event that you have *gilliganitis*
NG = the event that you don't have *gilliganitis*
Pos = the event that the blood test says you have *gilliganitis*
Neg = the event that the blood test says you don't have *gilliganitis*

Then $P(pos|G)$ represents the probability that the test will correctly identify the presence of the disease, $P(neg|G)$ the probability that it will miss it, $P(pos|NG)$ the probability that the test will tell you that you have the disease when you don't, and $P(neg|NG)$ the probability that the test will correctly say that you are free of the disease.

The tree looks like this.

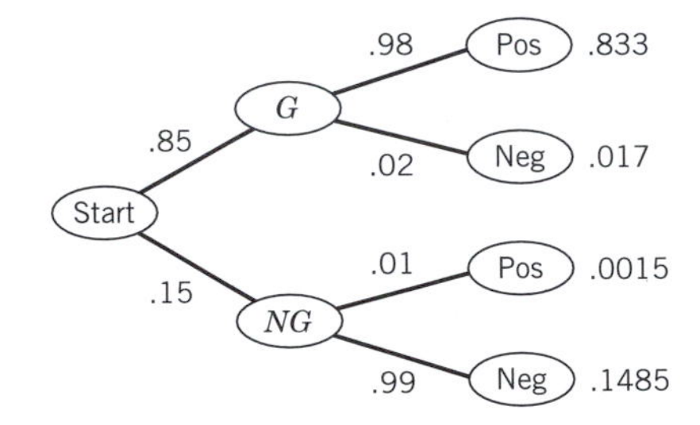

PRACTICE

All questions refer to the *gilliganitis* tree.

1. What does the number .1485 represent?

2. What is the probability that you have the disease and the test fails to detect it?

3. What is the probability that the test will say you don't have the disease given that you do have the disease?

We are now to the point where we can answer the question, ''What is the probability that you really have the disease when the test says that you do?''

In symbols, we are concerned with $P(G|pos)$. Notice that $P(pos|G)$ is on the tree (.98), but $P(G|\text{pos})$ isn't. The order or appearance is backwards. To get at this *backwards conditional,* we will use the conditional probability formula.

$$P(G|pos) = P(G \text{ and } pos)/P(pos)$$
$$= .833/P(pos)$$

What is $P(pos)$?

Good! $P(pos) = P(G \text{ and } pos) + P(NG \text{ and } pos) = .833 + .0015 = .8345$. So, $P(G|pos) = .833/.8345 = .998$.

PRACTICE

The congregation of a large church is 60% female and 40% male. The preacher has noticed that males have .77 probability of falling asleep during his sermons, but females only .40 probability of falling asleep.

4. Draw a probability tree for this situation.

5. What is the probability that a person selected at random will stay awake during the sermon?

6. If you hear someone sleeping during the sermon, what is the probability that it is a male?

7. In a comparative study of TV weather forecasters, it was discovered that, up to July 23, the TV-9 forecaster made accurate predictions 52% of the time and the TV-8 forecaster made accurate predictions 48% of the time. On July 23, the TV-9 forecaster said that there was a 70% chance of rain for July 24. The TV-8 forecaster predicted a 75% chance of rain. It rained on July 24. What is the probability that the TV-8 forecaster made an accurate prediction?

Situation 15

Quality Control

You are the quality inspector for Precision Bolts, Ltd. Your company produces small bolts that are used by the aerospace industry. The supervisor in charge of production says that the bolt-making machinery will turn out only one defective bolt per thousand bolts produced. You open a box of 50 bolts. If the supervisor is right about the rate of defects, what is the probability that you will find at least one defective bolt in that box?

4.12 BERNOULLI TRIALS

Our solution to the problem posed in Situation 15 will depend on some simplifying assumptions. First, we will assume that there are only two kinds of bolts, perfect and defective. Second, we assume that the supervisor is correct in her estimate about the defective rate. Third, we assume that the quality of a bolt is independent of the quality of any bolt preceding it in the production process. This last assumption may not necessarily hold in all cases, but it is necessary for our solution.

The bolt-manufacturing process is an experiment that repeats itself over and over again. Each time the experiment is performed, there are two possible outcomes, good bolt or bad bolt. We know the probability of getting a bad bolt (1/1000) and a good bolt (999/1000). And, we know that past history has no effect on the current experiment. Experiments that have the following three properties are called **Bernouilli experiments**:

1. Only two outcomes
2. Probability of one of the outcomes is known
3. If the experiment is repeated, the probability of each outcome is unaffected by the results of previous experiments

When the experiments are repeated over and over, they are referred to as **Bernoulli trials.**

Here are some other examples of Bernoulli trials.

Repeated-toss-a-coin experiment: Toss a fair coin 20 times in succession and keep track of what you get.

Repeated-throw-a-die experiment: Throw a fair die five times in succession and keep track of whether or not you get a 2. Notice that we are considering only two outcomes: 2 and *other*.

Consider the repeated-throw-a-die experiment. The fundamental counting principle tells us that we have $6 \cdot 6 \cdot 6 \cdot 6 \cdot 6 = 7776$ possible outcomes to this experiment. If we let T stand for the event that we roll a 2 (we'll call this a success) and O for the event that we roll something else (a failure), we can write the sample space for this experiment:

{*TTTTT*, *TTTTO*, *TTTOT*, *TTOTT*, *TOTTT*, *OTTTT*, *TTTOO*, *TTOTO*, *TTOOT*,

TOOTT, *OTTTO*, *TOTOT*, *TOTTO*, *OOTTT*, *OTTOT*, *OTOTT*, *TTOOO*, *TOTOO*,

TOOTO, *TOOOT*, *OTTOO*, *OTOTO*, *OTOOT*, *OOTTO*, *OOTOT*, *OOOTT*, *TOOOO*, *OTOOO*, *OOTOO*, *OOOTO*, *OOOOT*, *OOOOO*}

Keep in mind that *TTTTO* is shorthand for this event {22221, 22223, 22224, 22225, 22226}.

When we look at the probability of each of these 32 outcomes, we see that they are not equally likely. For example,

$$P(TTTTT) = 1 \cdot 1 \cdot 1 \cdot 1 \cdot 1/(6 \cdot 6 \cdot 6 \cdot 6 \cdot 6) = 1/7776$$
$$P(TOOTT) = 1 \cdot 5 \cdot 5 \cdot 1 \cdot 1/7776 = 25/7776$$

PRACTICE

Calculate

1. $P(TTOOO)$

2. $P(TOOOT)$

3. $P(TTOTT)$

What is the probability that we will roll exactly three 2's in our five rolls? Well, we can get three 2's in 10 different ways: *TTTOO*, *TTOOT*, *TTOTO*, *TOOTT*, *TOTOT*, *TOTTO*, *OTTTO*, *OTTOT*, *OTOTT*, and *OOTTT*. The probability of each of these is 25/7776. Why?

____

____

____

Right! In each case, we have three 1's and two 5's in the numerator: $P(TTTOO) = 1 \cdot 1 \cdot 1 \cdot 5 \cdot 5/7776$, $P(TTOOT) = 1 \cdot 1 \cdot 5 \cdot 5 \cdot 1/7776$, $P(TTOTO) = 1 \cdot 1 \cdot 5 \cdot 1 \cdot 5/7776$, and so on.

Thus, the probability of getting exactly three 2's (three successes) is $10 \cdot (25/7776) = 250/7776 \approx .321$.

You are probably wondering how I can be sure that I counted all the ways three *T*'s and two *O*'s can be arranged? Look at it this way.

There are five spots to be filled in our sequence of letters. Let's number them:

$$\underset{1}{\_}\ \underset{2}{\_}\ \underset{3}{\_}\ \underset{4}{\_}\ \underset{5}{\_}$$

Now three of those spaces are to be occupied by 2's. So, we have to select three of the five spaces to be *awarded* 2's. Suppose I select spaces 3, 1, and 4. Then I've formed the sequence *TOTTO*. Does this sound familiar? Yes, we are counting combinations. So, the number of different sequences with exactly three 2's is the number of ways I can select three places from five, or $C(5, 3) = 10$.

PRACTICE

Use combinations to count the following sequences in the repeated-throw-a-die experiment:

4. The number of sequences with exactly four 2's.

5. The number of sequences with exactly one 2.

6. The number of sequences with more than three 2's.

7. The number of sequences with at least one 2.

Consider $P(TOOTT)$ again. We found that $P(TOOTT) = 1 \cdot 5 \cdot 5 \cdot 1 \cdot 1/(6 \cdot 6 \cdot 6 \cdot 6 \cdot 6)$. This can be written as a product of fractions (1/6) (5/6) (5/6) (1/6) (1/6) and then regrouped to obtain (1/6) (1/6) (1/6) (5/6) (5/6). Finally, writing the last product using exponents gives $(1/6)^3(5/6)^2$. Now $P(TTOOT)$, $P(TOTOT)$, and all the other outcomes with exactly three 2's have the same probability as *TOOTT*. So, we can write the probability of getting exactly three 2's as

$$C(5, 3)\ (1/6)^3(5/6)^2$$

This is a very useful expression. Let's examine its parts carefully:

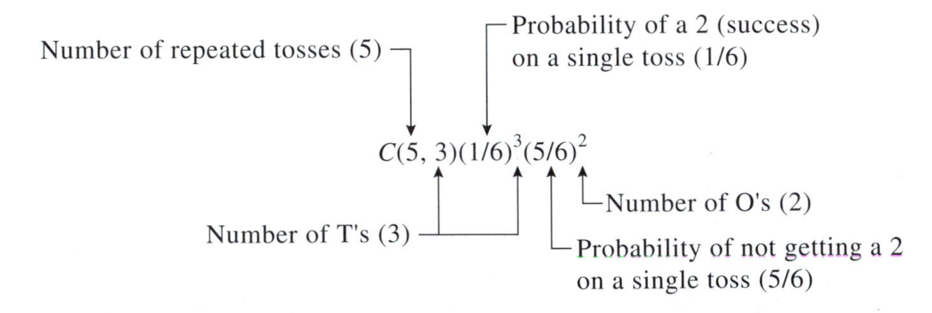

The expression for the probability of getting exactly three 2's in five tosses is a special case of the very useful **binomial formula.**

The probability of obtaining exactly k successes in n Bernoulli trials when the probability of a success in a single trial is p is

$$C(n, k) \cdot p^k \cdot (1 - p)^{n-k}$$

In the case of getting exactly three 2's in five tosses, $n = 5$, $k = 3$, and $p = 1/6$.

PRACTICE

Use the binomial formula to calculate the following probabilities:

8. The probability of getting exactly four 2's when a die is tossed five times in succession.

9. The probability of getting exactly three heads when a fair coin is tossed 10 times in succession.

If we want to find the probability of getting at least or at most a certain number of successes, we simply add the appropriate probabilities or subtract the appropriate probabilities from 1.

Thus, in the repeated-throw-a-die experiment,

$$P(\text{at least two 2's}) = P(\text{exactly two 2's}) + P(\text{exactly three 2's}) + P(\text{exactly four 2's}) + P(\text{exactly five 2's})$$

or

$$P(\text{at least two 2's}) = 1 - P(\text{exactly no 2's}) - P(\text{exactly one 2})$$

We can now answer the question posed in Situation 15. We want P(at least one defective in 50 trials) $= 1 - P$(exactly no defectives in 50 trials). We have $n = 50$, $k = 0$, and $p = .001$. So,

$$1 - P(\text{exactly no defectives}) = 1 - C(50, 0) \cdot (.001)^0 \cdot (.999)^{50} = .049$$

We would be very surprised to find a defective bolt in a box of 50!

PRACTICE

10. What is the probability of getting exactly four heads when a fair coin is tossed six times in succession?

11. What is the probability of getting at most two heads when a fair coin is tossed 10 times in succession?

12. What is the probability of getting at least five 5's when a fair die is tossed six times in succession?

Situation 16

Does the Cardinal Prefer the Cedar House?

Mei Li has four bird feeders on her apartment balcony. She fills them each day with the same bird seed mixture that she buys at a local hardware store. The first bird to visit Mei Li's feeder each morning is a brilliant red cardinal whom Mei Li has named Fred. Mei Li works as an actuary for an insurance company, so she is naturally interested in probability. She has noticed that Fred feeds at the cedar house bird feeder more often than at the others. In fact, in 40 observations, Mei Li saw Fred at the cedar house 17 times, at the pine house feeder 8 times, at the redwood house 7 times, and at the plastic house 8 times.

Does Fred really prefer the cedar house or are the variations in feeding locations just the result of chance?

4.13 THE CHI-SQUARE STATISTIC

The answer to the question posed in Situation 16, and to many similar questions, can be found in a comparison between the observed feeding habits of the cardinal and what we would expect to see if each feeder had an equally likely chance of being visited. Such a comparison should point us toward deciding whether or not the cardinal's behavior is not random.

Suppose that the cardinal's behavior was completely random. In 40 observations, how many times would we expect him to visit the cedar house? What about the other feeders?

If you said that the cardinal should visit each feeder 10 times, then you were correct. The probability of visiting each feeder is 1/4, so each feeder should be visited 1/4 of 40 visits or 10 times.

Let's put the observed visits of Fred and the visits expected from a randomly feeding cardinal in a table:

OBSERVED AND EXPECTED FEEDER VISITS

	Observed	Expected
Cedar house	17	10
Pine house	8	10
Redwood house	7	10
Plastic feeder	8	10

Notice that the observed values may be greater or smaller than expected. A statistic proposed by Karl Pearson in 1900 summarizes the variation between the observed and expected values. It is called the **chi-square** (or χ^2) **statistic** and computed using this formula:

$$\chi^2 = \sum \frac{(O - E)^2}{E}$$

The Σ means *sum up*, the O stands for the observed value, and the E stands for the corresponding expected value. This calculation can be accomplished readily using the table of observations we have already created. We will just add two new columns and add up the last one. Here it is:

OBSERVED AND EXPECTED FEEDER VISITS

	Observed O	Expected E	$(O - E)^2$	$(O - E)^2/E$
Cedar house	17	10	$(17 - 10)^2$	$(17 - 10)^2/10$
Pine house	8	10	$(8 - 10)^2$	$(8 - 10)^2/10$
Redwood house	7	10	$(7 - 10)^2$	$(7 - 10)^2/10$
Plastic feeder	8	10	$(8 - 10)^2$	$(8 - 10)^2/10$

And here is the same table after a little arithmetic:

OBSERVED AND EXPECTED FEEDER VISITS

	Observed O	Expected E	$(O - E)^2$	$(O - E)^2/E$
Cedar house	17	10	49	4.9
Pine house	8	10	4	.4
Redwood house	7	10	9	.9
Plastic feeder	8	10	4	.4
				$\chi^2 = 6.6$

Make sure that you see where each entry in the preceding tables came from.

PRACTICE

Compute the χ^2 statistic for each of the following sets of data:

1.

30 COIN FLIPS

	Observed	Expected
Heads	20	15
Tails	10	15

2.

42 DIE TOSSES

	Observed	Expected
One	7	7
Two	5	7
Three	20	7
Four	0	7
Five	5	7
Six	5	7

3. A single card is drawn from a deck, its suit is recorded, and then it is put back into the deck. The deck is then shuffled. This is repeated 100 times. It is found that a heart is drawn 24 times, a spade 27 times, a club 23 times, and a diamond 26 times.

4. Look at the χ^2 values for the preceding sets of data. What can you say about the relationship between the size of χ^2 and the closeness of the observed values to the expected values.

Your answer to the last Practice Exercise gave a rough description of the relationship between the χ^2 statistic and the appearance of a set of data. We can get a more precise relationship between the χ^2 statistic and the variation between observed and expected values by using a statistical test called the **chi-square test.** However, before we describe this test, we need to talk a little about a class of important statistical processes called **tests of significance.**

4.14 TESTS OF SIGNIFICANCE

The question Mei Li had about the feeding behavior of the cardinal is an example of a vast group of questions that ask, "Is it chance or something else?" A **test of significance** is a mathematical way of answering this kind of question. In this section, we will look at what makes up a test of significance and then in the next section we will use a particular test (the chi-square test) to answer the question about Mei Li's cardinal.

The four characteristics of a test of significance are

1. The **null hypothesis** H_0
2. The **alternative hypothesis** H_a
3. The **significance level** α
4. The **test statistic**

The two hypotheses, H_0 and H_a, represent the possible answers to the question, "Is it chance, or something else?" The null hypothesis H_0 says that the observed variation is just due to chance. The alternative hypothesis H_a says that the observed variation results from something other than chance; it is real.

For Mei Li and her cardinal, the two hypotheses are

H_0: The cardinal's feeding patterns are just due to chance.

H_a: The cardinal's feeding patterns are not just due to chance.

These hypotheses can also be expressed mathematically in terms of probabilities. Let

C = the event that the cardinal feeds at the cedar house
I = the event that he feeds at the pine house
R = the event that he feeds at the redwood house
L = the event that he feeds at the plastic feeder.

Then

H_0: $P(C) = P(I) = P(R) = P(L) = 1/4$

H_a: At least one of the probabilities $P(C)$, $P(I)$, $P(R)$, or $P(L)$ is not 1/4.

PRACTICE

Suppose that you are conducting a test of significance for each of the following situations. State the null and alternative hypotheses as statements about the situation and as probabilities.

1. A coin is tossed 30 times with the results shown below. Is this a fair coin?

30 COIN FLIPS

	Observed	Expected
Heads	20	15
Tails	10	15

2. A die is tossed 42 times with the results shown below. Is this a fair die?

42 DIE TOSSES

	Observed	Expected
One	7	7
Two	5	7
Three	20	7
Four	0	7
Five	5	7
Six	5	7

3. A single card is drawn from a deck, its suit is recorded, and then it is put back into the deck. The deck is then shuffled. This is repeated 100 times. It is found that a heart is drawn 24 times, a spade 27 times, a club 23 times, and a diamond 26 times. Do the results suggest something *fishy*?

If we perform a test of significance and then act on the results of that test, we could make one of two different errors. On the one hand, we could decide that the null hypothesis is incorrect and reject it. But, if the null hypothesis is really true, we made a mistake. This kind of error, in which we reject the null hypothesis when it is true, is called a **type I error.** On the other hand, we could decide that the null hypothesis is correct and not reject it. But, if the null hypothesis is really false, we made a mistake. This kind of error, in which we accept (do not reject) the null hypothesis when it is false, is called a **type II error.**

The box below summarizes the possible errors and correct decisions that we could make using a test of significance.

	NULL HYPOTHESIS	
Decision	**True**	**False**
Reject H_0	Type I error	Correct decision
Accept H_0	Correct decision	Type II error

The **level of significance** of a test of significance is the probability of making a type I error. In Situation 16, for example, Mei Li could decide that the cardinal's behavior is not just due to chance. This decision would reject the null hypothesis, H_0. But, statistical tests are not perfect! If the cardinal's behavior is truly just due to chance then Mei Li would be rejecting a true null hypothesis. She would be making a type I error. She cannot know for sure if she is right or wrong in rejecting the null hypothesis, but she can estimate the probability that she is wrong. This probability is the level of significance of her test.

Obviously, the lower the level of significance the more confident we can be about rejecting the null hypothesis. Why?

____

____

Right! The lower the level of significance the lower the probability of erroneously rejecting the null hypothesis.

If the level of significance is .05 (i.e., our probability of a type I error is .05) and we reject the null hypothesis, then we say that we have rejected H_0 at the .05 level of significance.

Certain small levels of significance are given special names that emphasize the fact that the probability of rejecting the null hypothesis is small. If we can reject H_0 at the .05 level of significance (i.e., our probability of a type I error is .05), then we will say along with many statisticians that the variation in our data is *statistically significant.* If we can reject H_0 at the .01 level of significance or lower (i.e., our probability of a type I error is .01 or less), then we will say that the variation is *highly significant.*

The **test statistic** of a test of significance is a numerical measure of the difference between the observed data and the expected outcome according to the null hypotheses. Each test of significance has its own test statistic. The test of significance that uses the chi-square statistic as its test statistic is called the **chi-square test.**

The test statistic, when compared to a standard value given in a statistical table, tells you whether or not you may reject the null hypothesis with an α probability of error. We will use a specific table in the next section, but for now just assume that some appropriate table value is available.

For example, if $\alpha = .05$, the test statistic has a value of 4.8, and the table value for the test of significance is 4.5, then we may reject the null hypothesis with a probability of error of, at most, .05. On the other hand, if $\alpha = .05$, the test statistic has a value of 4.0, and the table value for the test of significance is 4.5, then we may not reject the null hypothesis if we want our probability of error to be no more than .05.

The chi-square statistic is one of many possible test statistics. (For a discussion of the strengths and weaknesses of the various possible test statistics, consult a statistics textbook.) In this brief discussion, we will assume that the chi-square statistic is the appropriate one to use.

In making a test of significance, we generally proceed as follows:

1. Select a test of significance.
2. State the null and alternative hypotheses.
3. Specify the level of significance α that we are willing to accept.
4. Compute the test statistic.
5. Reject or do not reject the null hypothesis based on the test statistic and α.

In Situation 16, Mei Li is a cautious statistician. Thus, she decides that she wants a .01 level of significance. Her test of significance then proceeds as follows:

1. Select a test of significance: Mei Li uses the chi-square test.
2. State the null and alternative hypotheses:

   H_0: The cardinal's feeding patterns are just due to chance.
   H_a: The cardinal's feeding patterns are not just due to chance.

3. Specify the level of significance α that we are willing to accept: $\alpha = .01$.
4. Compute the test statistic: $\chi^2 = 6.6$.
5. Reject or do not reject the null hypothesis based on the test statistic and α: To complete this step, we need only to see how to interpret the chi-square statistic in light of the data and α. We do that in the next section.

4.15 THE CHI-SQUARE TEST

In Appendix C, you will find a table entitled "χ^2 Critical Points." A portion of that table is reproduced below.

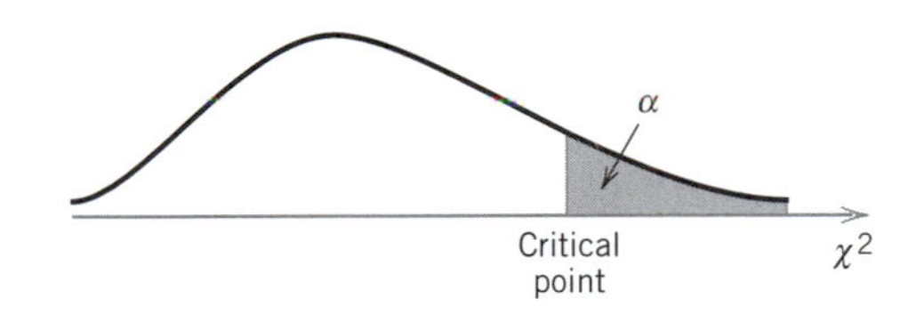

df\α	.250	.100	.050	.025	.010
1	1.32	2.71	3.84	5.02	6.63
2	2.77	4.61	5.99	7.38	9.21
3	4.11	6.25	7.81	9.35	11.30
4	5.39	7.78	9.49	11.10	13.30

The top row of the table lists various values for α. Mei Li, who is seeking a .01 level of significance, would be concerned with the column headed by .010.

The first column, marked *df*, lists possible **degrees of freedom** of the test of significance. Degrees of freedom refers to the number categories in your data that can be freely chosen if you choose one at a time. For example, the cardinal at Mei Li's balcony initially has a choice of four bird feeders. Once he chooses one of these feeders, for his next choice he has only three to pick from (for the purposes of calculation, he is not allowed to repeat a feeder). Next, he has two to choose from. Finally, only one feeder remains and the cardinal has no freedom of choice. Thus, overall the cardinal had three free choices.

In the tests of significance under consideration in this book, the number of degrees of freedom will be 1 less than the number of possible outcomes. How many degrees of freedom are there in the coin flip, die toss, and card draw examples?

(Answers: coin flip: $df = 2 - 1 = 1$; die toss: $df = 6 - 1 = 5$; card draw (suits!): $df = 4 - 1 = 3$)

Returning to the table, Mei Li would look down the *df* column until she found 3 (the number of degrees of freedom for her data). Next, she would read across until she was under .010 (the α value for her test of significance). The number there is 11.3.

Check it yourself, now!

This value, 11.3, is the critical point for Mei Li's test. If her test statistic is greater than 11.3, then she may reject the null hypothesis at the .01 level of significance. If her test statistic is less than or equal to 11.3, then she may not reject the null hypothesis at the .01 level of significance.

The computed χ^2 is 6.6. So, Mei Li cannot reject the null hypothesis at the .01 level of significance. She cannot conclude with a probability of error .01 or less than the cardinal's behavior at the feeders is not random.

However, because 6.6 is greater than 6.25, Mei Li can reject the null hypothesis at the .10 level of significance. She can conclude that the cardinal's behavior is not random, but the probability that she is wrong is at least .10. Still we would not call the variation from chance statistically significant because for that we would like the probability of a type I error to be no more than .05.

It is important to remember that the chi-square test of significance provides a *probability* that variations from expected distributions are not due to chance. It says nothing about *why* the variations occur.

PRACTICE

Use the chi-square test as a test of significance for each of the following sets of data. Use $\alpha = .1$, $\alpha = .05$, and $\alpha = .01$.

1. A coin is tossed 30 times with the results shown below. Is this a fair coin?

30 COIN FLIPS

	Observed	Expected
Heads	20	15
Tails	10	15

2. A die is tossed 42 times with the results shown below. Is this a fair die?

42 DIE TOSSES

	Observed	**Expected**
One	7	7
Two	5	7
Three	20	7
Four	0	7
Five	5	7
Six	5	7

3. A single card is drawn from a deck, its suit is recorded, and then it is put back into the deck. The deck is then shuffled. This is repeated 100 times. It is found that a heart is drawn 24 times, a spade 27 times, a club 23 times, and a diamond 26 times. Do the results suggest something *fishy*?

4.16 WHAT DO YOU KNOW?

If you have worked carefully through this chapter, then you can now compute the probability of a variety of experiments, calculate odds, and find the expected value of games of chance. The exercises in this section are designed to test and refine your ability to work with probability.

Fred, Mel, Polly, and Kate draw straws to see who will shampoo the carpet.

1. Write the sample space for this experiment.
2. What is the probability that Kate will shampoo the carpet?
3. What is the probability that Fred will not shampoo the carpet?
4. What is the probability that a man will shampoo the carpet?

A fair coin is tossed three times in succession and the result noted each time.

5. Write the sample space for this experiment.
6. What is the probability that the coin will come up heads three times?
7. What is the probability that we will get a tail followed by two heads?
8. What is the probability that we will get no tails?
9. What is the probability that we will get more tails than heads?

Sade selects a soup and salad from a menu listing bean soup, chicken soup, Caesar salad, spinach salad, and shrimp salad.

10. Write the sample space for this experiment.
11. What is the probability that Sade will select bean soup and shrimp salad?
12. What is the probability that Sade will select bean soup?

A bucket contains three apples and two pears. A piece of fruit is selected at random and given to Fern.

13. Write the sample space for this experiment.
14. What is the probability that Fern will get a pear?
15. What are the odds against Fern receiving an apple?
16. The odds against winning a certain game are 40 to 1. What is the probability of winning this game? What is the probability of not winning this game?
17. The odds in favor of winning a certain game are 20 to 3. What is the probability of not winning this game? What is the probability of winning this game?
18. What are the odds against drawing an ace when you draw a single card from an ordinary deck?
19. What is the probability of drawing a jack when you draw a single card from an ordinary deck?
20. What is the probability that you will not draw a king when you draw a single card from an ordinary deck?

Two fair dice are tossed.

21. Find $n(S)$.
22. What is the probability of rolling a sum of 3?
23. What is the probability of rolling a sum of 7?
24. What are the odds in favor of rolling a sum of 6?
25. If you receive \$2 for rolling a sum of 2, \$3 for rolling a sum of 3, \$4 for rolling a sum of 4, and so on, what is the expected value of this game?
26. How many different permutations of the letters in the word *oryx* are possible?
27. How many different permutations of the letters in the word *chain* are possible?
28. How many different three-letter code words begin with the letter "B"?
29. How many different three-letter codes end with the letter "G"?
30. How many different three-letter codes consist of the same letter repeated three times?
31. How many different three-letter codes begin and end with the letter "Y"?
32. How many different three-letter codes have no repeated letters?
33. How many different three-letter codes have at least one repeated letter?
34. In how many different ways can five people line up for a group photograph?
35. In how many different ways can five people line up for a group photograph if one of them must be in the middle of the picture?

Andrew, Becky, Carlos, Denzel, Everett, and Fern are finalists in a contest. Two of them will win a free trip to Bhutan.

36. Write all the possible pairs of winners.
37. What is the probability that Andrew and Becky will win the free trip?
38. What is the probability that Denzel will win a free trip?
39. What is the probability that Denzel will not win a free trip?
40. What are the odds against Denzel winning a free trip?
41. Hong wants to open his front door. It is dark and so he cannot see well enough to find his keys. So, he picks a key at random. He has seven keys on his key ring. What are the odds against Hong finding the right key on his first try?

A bucket contains a red orb, blue orb, white orb, green orb, and golden orb. The orbs are all the same size and small enough so that two can be grabbed with one hand. A blindfolded Mario reaches into the bucket and grabs two orbs. He puts the orbs on a table and removes his blindfold. Mario then notes the colors of his two orbs.

42. Write the sample space for this experiment.

43. What is the probability that Mario will get the red orb and blue orb?

44. What is the probability that Mario will not get the golden orb?

45. What is the probability that Mario will get the green orb?

46. What are the odds against Mario getting the red orb?

47. Suppose that the red orb is worth $5 and the golden orb $100. The other orbs have no monetary value. Mario paid $5 to play the game. What is Mario's expected value?

Ms. DeRight's third grade class consists of 25 students. Fifteen of the students are girls. Three of the students are to be chosen at random to take part in the annual Arbor Day parade.

48. How many different groups of three students are possible?

49. What is the probability that only girls will be chosen to take part in the parade?

50. What is the probability that you will be dealt three aces when you are dealt three cards from a standard deck?

51. You toss a fair die. If you get an even number, then you win $12; otherwise, you must pay $12. Find the expected value of this game.

A bucket contains a red orb, blue orb, white orb, golden orb, and green orb. The red orb is worth $5 and the golden orb $100. The other orbs have no monetary value. Rachel selects one orb at random from the bucket. If that orb is red, golden, or green, then she keeps it and the game is over. If the orb she selects is blue or white, then it is thrown away and Rachel gets to select another orb at random from the bucket. Rachel gets to keep the second orb she draws and the game is over.

52. Draw a probability tree for this game.

53. What is the probability that Rachel will get to keep the golden orb?

54. If it costs Rachel $5 to play this game, what is her expected value?

A single card is drawn from a standard deck.

55. What is the probability of drawing the queen of hearts?

56. What is the probability of drawing the queen of hearts given that a heart was drawn?

57. What is the probability of drawing the queen of hearts given that a queen was drawn?

58. What is the probability of drawing a heart given that a queen was drawn?

59. What is the probability of drawing a queen given that a heart was drawn?

60. A medical test correctly identifies the presence of a virus 95% of the time. However, the test will falsely say that the virus is present when it really isn't 2% of the time. It is estimated that 80% of the people in a certain town have the virus. A person from this town is tested and the test results show that she doesn't have the virus. What is the probability that she does have the virus?

Bouncy Autos, Inc. has two manufacturing plants. The Rolling Hills plant produces 40% of Bouncy's cars. The remainder of the cars are produced at the Desert Springs plant. One percent of the cars produced by the Rolling Hills plant have defective dome lights, whereas only 0.7% of the Desert Springs cars have defective dome lights. A Bouncy auto selected at random has a defective dome light.

61. What is the probability that this car was made at the Rolling Hills plant?

62. What is the probability that this car was manufactured at the Desert Springs plant?

In a human population, 49% are males and 51% females. Five percent of the males are color-blind and 0.4% of the females are color-blind. A person is selected at random from this population.

63. What is the probability this person is female given that this person is color-blind?

64. What is the probability this person is color-blind given that this person is female?

65. What is the probability that this person is female and color-blind?

Suppose that 5% of the items produced by a factory are defective. Also suppose that the production of an item is independent of the production of any previous item. Six items are produced by this factory.

66. What is the probability that exactly two items are defective?

67. What is the probability that exactly five items are defective?

68. What is the probability that at least one item is defective?

69. What is the probability that at most one item is defective?

70. In a 10-question true–false test, what is the probability of randomly guessing the answers to at least six questions correctly?

Bouncy Autos, Inc. has three manufacturing plants. The Rolling Hills plant produces 40% of Bouncy's cars. The Rock Island plant produces 20% of Bouncy's cars. The remainder of the cars are manufactured at the Desert Springs plant. One percent of the cars produced by the Rolling Hills plant have defective dome lights, 1.3% of the cars produced at the Rock Island plant have defective dome lights, and 0.7% of the Desert Springs cars have defective dome lights. A Bouncy auto selected at random has a defective dome light.

71. What is the probability that this car was made at the Rolling Hills plant?

72. What is the probability that this car was manufactured at the Rock Islant plant?

73. What is the probability that this car was produced at the Desert Springs plant?

Two cards are dealt in succession from a standard deck. The first card is dealt face down and the second face up.

74. Draw a probability tree for this experiment. Consider only two outcomes for each card dealt: queen or not queen.

75. If the second card dealt is a queen, what is the probability that the first card dealt is a queen?

76. What are the odds against being dealt two queens?

77. You win $2600 if you get two queens; otherwise, you must pay $26. What is your expected value?

In a popular game, you toss five dice in an attempt to get all five of the dice to come up with the same number. Consider the tossing of the five dice to be equivalent to tossing a single die five times in succession.

78. What is the probability that you will roll five 6's?

79. What is the probability that you will roll five of the same number?

80. Five fair dice are tossed simultaneously. Are you more likely to get five of the same number or a sequence of five numbers (i.e., 1, 2, 3, 4, 5 or 2, 3, 4, 5, 6)?

A European Roulette wheel has 37 compartments: 18 black, 18 red, and 1 green. To win, a player must try to pick the color of the compartment that the silver ball will come to rest in.

81. What is the probability of winning if the player picks green?

82. What is the probability of winning if the player picks red?

83. What is the probability of winning if the player picks black?

84. What is the probability of winning?

85. A player bets $1 and picks the color black. If the ball lands in black, the player will receive $2 (this includes her $1 bet). What is the player's expected value?

86. A player bets $1 and picks the color green. If the ball lands in green, the player will receive $2 (this includes her $1 bet). What is the player's expected value?

The probability that a transplanted clump of prairie grass will survive is .70 if it rains in the next 36 hours. If it doesn't rain in the next 36 hours, the survival probability is .40.

87. What is the probability that the clump will survive?

88. The clump is alive after 36 hours. What is the probability that it rained?

89. In a small town in Iowa, of 100 babies born in the last year, 65 were female. Is this statistically significant?

90. Fred was throwing dice. In 10 tosses, he threw *doubles* 8 times. Is this statistically significant?

91. In Dr. Koopa's geography class, there are 14 men and 14 women. In three consecutive days of class, Dr. Koopa called on a man 25 times and a woman 10 times. Was this statistically significant?

92. A child's game has a spinner. The spinner is divided into four equal-sized regions: red, yellow, green, and blue. In 64 spins, red was pointed to 20 times, yellow 30 times, green 10 times, and blue 4 times. Is the spinner balanced fairly?

93. A basketball player usually hits 55% of her jump shots. In a recent game, she hit 70% of her jump shots. Can you explain this increase in performance using probabilities?

94. Design a project for which a χ^2 test would be appropriate. Collect the data, carry out the test, and discuss the results.

ANSWERS TO PRACTICE EXERCISES

Section 4.1

1. Lions, tigers, bears, manatees; 4
2. Bottle number 1, bottle number 2, . . . bottle number 99; 99
3. {Lions, tigers, bears, manatees}
{Bottle number 1, bottle number 2, . . . bottle number 99}
4. {blue, red, yellow}; 3
5. A female baby is born.
6. {Lions, tigers}
7. The bottle of beer chosen has a number that is divisible by 11.

Section 4.2

1. 1/6
2. 2/6 = 1/3
3. 3/6 = 1/2
4. 2/4 = 1/2
5. 1/4
6. 3/4
7. 1 − 1/2 = 1/2
8. 1 − 4/52 = 48/52 = 12/13
9. 1 − 1/1000 = 999/1000 = .999

Section 4.3

1. 51 to 1
2. 999 to 1
3. 11 to 1
4. 44 to 1
5. 2 to 13
6. 1 to 2
7. 1/15
8. 1/101
9. 2/25
10. 7/52

Section 4.5

1. $12 \cdot 5 \cdot 6 \cdot 15 \cdot 1 = 5400$ Different outfits
2. $26 \cdot 26 \cdot 26 \cdot 26 = 456{,}976$ Different codes
3. $26 \cdot 26 \cdot 10 \cdot 10 \cdot 10 = 676{,}000$ Different license plates
4. $1 \cdot 26 \cdot 26 \cdot 1 = 676$ Different codes
5. All codes − codes with all letters the same = $26 \cdot 26 \cdot 26 \cdot 26 - 26 \cdot 1 \cdot 1 \cdot 1 = 456{,}950$ Different codes
6. $1 \cdot 26 \cdot 26 \cdot 1 = 676$ Different codes

Section 4.6

1. $7 \cdot 6 \cdot 5 \cdot 4 \cdot 3 \cdot 2 \cdot 1 = 5040$

2. $7 \cdot 6 \cdot 5 = 210$

3. $5 \cdot 4 \cdot 3 \cdot 2 = 120$

Section 4.7

1. 362,880

2. 14

3. $77 \cdot 76 \cdot 75 = 438{,}900$

Section 4.8

1. 15

2. 10

3. 70

4. 56

5. 28

6. 1

7. $C(13, 3)/C(52, 3) = 11/850$

8. $C(17, 3)/C(20, 6) = 1/57$

Section 4.9

1. $(1/5) \cdot (\$5) + (4/5) \cdot (-\$1) = \$0.20$

2. $(1/6) \cdot (\$5) + (5/6) \cdot (-\$1) = \$0.00$

3. $(1/7) \cdot (\$5) + (6/7) \cdot (-\$1) = -\$0.14$

4. $(1/2) \cdot (\$5) + (1/2) \cdot (-\$1) = \$2.00$

5. Let E_1 = you roll doubles, and E_2 = you don't roll doubles.

Event	Probability of Event	Payoff for Event
E_1	6/36	\$72
E_2	30/36	−\$6

$$E = (6/36) \cdot (\$72) + (30/36) \cdot (-\$6) = \$7.00$$

This game is not fair.

6. Let E_1 = you win the lottery, and E_2 you don't win the lottery.

Event	Probability of Event	Payoff for Event
E_1	1/25,827,165	\$50 million − \$1
E_2	25,827,164/25,827,165	−\$1

$$E = (1/25{,}827{,}165) \cdot (\$49{,}999{,}999) + (25{,}827{,}164/25{,}827{,}165) \cdot (-\$1) = \$0.94$$

7. Let E_1 = you draw a heart, E_2 = you draw a club, and E_3 = you draw a spade.

Event	Probability of Event	Payoff for Event
E_1	1/4	\$52
E_2	1/4	\$26
E_3	1/4	−\$78

$$E = (1/4) \cdot (\$52) + (1/4) \cdot (\$26) + (1/4) \cdot (-\$78) = \$0.00$$

Section 4.10

1. $P(\text{yellow second}) = P(\text{red first and yellow second}) + P(\text{yellow first and yellow second})$

$$= \frac{3 \cdot 4}{7 \cdot 10} + \frac{4 \cdot 6}{7 \cdot 9} = \frac{12}{70} + \frac{24}{63} = \frac{6}{35} + \frac{8}{21}$$

$$= 406/735 \approx .553$$

2. $6/10 = .60$

3. $3/9 \approx .33$

4. .447

5. 4/7

6. Y_1

7. 6/10

8. R_2

9.

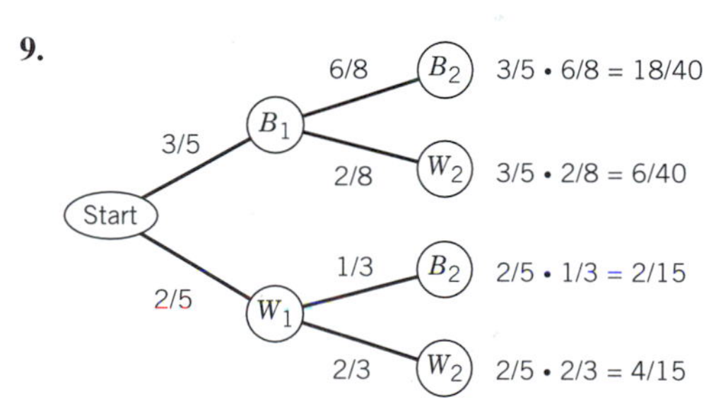

10. 2/8

11. 6/40

12. $6/40 + 4/15 = 5/12$

Section 4.11

1. The probability that you don't have the disease and the test says you don't.

2. .017

3. 0.02

4.

Start
.4 → M: .77 → S .308; .23 → NS .092
.6 → F: .4 → S .24; .6 → NS .36

5. $.092 + .36 = .452$

6. $P(M|S) = P(M \text{ and } S)/P(S) = .308/(.308 + .24) = .308/.548 \approx .562$
7. $P(\text{TV-8}|\text{rain}) = (.48) \cdot (.75)/((.48) \cdot (.75) + (.52) \cdot (.70)) \approx .50$

Section 4.12

1. 125/7776
2. 125/7776
3. 5/7776
4. $C(5, 4) = 5$
5. $C(5, 1) = 5$
6. $C(5, 4) + C(5, 5) = 5 + 1 = 6$
7. $C(5, 1) + C(5, 2) + C(5, 3) + C(5, 4) + C(5, 5) = 31$ or $32 - C(5, 0) = 31$
8. $C(5, 4) \cdot (1/6)^4 \cdot (5/6)^1 = 25/7776$
9. $C(10, 3) \cdot (1/2)^3 \cdot (1/2)^7 = 120/1024$
10. $C(6, 4) \cdot (1/2)^4 \cdot (1/2)^2 = 15/64$
11. $C(10, 0) \cdot (1/2)^0(1/2)^{10} + C(10, 1) \cdot (1/2)^1(1/2)^9 + C(10, 2) \cdot (1/2)^2(1/2)^8$
12. $C(6, 5) \cdot (1/6)^5 \cdot (5/6)^1 + C(6, 6) \cdot (1/6)^6 + (5/6)^0$

Section 4.13

1.

30 COIN FLIPS

	O	E	$(O - E)^2$	$(O - E)^2/E$
Heads	20	15	25	25/15
Tails	10	15	25	25/15

$\chi^2 = 50/15 = 3.333.$

2.

42 DIE TOSSES

	O	E	$(O - E)^2$	$(O - E)^2/E$
One	7	7	0	0
Two	5	7	4	4/7
Three	20	7	169	169/7
Four	0	7	49	49/7
Five	5	7	4	4/7
Six	5	7	4	4/7

$\chi^2 = 230/7 = 32.857.$

3.

100 CARD DRAWS

	O	E	$(O - E)^2$	$(O - E)^2/E$
Heart	24	25	1	1/25
Spade	27	25	4	4/25
Club	23	25	4	4/25
Diamond	26	25	1	1/24

$\chi^2 = 10/25 = .400.$

4. The card-drawing data had the lowest χ^2 and the observed card draws differed very little from the expected values. The die-tossing data gave the largest χ^2 and also showed the greatest variation between observed and expected values. The coin-tossing data fell somewhere in between. It looks like a larger χ^2 corresponds to a larger variation between observed and expected values.

Section 4.14

1. H_0: The results of the coin flips are due to chance.
   H_a: Something other than chance is causing the variation.
   H_0: $P(H) = P(T) = 1/2$ (the coin is fair).
   H_a: $P(H) \neq 1/2$ (the coin is not fair).
2. H_0: The results of the die tosses are due to chance.
   H_a: Something other than chance is causing the variation.
   H_0: $P(1) = P(2) = P(3) = P(4) = P(5) = P(6) = 1/6$ (the die is fair).
   H_a: At least one of the probabilities $P(1), P(2), \ldots P(6)$ is not 1/6 (the die is not fair).
3. H_0: The results of the card draws are due to chance.
   H_a: Something other than chance is causing the variation.
   H_0: $P(\heartsuit) = P(\spadesuit) = P(\clubsuit) = P(\diamondsuit) = 1/4$ (the shuffling is not fishy).
   H_a: At least one of the probabilities $P(\heartsuit), P(\diamondsuit), P(\clubsuit), P(\spadesuit)$ is not 1/4 (something is fishy).

Section 4.15

1. $\alpha = .10$; $3.333 > 2.71$; reject H_0 (the coin is not fair).
   $\alpha = .05$; $3.333 < 3.84$; do not reject H_0; not statistically significant.
   $\alpha = .01$; $3.333 < 6.63$; do not reject H_0; not highly significant.
2. $\alpha = .10$; $32.857 > 9.24$; reject H_0.
   $\alpha = .05$; $32.857 > 11.1$; reject H_0; statistically significant.
   $\alpha = .01$; $32.857 > 15.1$; reject H_0; highly significant (the die is not fair).
3. $\alpha = .10$; $.4 < 6.25$; do not reject H_0.
   $\alpha = .05$; $.4 < 7.81$; do not reject H_0; not statistically significant.
   $\alpha = .01$; $.4 < 11.3$; do not reject H_0; not highly significant.

Chapter Five

Mathematics of Relationships

What's the Point? Much of the work of science involves the defining of relationships between quantities. How many hawks are needed to control the rodents in a corn field? How long will it take the world population to double? How much power is needed to run a room air conditioner?

Scientific relationships are usually expressed in mathematical terms. This chapter will help you to understand some of the basic kinds of mathematical relationships that you are likely to encounter in science and business.

Situation 17 If This Trend Continues . . .

Your automobile-tire-installing business is doing well. In February, your business earned a profit of $1980, in March a profit of $2000, and in May a profit of $2040. If this trend continues, how much more will your profit from the tire-installing business be in June than it was in May? _____ Why?

____

____

____

If this trend continues, will your profit be greater in November or December? Why?

____

____

____

The profit amounts given in Situation 17 show an increase each month. Thus, if this trend continues, we should expect to have greater and greater profits as each month goes by.

We can see the *trend* in a graph like the one below. The boxes (■) indicate the profit amounts for the given months of February, March, and May. The line indicates that we expect the trend of increasing profits to continue. Because the trend seems to be a straight line, the relationship between time and profit in Situation 17 is said to be **linear.**

Notice that we can estimate our profits for the coming months from this graph.

What will the profit be in September?

____

____

____

(Answer: $2120)

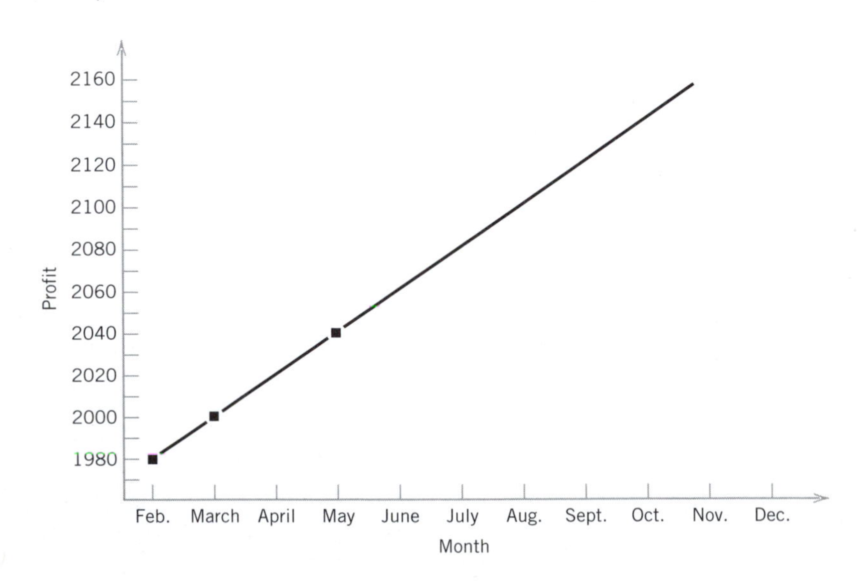

We obtained the estimate for September by reading straight up from September on the month line to the trend line and then reading straight over to the profits. However, this method of estimation is really very dependent on how good our graph is. Fortunately, there are mathematical techniques for making the September profit estimate and even predicting when we will make a profit of $3000. Let's look at these techniques and some other properties of linear relationships.

5.1 LINEAR RELATIONSHIPS

If the graph of the relationship between two things is a straight line, then we say that the relationship is **linear** and the two things are **linearly related.**

In Situation 17, profit and time are linearly related. Here are two more examples of linear relationships:

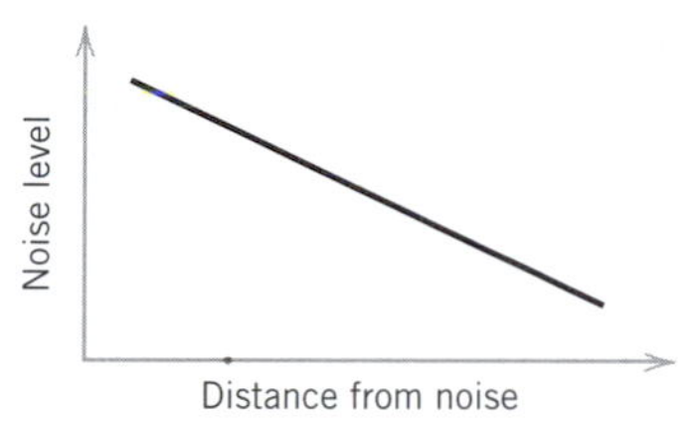

This graph says that the farther away we get from the source of a noise (say, a jackhammer), the lower the noise level of that source.

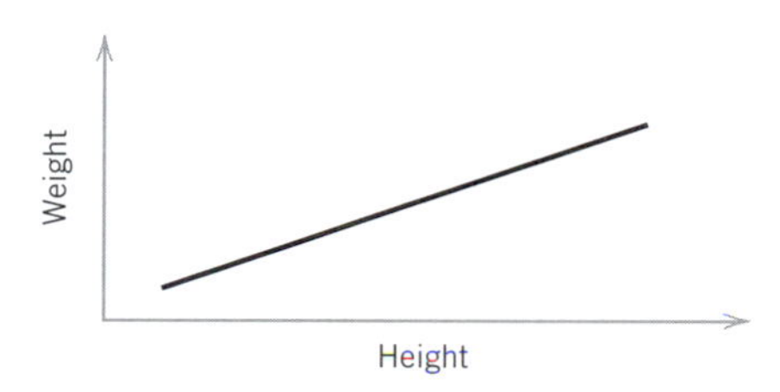

This graph says that tall people tend to weigh more than short people.

What difference do you see between the lines in these two graphs?

____

____

____

When the line of a linear relationship goes down to the right, we say that the line has **negative slope.** The line that shows the relationship between noise level and distance has negative slope.

When the line of a linear relationship goes up to the right, we say that the line has **positive slope.** The line that shows the relationship between height and weight has positive slope.

Not all relationships are linear. Here is the relationship between time and the number of bacteria in a petri dish.

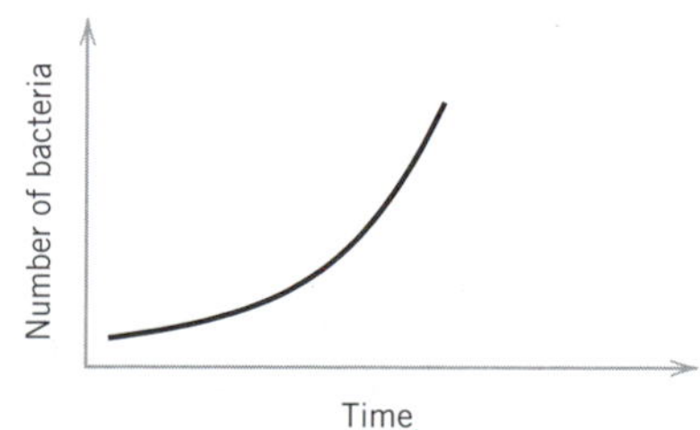

PRACTICE

Which of the following relationships are linear? For those that are linear, tell if they have positive or negative slope.

1. 

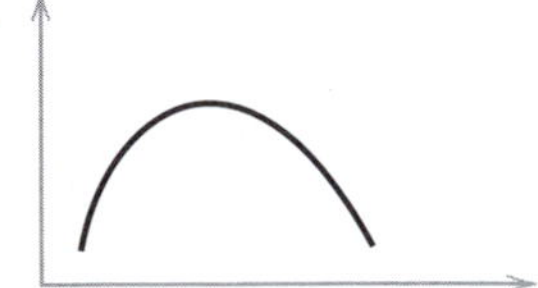

2. 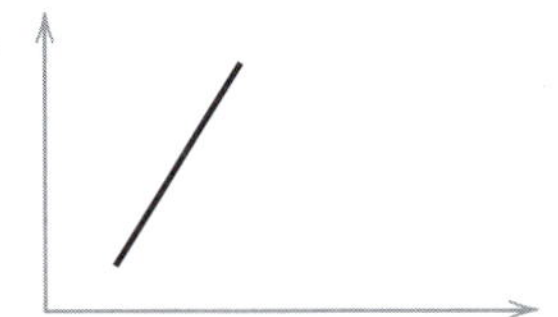

3. 

4. 

5. 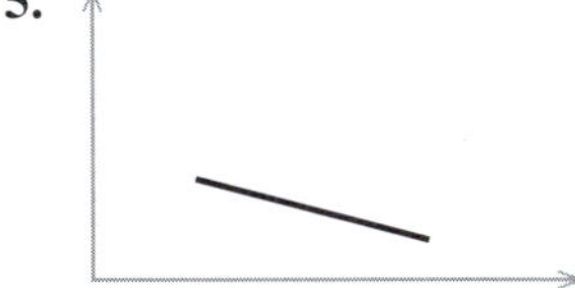

6. 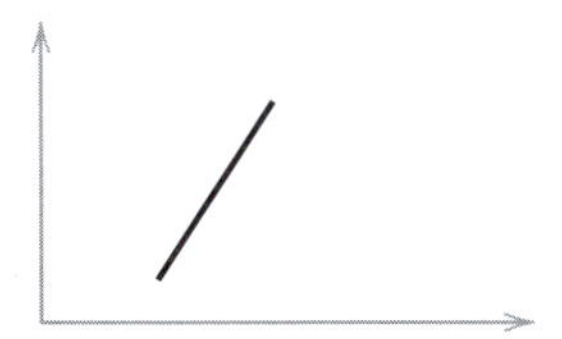

Before we get to the mathematical description of a linear relationship, we need to introduce some terminology about the graphs of linear relationships. Here's a linear relationship between two quantities A and B.

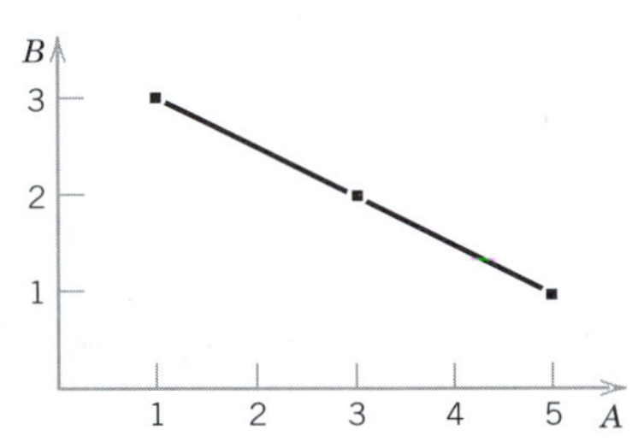

The horizontal line that is labeled with the possible values of quantity A is called the ***A*-axis** or **horizontal axis.** The vertical line that is labeled with the possible values of quantity B is called the ***B*-axis** or **vertical axis.**

Both the horizontal and vertical axes can be extended in either direction if the values of the quantities they represent include negative values. For example, the linear relationship between temperature given in degrees Fahrenheit and temperature in degrees Celsius looks like this.

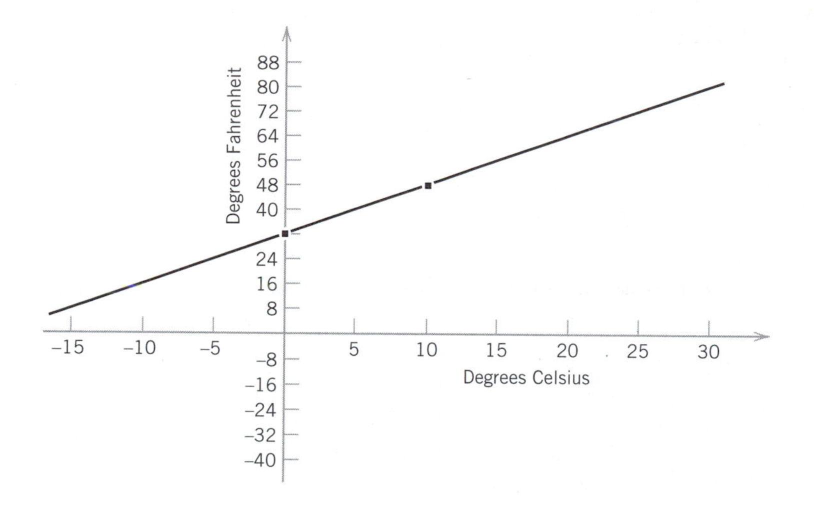

Here the horizontal axis gives the values for temperature in degrees Celsius and the vertical axis the values for temperature in degrees Fahrenheit. Since temperatures can be negative (i.e., below zero), both axes have some negative labels. It is customary to label both axes with the positive numbers increasing to the right (horizontally) and up (vertically) and the negative numbers decreasing in magnitude to the left and down. Unless otherwise labeled, the point where the two axes cross is the zero point on both. This zero point is called the **origin.**

Notice that for any graph of a linear relationship, any location on the line is completely specified by two values; the distance that location is horizontally from the vertical axis and the distance that location is vertically from the horizontal axis.

In our A–B relationship (redrawn below), you can see that the position of the point on the line marked with a heart is a distance of 3 horizontally from the origin (shown by the value on the A-axis) and 2 vertically from the origin (shown by the value on the B-axis).

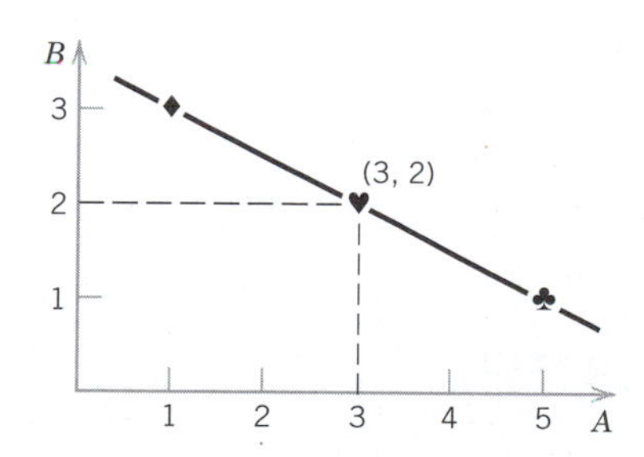

We use the notation (3, 2) to identify a point located a distance of 3 horizontally from the origin and a distance of 2 vertically from the origin. (3, 2) are called the **coordinates** of the heart point.

What are the coordinates of the point marked with a ♦? _____

What are the coordinates of the point marked with a ♣? _____

Right! The coordinates of the point marked with a ♣ are (5, 1) and those of the point marked with a ♦ are (1, 3). Remember that in the coordinate notation, the horizontal axis value always comes first!

PRACTICE

7. Find the coordinates of each of the points *A–D*.

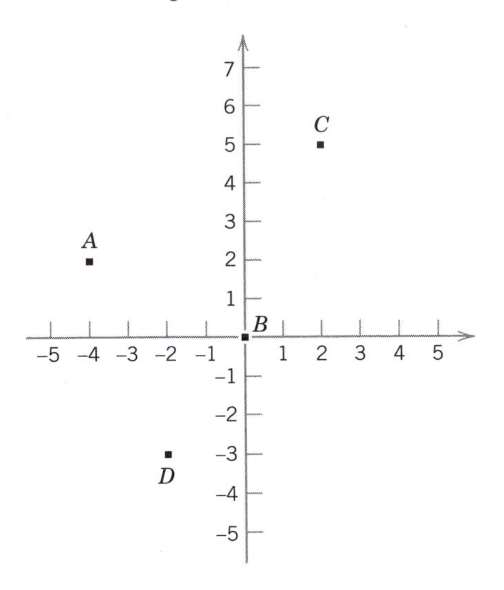

8. Locate the following points on the graph:
Point *E* with coordinates (4, 3)
Point *F* with coordinates (−2, 3)

Each of the points on the line of the *A–B* relationship (redrawn again below) can be given coordinates that describe its position with respect to the two axes. The coordinates of the boxes on the line are (1, 3), (3, 2), and (5, 1).

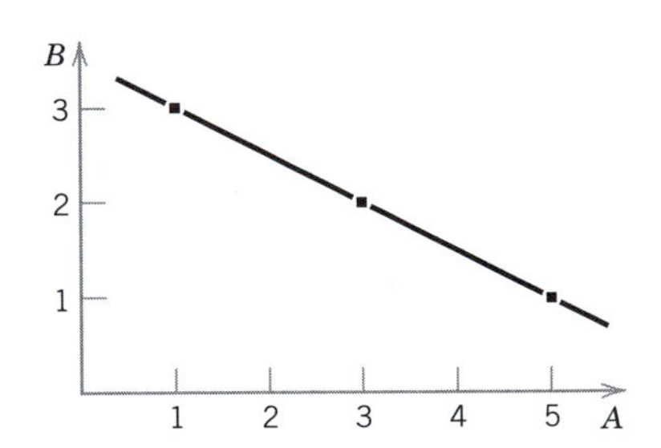

A mathematical description of this linear relationship is an equation that would give us these coordinates without the picture. Such an equation is called a **linear equation.**

5.2 LINEAR EQUATIONS

A **linear equation** is an equation with two variables that has as its only solutions the coordinates of a straight line. ''Fine,'' you say, ''but, what does that mean?''

Here is a linear equation

$$y = -0.5x + 3.5$$

This equation is true for some values of x and y, and false for others. For example, if we let $x = 5$ and $y = 7$ and then substitute those values into the equation for their respective letters, we get

$$7 = -0.5 \cdot 5 + 3.5$$

or

$$7 = -2.5 + 3.5$$

or

$$7 = 1$$

which is certainly false.

On the other hand, if we let $x = 3$ and $y = 2$ and then substitute, we get

$$2 = -0.5 \cdot 3 + 3.5$$

or

$$2 = 2$$

which is true.

If a pair of values makes a linear equation true, then we say that such a pair is a **solution** to the linear equation.

Complete the following table:

THE EQUATION
$y = -0.5x + 3.5$

x	y	A solution?
5	7	No
1	3	___
3	2	Yes
4	5	___
0	3.5	___
−2	1	___

(Answers: yes, no, yes, no)

It is customary in linear equations to let x represent the horizontal axis values and y the vertical axis values. The variable that represents the horizontal axis values is called the **independent variable,** and the variable that represents the vertical axis values is called the **dependent variable.**

We can obtain as many solutions to a linear equation as we want by picking any value for the independent variable and then computing the corresponding value for the dependent variable. Here's how that works for our example equation $y = -0.5x + 3.5$:

Pick x	Compute y
1	$-0.5 \cdot 1 + 3.5 = 3$
2	$-0.5 \cdot 2 + 3.5 = 2.5$
9	$-0.5 \cdot 9 + 3.5 = -1$
−1000	$-0.5 \cdot (-1000) + 3.5 = 503.5$

Furthermore, each of these solutions represents the coordinates of a point. So, using the solutions just obtained, we have the coordinates (1, 3), (2, 2.5), (9, −1), (−1000, 503.5). If we locate the first three of these points in a picture, we get

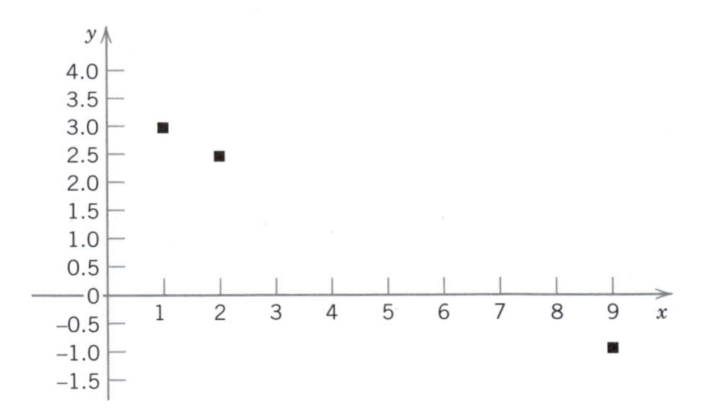

and if we connect these dots, we obtain a straight line.

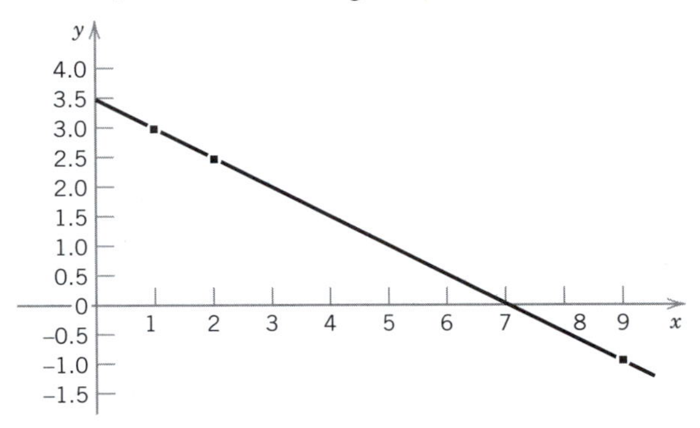

The moral of this rather long story is that the solutions of a linear equation are coordinates on a line. That line depicts the linear relationship between the x values and y values.

PRACTICE

For each of the following linear equations, find two solutions, interpret the solutions as coordinates, and draw the line produced by these two points. This process is called **graphing a linear equation.**

1. $y = 2x + 1$

2. $y = -3x + 5$

3. $y = 3x - 1$

4. $y = -4x - 5$

5. Which of the lines in Practice Exercises 1–4 have positive slope? Which have negative slope? Can you tell what number in the equation determines whether the slope of the line will be positive or negative?

You have seen how to graph a line from a linear equation. We also have seen how to produce the line from given coordinates. Our next step, and the one that will return us back to Situation 17, is to produce the linear equation from the given coordinates.

5.3 CONSTRUCTING LINEAR MODELS

If a relationship is presumed to be linear and you have at least two pairs of values of the relationship, then you can construct a linear equation whose graph will be the line that depicts such a linear relationship. You will need to follow a few short steps to construct the equation, but they will work in every situation. I'll tell you the steps first and then work through a few examples.

How to Construct a Linear Equation

For this process to work successfully, you must have two pairs of values of the relationship that you are presuming to be linear.

Step 1: Decide which quantity in the relationship is to be your independent variable (horizontal axis) and which is to be your dependent variable (vertical axis).

Step 2: Write your given pairs of values as coordinates. For the sake of discussion, call them (x_1, y_1) and (x_2, y_2).

Step 3: Compute $(y_2 - y_1)/(x_2 - x_1)$. Call this value m.

Step 4: Write the equation

$$y = mx + y_1 - mx_1$$

Step 5: Simplify the right-hand side of the equation you wrote in Step 4. The result is the linear equation whose line includes the points (x_1, y_1) and (x_2, y_2).

The equation that you get from the five steps outlined above is called a **linear model** of the relationship between the two quantities in question. I'll conclude this section with several linear models, including the one posed in Situation 17, that I get by following the five-step procedure.

LINEAR MODEL 1: FAHRENHEIT AND CELSIUS

In Canada and Europe, temperatures are given in degrees Celsius (C). When it is 0°C, it is 32° Fahrenheit (F), and when it is 100°C, it is 212°F. The relationship between degrees Celsius and degrees Fahrenheit is linear. Here is how we can construct a linear equation to show this relationship.

Step 1: I'll let degrees Celsius be my independent variable and degrees Fahrenheit be my dependent variable.

Step 2: The given information supplies me with these two sets of coordinates (0, 32) and (100, 212). So, from now on, $x_1 = 0$, $y_1 = 32$, and $x_2 = 100$, $y_2 = 212$.

Step 3: $m = (y_2 - y_1)/(x_2 - x_1) = (212 - 32)/(100 - 0) = 180/100 = 1.8$.

Step 4:

$$
\begin{aligned}
y &= mx + y_1 - mx_1 \\
&= 1.8x + 32 - 1.8 \cdot 0
\end{aligned}
$$

Step 5:

$$y = 1.8x + 32$$

We now have an equation that gives the relationship between degrees Celsius (x) and degrees Fahrenheit (y). It is often helpful to replace the x and y by more meaningful letters once you have your equation. In this case, we may want to write the equation this way

$$\mathrm{F} = 1.8\mathrm{C} + 32$$

so that the variables standing for degrees Celsius and Fahrenheit are more obvious.

LINEAR MODEL 2: HEIGHT AND WEIGHT

The height and weight of women in the age range of 20 to 24 years old are linearly related. A woman who is 5 feet tall weighs 112 pounds. A woman who is 5 feet, 4 inches tall weighs 128 pounds. Here is how we can construct a linear equation to show this relationship.

Step 1: I'll let a woman's height, in inches, be my independent variable and her weight, in pounds, my dependent variable. I chose to represent the height in inches to avoid fractions. Thus, 5 feet, 4 inches will be written as 64 inches instead of 5.3333 feet.

Step 2: The provided information gives me these two sets of coordinates (60, 112) and (64, 128). So, $x_1 = 60$, $y_1 = 112$, and $x_2 = 64$, $y_2 = 128$.

Step 3: $m = (y_2 - y_1)/(x_2 - x_1) = (128 - 112)/(64 - 60) = 16/4 = 4$.

Step 4:

$$
\begin{aligned}
y &= mx + y_1 - mx_1 \\
&= 4x + 112 - 4 \cdot 60
\end{aligned}
$$

Step 5:

$$y = 4x - 128$$

Again, we may want to rename the variables and get the linear equation $\mathrm{W} = 4\mathrm{H} - 128$.

LINEAR MODEL 3: PRICE OF BANANAS

Suppose that the price (per pound) of bananas is linearly related to the mean July temperature in Quirigua, Guatemala. When the mean July temperature is 90°F, bananas cost $1.29 per pound. When the mean July temperature is 93°F, bananas cost $1.34 per pound.

PRACTICE

Complete the steps in the construction of linear model 3, the price of bananas.

Step 1: I'll let the mean July temperature in Quirigua, Guatemala be my independent variable and the price per pound of bananas my dependent variable.

Step 2: The provided information gives me these two sets of coordinates

1. (_____, _____) and (_____, _____). So,

2. $x_1 =$ _____, $y_1 =$ _____ and $x_2 =$ _____, $y_2 =$ _____

Step 3:

3. $m = (y_2 - y_1)/(x_2 - x_1) =$ _________

Step 4:

$$y = mx + y_1 - mx_1$$

4. $y =$ _________

Step 5:

5. $y =$ _________

LINEAR MODEL 4: SITUATION 17

Recall that your tire-installing business made a profit of \$1980 in February and a profit of \$2040 in May. We also are assuming that profit and time are linearly related. Here is how we can construct a linear equation to show this relationship.

Step 1: I'll let time be the independent variable and profit the dependent variable.

Step 2: The given information supplies me with these two sets of coordinates (February, 1980) and (May, 2040). Now we have a slight difficulty.

We can't put the names of months in our formulas. So, instead we will assign numerical values to the months. This can be done any way you like, but perhaps using 2 for February (the second month of the year) and 5 for May (the fifth month of the year) is the most natural. Our coordinates then are (2, 1980) and (5, 2040). So, from now on, $x_1 = 2$, $y_1 = 1980$, and $x_2 = 5$, $y_2 = 2040$.

Step 3: $m = (y_2 - y_1)/(x_2 - x_1) = (2040 - 1980)/(5 - 2) = 60/3 = 20$.

Step 4:

$$
\begin{aligned}
y &= mx + y_1 - mx_1 \\
&= 20x + 1980 - 20 \cdot 2
\end{aligned}
$$

Step 5:

$$y = 20x + 1940$$

PRACTICE

Construct linear equations to model each of the following:

6. The price of grapefruit is linearly related to the quantity you buy. One grapefruit costs 69¢. Four grapefruits cost 60¢ each. Let the independent variable be the number of grapefruit purchased and the dependent variable the price per grapefruit.

7. The number of robins hopping happily in Howard's yard is linearly related to the number of dandelions in Howard's yard. When there are 30 dandelions, there are four robins. When there are 40 dandelions, there are five robins. Let the number of dandelions be the independent variable.

Once you have constructed a linear equation to model a situation, you can use that equation to obtain estimates of one quantity in the relationship from the other.

For example, the linear equation for Situation 17 is

$$y = 20x + 1940$$

where y is the profit in dollars and x the month with $x = 2$ representing February. Now, if we wish to estimate our profit for the month of June, we simply let $x = 6$, substitute into the equation, and compute y.

$$y = 20 \cdot 6 + 1940 = \$2060$$

The linear model estimates that our profit for June will be \$2060.

On the other hand, if we wish to know when our profits will reach \$3000, we can let $y = 3000$, substitute into the equation, and solve for x.

$$3000 = 20x + 1940$$

$$3000 - 1940 = 20x$$

$$1060 = 20x$$

$$1060/20 = x$$

$$x = 53$$

The linear model estimates that we will make a profit of \$3000 in the 53rd month (or about 4 years from May).

In the Fahrenheit–Celsius model, we obtained the linear equation

$$F = 1.8C + 32$$

If the temperature is 7°C, then we can compute the Fahrenheit equivalent like this:

$$F = 1.8 \cdot 7 + 32 = 44.6°$$

If the temperature is 88°F, then we compute the Celsius equivalent like this:

$$88 = 1.8C + 32$$
$$88 - 32 = 1.8C$$
$$56 = 1.8C$$
$$56/1.8 = C$$
$$C = 31.1°$$

PRACTICE

8. Refer to linear model 2.

(a) Estimate the weight of a woman who is 6 feet tall.

(b) Estimate the height of a woman who weighs 100 pounds.

9. Refer to linear model 3.

(a) Estimate the price of bananas when the mean July temperature in Quirigua, Guatemala is 96°F.

(b) Estimate the mean July temperature in Quirigua, Guatemala when the price of bananas is $1.50 per pound.

5.4 LINEAR REGRESSION

In all the examples we've looked at so far, we have assumed a perfect linear relationship: a straight line connects all the pairs of points. However, sometimes two sets of measurements are only approximately linearly related. As an example, let's look again at Situation 17.

Suppose that your automobile-tire-installing business earned a profit of $2090 in July, rather than the $2080 predicted by the linear model. Also suppose that the profit for September was $2100. What does the linear model predict for September's profit?

(Answer: $2120)

The actual profits for July and September are fairly close to those predicted by the linear model, but they are not perfect. You can see this on the graph.

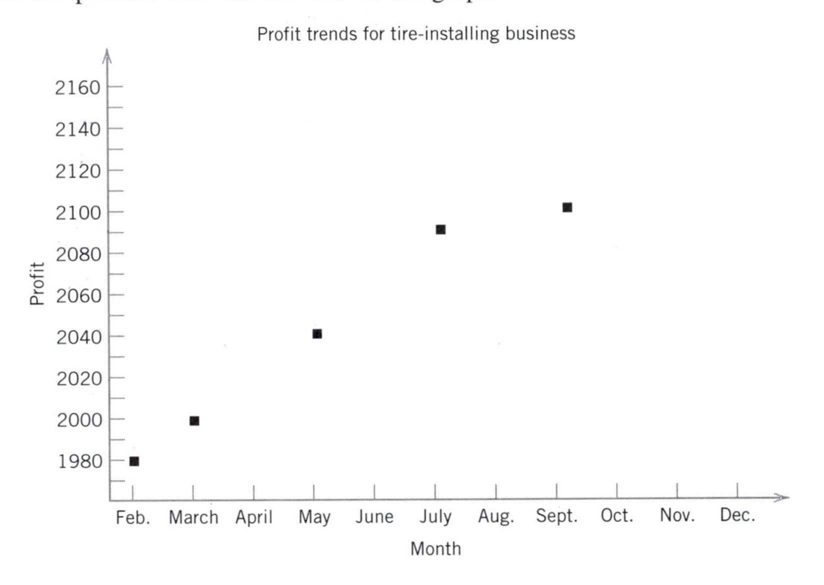

Your profits are still increasing each month and the points on the graph are approximately lined up. In this type of situation, we can construct a linear model that approximates the actual situation. We call such a model a **regression line** and the process of constructing the model **linear regression.**

The goal of linear regression is to find a linear model that comes as close as possible to connecting the points on our graph. A line is *as close as possible* to connecting the points on our graph when the sum of distances between the points on our graph and the line is as small as possible.

Does this seem like a reasonable way of defining *as close as possible*? Can you see any problems with this definition?

In order to find the regression line, we make use of the **regression formulas** shown below. The justifications for these formulas involve differential calculus, so I'll need to ask that you just accept them.

Regression Formulas

In order to find the linear model

$$y = mx + b$$

that best fits a set of data, use the following formulas in the order given:

$$m = \frac{\Sigma xy - \Sigma x \cdot \Sigma y/n}{\Sigma x^2 - \Sigma x \cdot \Sigma x/n}$$

$$b = \bar{y} - m \cdot \bar{x}$$

Don't Panic!! These are truly monstrous formulas, but they are actually not so bad to work with if you are careful. Let's first look at what all the symbols represent and then we'll work carefully through an example.

The regression formulas assume that you have a collection of points to which you wish to fit a regression line. Each of these points has an x-coordinate and a y-coordinate. In the formulas, $\bar{x}$ represents the mean of all the x-coordinates and $\bar{y}$ the mean of all the y-coordinates. The letter n in the formulas represents the number of points you have.

In our current example, we have the points (2, 1980), (3, 2000), (5, 2040), (7, 2090), and (9, 2100). (Recall that, in this model, February = 2, March = 3, etc.) We have five points, so $n = 5$. Also

$$\bar{x} = \text{(the mean of 2, 3, 5, 7, and 9)} = 5.2$$

and

$$\bar{y} = \text{(the mean of 1980, 2000, 2040, 2090, and 2100)} = 2042$$

PRACTICE

Find n, $\bar{x}$, and $\bar{y}$ for each set of data points.

1. (1, 2), (2, 4), (3, 5)

2. (10, 200), (12, 400), (13, 500), (14, 700), (16, 700), (18, 800)

The Greek letter sigma, Σ, stands for *sum*. So, Σx means "sum up the x values," Σy means "sum up the y values," Σxy means "sum up the products of the x values with the y values," and Σx^2 means "sum up the squares of the x values." These sums can be conveniently calculated and kept track of by using a table.

Make a table column heading for each of the quantities x, y, x^2, and xy.

x	y	x^2	xy

Now fill in the table one line at a time using your given data points. The first line of the table for the example we are working looks like this:

x	y	x^2	xy
2	1980	4	3960

Notice that the entry in the x^2 column (4) is the square of the entry in the x column, and the entry in the xy column (3960) is the product of the entries in the x and y columns ($2 \cdot 1980 = 3960$). The complete table looks like this:

x	y	x^2	xy
2	1980	4	3960
3	2000	9	6000
5	2040	25	10,200
7	2090	49	14,630
9	2100	81	18,900

Now add up each column. The table now looks like this:

	x	y	x^2	xy
	2	1980	4	3960
	3	2000	9	6000
	5	2040	25	10,200
	7	2090	49	14,630
	9	2100	81	18,900
Sums	26 Σx	10,210 Σy	168 Σx^2	53,690 Σxy

We'll call a table constructed in this fashion a **table of sums.**

PRACTICE

Make a table of sums for each of the sets of points.

3. (1, 2), (2, 4), (3, 5)

4. (10, 200), (12, 400), (13, 500), (14, 700), (16, 700), (18, 800)

Now that we have our table of sums, we can find the regression line using the regression formulas:

$$m = \frac{\Sigma xy - \Sigma x \cdot \Sigma y/n}{\Sigma x^2 - \Sigma x \cdot \Sigma x/n} = \frac{53{,}690 - 26 \cdot 10{,}210/5}{168 - 26 \cdot 26/5} = \frac{598}{32.8} \approx 18.23$$

$$b = \bar{y} - m \cdot \bar{x} = 2042 - 18.23 \cdot 5.2 = 1947.20$$

Thus, the linear regression model is

$$y = 18.23x + 1947.20$$

What does this linear regression model predict for March?

(Answer: \$2001.89)

The regression model's prediction is off by $1.89. Is that OK?

Correct! The linear regression model is a linear model that comes as close to all of the points as possible. In doing this, it may not actually hit any of the points! Here is a graph of the linear regression line and given points:

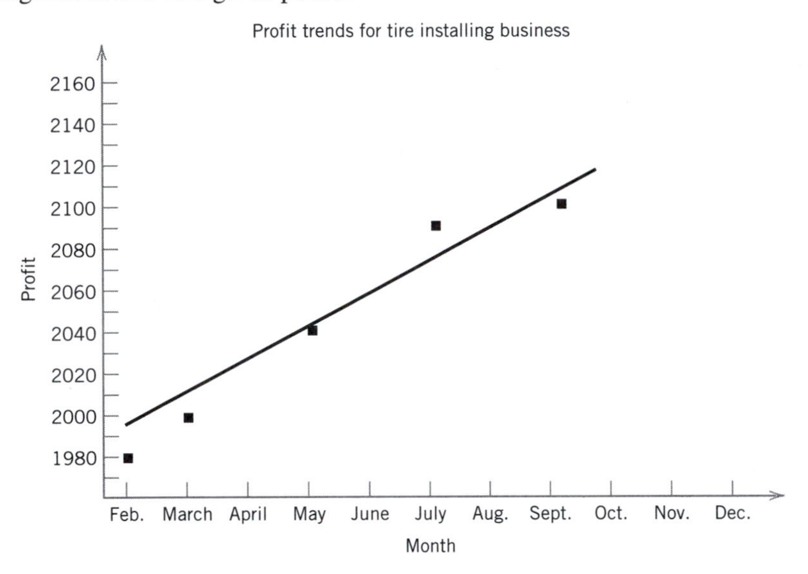

The linear regression line gives us an approximation that allows us to estimate future points based on general trends.

PRACTICE

Compute a linear regression model for each of the following sets of points. You already have the tables of sums for Practice Exercises 5 and 6.

5. (1, 2), (2, 4), (3, 5)

6. (10, 200), (12, 400), (13, 500), (14, 700), (16, 700), (18, 800)

7. (3, 4), (4, 3), (5, 1), (6, −1)

The linear regression model can be constructed for any collection of points whether or not the points look like they are arranged linearly. We could even construct a linear regression model for a set of points arranged in a circle! So, how can we tell if our regression line is really a good model for the data? One way is to compute a statistic called the **Pearson product moment coefficient** or ***r*-value.**

The r-value gives us an idea of how linearly related our data really are. The r-value is a number in the range -1 to 1. If the r-value for a set of points is 1, then the points line up perfectly with positive slope. If the r-value is -1, then the points line up perfectly with negative slope. If the r-value is 0, then the points don't line up at all. The technical term for *lining up* is **correlation.** We can summarize the meaning of the r-value in a small table:

PEARSON PRODUCT MOMENT COEFFICIENT (r-VALUE)

Value	Correlation
1	Positive
0	None
−1	Negative

r-values other than $1, -1$, and 0 represent degrees of correlation. The closer the r-value is to 1 or -1, the better the correlation. A set of data points with an r-value of 0.86 line up better than a set of points with an r-value 0.43. We say that the set with r-value 0.86 is *better correlated* than the set with r-value 0.43.

A set of data points with r-value -0.94 is better correlated than a set of points with r-value 0.86 because the first r-value is closer to a perfect correlation than the second.

Arrange the following r-values in order of increasing correlation: 0.34, −0.90, 1, −0.78, 0, 0.45, −1.

(Answer: 0, 0.34, 0.45, −0.78, −0.90, −1, 1 (the last two are tied))

The r-value may be calculated using the table of sums and one additional column. Here is the formula for the r-value:

$$r = \frac{\Sigma xy - \Sigma x\,\Sigma y/n}{\sqrt{(\Sigma x^2 - \Sigma x\,\Sigma x/n)(\Sigma y^2 - \Sigma y\,\Sigma y/n)}}$$

Can you tell what additional column is needed in the table of sums?

Correct! We need a y^2 column.

Here is the modified table of sums for our tire-installing business:

	x	y	x^2	y^2	xy
	2	1980	4	3,920,400	3960
	3	2000	9	4,000,000	6000
	5	2040	25	4,161,600	10,200
	7	2090	49	4,368,100	14,630
	9	2100	81	4,410,000	18,900
Sums	26	10,210	168	20,860,100	53,690
	Σx	Σy	Σx^2	Σy^2	Σxy

The r-value for our tire-installing business data is

$$\begin{aligned} r &= \frac{\Sigma xy - \Sigma x\,\Sigma y/n}{\sqrt{(\Sigma x^2 - \Sigma x\,\Sigma x/n)(\Sigma y^2 - \Sigma y\,\Sigma y/n)}} \\ &= \frac{53{,}690 - 26 \cdot 10{,}210/5}{\sqrt{(168 - 26 \cdot 26/5)(20{,}860{,}100 - 10{,}210 \cdot 10{,}210/5)}} \\ &= \frac{598}{\sqrt{32.8 \cdot 11{,}280}} \\ &= \frac{598}{\sqrt{369{,}984}} \\ &= \frac{598}{608.26} = 0.98 \end{aligned}$$

Thus, we have a fairly good positive correlation between the time and profits.

A more accurate reading of the r-value (more precise than *fairly good*) can be statistically determined. To learn more about regression and correlation, you should take a statistics course or read a good statistics book. I recommend *Introduction to Probability and Statistics* by William Mendenhall and Robert Beaver (Boston: PWS-Kent, 1991) and *How to Lie with Statistics* by Darrell Huff (NY: Norton, 1954).

PRACTICE

Compute the r-value for each of the sets of data points. These are the same sets you investigated in Practice Exercises 5–7.

8. (1, 2), (2, 4), (3, 5)

9. (10, 200), (12, 400), (13, 500), (14, 700), (16, 700), (18, 800)

10. (3, 4), (4, 3), (5, 1), (6, −1)

Situation 18 Population Explosion

Dandelions are an ever-present, if not welcome, inhabitant of people's lawns. If left unpulled, dandelions seem to take over an area within a very short time. Suppose that it will take 2000 dandelion plants to completely cover Zork's backyard. Also, suppose that a single dandelion plant will produce two additional dandelion plants in 4 days. If there is a single dandelion plant growing in Zork's backyard on March 1, when will the dandelions completely cover Zork's yard if Zork does nothing to stop them and none of them die natural deaths?

____

____

____

We can view the growth of the dandelion population one step at a time. On March 1, there is one dandelion. On March 5, there will be three dandelions: the original one and its two offspring. On March 9, there will be nine dandelions: the three that were alive on March 5 and two offspring for each of them. How many dandelions will there be on March 13? _____

Right! On March 13, there will be 27 dandelions. Here's a table that shows the population explosion of dandelions.

NUMBER OF DANDELIONS IN ZORK'S BACKYARD

Date	Number of Dandelions
March 1	1
March 5	3
March 9	9
March 13	27
March 17	81
March 21	243
March 25	729
March 29	2187

You can see that Zork's yard will be full of dandelions on March 29. Look at the number of dandelions for each data. Can you see how each successive number is obtained from the previous number?

____

____

____

Every 4 days the number of dandelions triples!

I'd like to examine the explosion of dandelions in a little more detail. Here's the number of dandelions table again with some minor changes.

NUMBER OF DANDELIONS IN ZORK'S BACKYARD

Time	Number of Dandelions
0	$1 = 3^0$
1	$3 = 3^1$
2	$3 \cdot 3 = 3^2$
3	$3 \cdot 3 \cdot 3 = 3^3$
4	$3 \cdot 3 \cdot 3 \cdot 3 = 3^4$
5	$3 \cdot 3 \cdot 3 \cdot 3 \cdot 3 = 3^5$
6	$3 \cdot 3 \cdot 3 \cdot 3 \cdot 3 \cdot 3 = 3^6$
7	$3 \cdot 3 \cdot 3 \cdot 3 \cdot 3 \cdot 3 \cdot 3 = 3^7$

In this version of the table, I renamed March 1 as time = 0. Then I numbered the remaining dates accordingly. I also rewrote the number of dandelions as products and powers of 3. Do you see a relationship between the time and the number of dandelions?

____

____

____

Because the time value and the number of dandelions are related, we can write an equation that tells us the number of dandelions N, given the time t. That equation is

$$N = 3^t$$

This is an example of an **exponential relationship.**

5.5 EXPONENTIAL RELATIONSHIPS

Two quantities x and y are related exponentially if they are related by an equation of the form

$$y = c \cdot b^x$$

where c and b are specified numbers called **constants.** The value b is called the **base** of the exponential relationship. In our dandelion example, $c = 1$ and the base was 3.

If we let time be the independent variable and the number of dandelions the dependent variable, then we can obtain the following coordinates of points for the exponential relationship between dandelions and time: (1, 3), (2, 9), (3, 27), (4, 81), (5, 243), (6, 729), (7, 2187).

If we plot the first six of these points, we get a view of what the graph of an exponential function looks like.

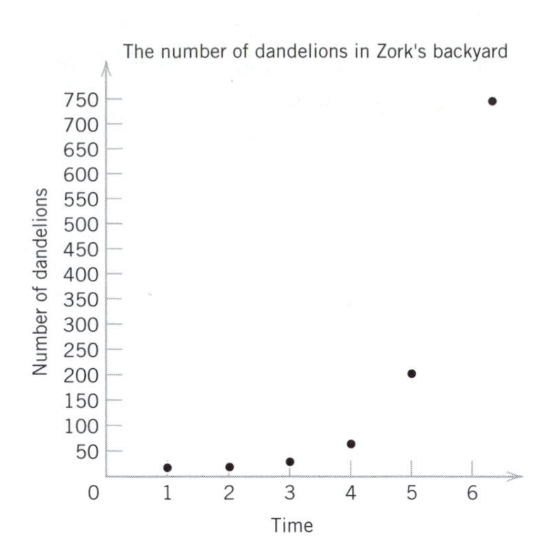

If we then connect the dots with a smooth curve, we can see the shape of an exponential relationship. Describe what you see.

____

____

____

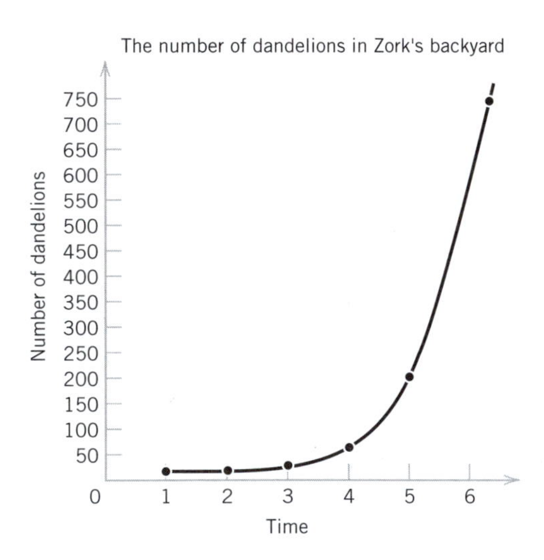

PRACTICE

1. How many dandelions would Zork have in his yard on April 6, if there were enough room for all of them?

2. The relationship between atmospheric pressure and elevation is an exponential relationship given by the equation

$$P = 14.7 \cdot 0.99996^{A}$$

where

P = the atmospheric pressure in pounds per square inch
A = the elevation in feet

(a) What is the atmospheric pressure at sea level ($A = 0$)?

(b) What is the atmospheric pressure at an elevation of 1 mile (5280 feet)?

(c) What is the atmospheric pressure at 10,000 feet?

(d) What is the atmospheric pressure at the top of Mt. Everest (29,028 feet)?

(e) Draw the graph of this relationship.

(f) Compare this graph to that of the dandelion population explosion. How are they the same? How are they different?

3. Below are three exponential relationships between quantities J and K.

$$R_1: \quad K = 5 \cdot 1^J$$

$$R_2: \quad K = 5 \cdot 2^J$$

$$R_3: \quad K = 5 \cdot 0.5^J$$

Complete the table.

J	K in R_1	K in R_2	K in R_3
0	5	5	5
1	5	10	2.5
2	____	____	____
3	____	____	____

4. Describe in general terms what happens to K as J gets larger in each of the exponential relationships R_1, R_2, and R_3. What part of the equation accounts for the differences between the relationships?

5. Draw the general shape of the graph of each of the following exponential relationships without computing any values:

(a) $y = 14 \cdot 9^x$

(b) $B = 1.3 \cdot 0.08^A$

Situation 19

What Goes Up, Comes Down

You've been watching a Professional Golfer's Association tournament on television and have noticed that the golf balls hit the highest do not necessarily go the farthest. Why not?

____

____

If you said that the height and distance of a driven golf ball depend on how hard and at what angle the ball is hit, then you have the basic idea. Although we cannot discuss the details of the flying golf ball's motion in this text, we can look at some of the general mathematical principles involved. (See Chris Dunley and Chris Pritchard, "The golf ball aerodynamics of Peter Guthrie Tait," in *The Mathematical Gazette* (Nov. 1993, pp. 298–313) for a detailed historical discussion.)

5.6 QUADRATIC RELATIONSHIPS

The path traced out by a golf ball (hit with a nine-iron) in flight may look something like this:

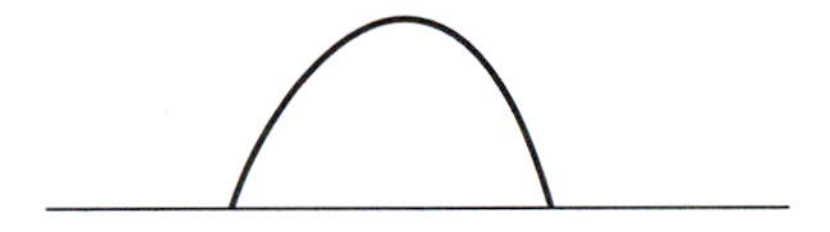

This kind of curve is called a **parabola** and produced by a **quadratic equation.** Quadratic equations are equations that have the form

$$y = ax^2 + bx + c$$

where a, b, and c are constants. We also require that the constant a be nonzero. What kind of relationship do we have if $a = 0$?

____

Here is a table of values for the quadratic equation

$$y = x^2 - 5x + 4$$

followed by the graph of the points generated in the table.

THE QUADRATIC EQUATION $y = x^2 - 5x + 4$

x	y
−1	10
0	4
1	0
2	−2
3	−2
4	0
5	4
6	10

The graph of this same equation looks like this:

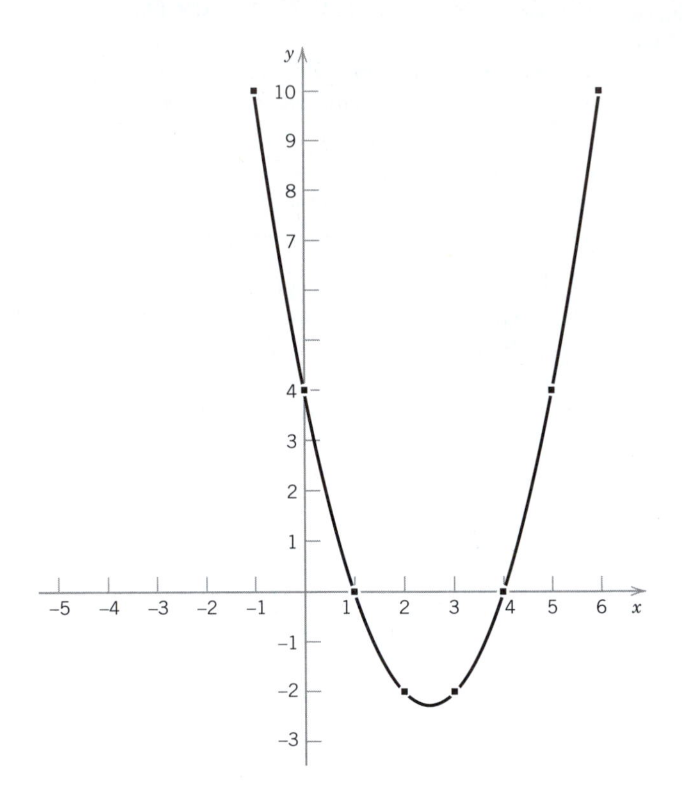

Describe the shape of the quadratic relationship and compare it to the linear and exponential relationships.

___

___

___

The graphs of quadratic relationships may also have a peak instead of a valley. For example,

$$y = -2x^2 + 5x + 1$$

THE QUADRATIC EQUATION
$\mathbf{y = -2x^2 + 5x + 1}$

x	y
−1	−6
0	1
1	4
2	3
3	−2
4	−11

The graph of this same equation looks like this:

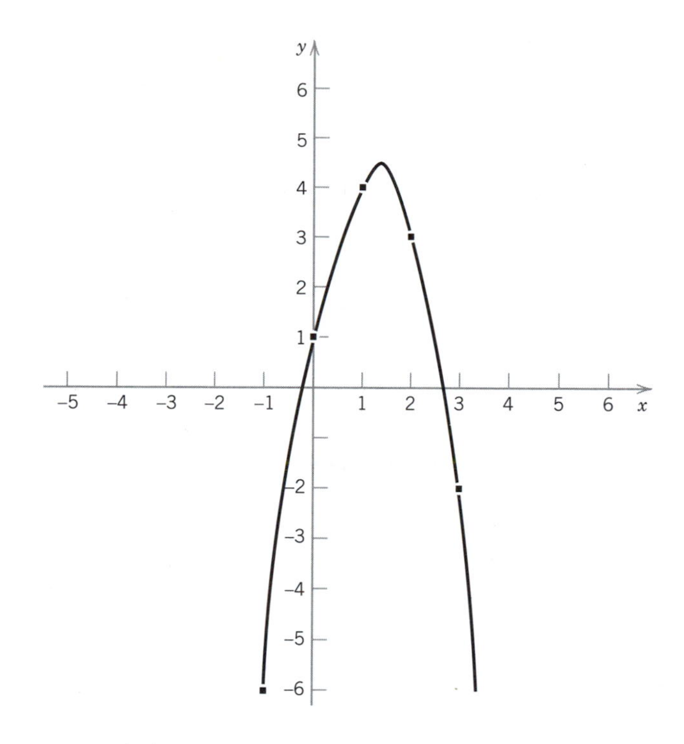

Notice that some parabolas open up, and others open down.

PRACTICE

Draw the graph of each of the following quadratic equations:

1. $y = x^2 + x - 6$

2. $y = -x^2 - x + 6$

3. $y = 2x^2 - 7x - 4$

4. $y = -2x^2 + 7x + 4$

5. Which of the curves in Practice Exercises 1–4 open up? Which open down? What part of the equation tells you which way the parabola will open?

5.7 TRAJECTORIES

The flight of a golf ball, or a baseball or missile, is a parabola that opens down. A detailed discussion of the flight paths, or **trajectories,** of these objects involves some physics and mathematics that are beyond the scope of this textbook. However, we can acquire some understanding of how the numbers in a quadratic equation affect the shape of a parabola by working with and selectively modifying a single equation.

Suppose that the flight of a golf ball is described by the quadratic equation

$$y = -x^2 + 6x,$$

where

x = the distance from the tee
y = the height of the ball at that distance

Complete the table of values for the equation $y = -x^2 + 6x$.

THE QUADRATIC EQUATION $y = -x^2 + 6x$

x	y	
0	0	(ball on tee)
1	5	
2	____	
3	9	(highest point)
4	____	
5	____	
6	0	(ball hits ground)

The graph of this same equation looks like this:

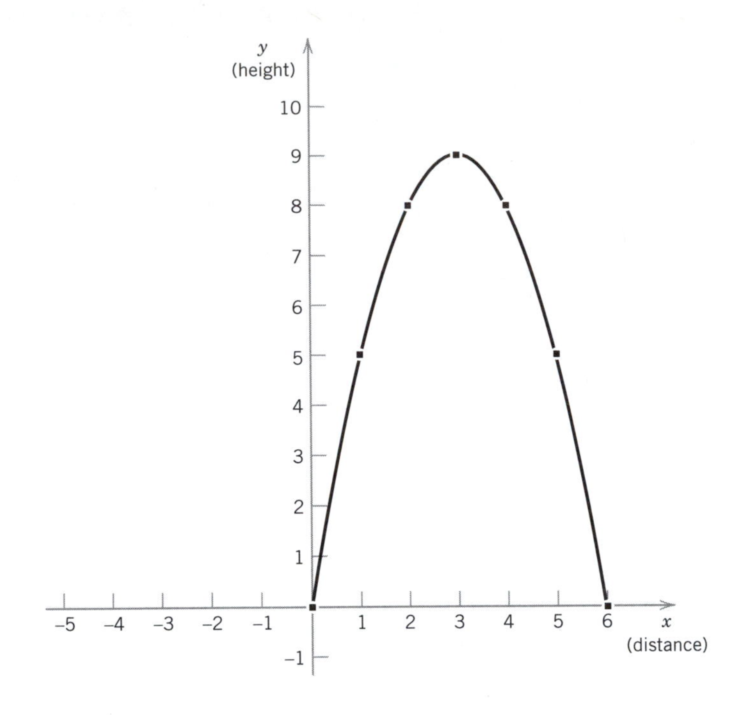

I've not drawn the curve below the x-axis or to the left of the y-axis because those areas would represent ball positions underground and behind the tee, respectively.

Now look at what happens when we modify the equation by multiplying $-x^2 + 6x$ by 0.5 to get the equation $y = -0.5x^2 + 3x$.

THE QUADRATIC EQUATION $y = -0.5x^2 + 3x$

x	y	
0	0	(ball on tee)
1	2.5	
2	4	
3	4.5	(highest point)
4	4	
5	2.5	
6	0	(ball hits ground)

The graph of this same equation looks like this. I've marked the new one with ● to distinguish it from the first graph.

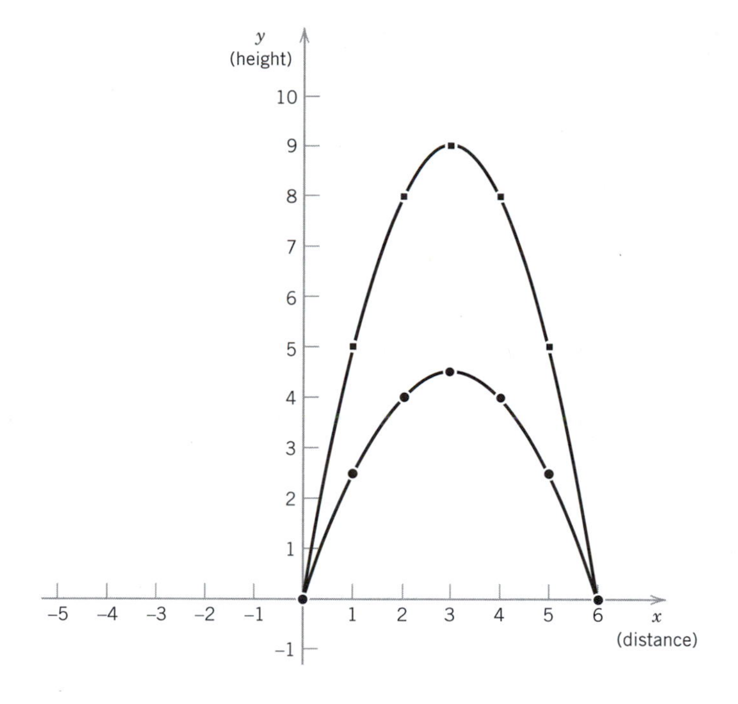

What is the difference between the two graphs?

____

____

____

Right! The height of the trajectory has been cut in half, but the distance traveled is still the same.

PRACTICE

1. Draw the graph of the quadratic equation $y = -2x^2 + 12x$ and compare to that of $y = -x^2 + 6x$.

2. Draw the graph of the quadratic equation $y = -0.25x^2 + 1.5x$ and compare to that of $y = -x^2 + 6x$.

Finally, let's modify our original quadratic equation by increasing the $6x$ term to $8x$. Thus, we have $y = -x^2 + 8x$.

THE QUADRATIC EQUATION $y = -x^2 + 8x$

x	y	
0	0	(ball on tee)
1	7	
2	12	
3	15	
4	16	(highest point)
5	15	
6	12	
7	7	
8	0	(ball hits ground)

The graph of this same equation looks like this. I've marked the new one with ❍ to distinguish it from the first two graphs.

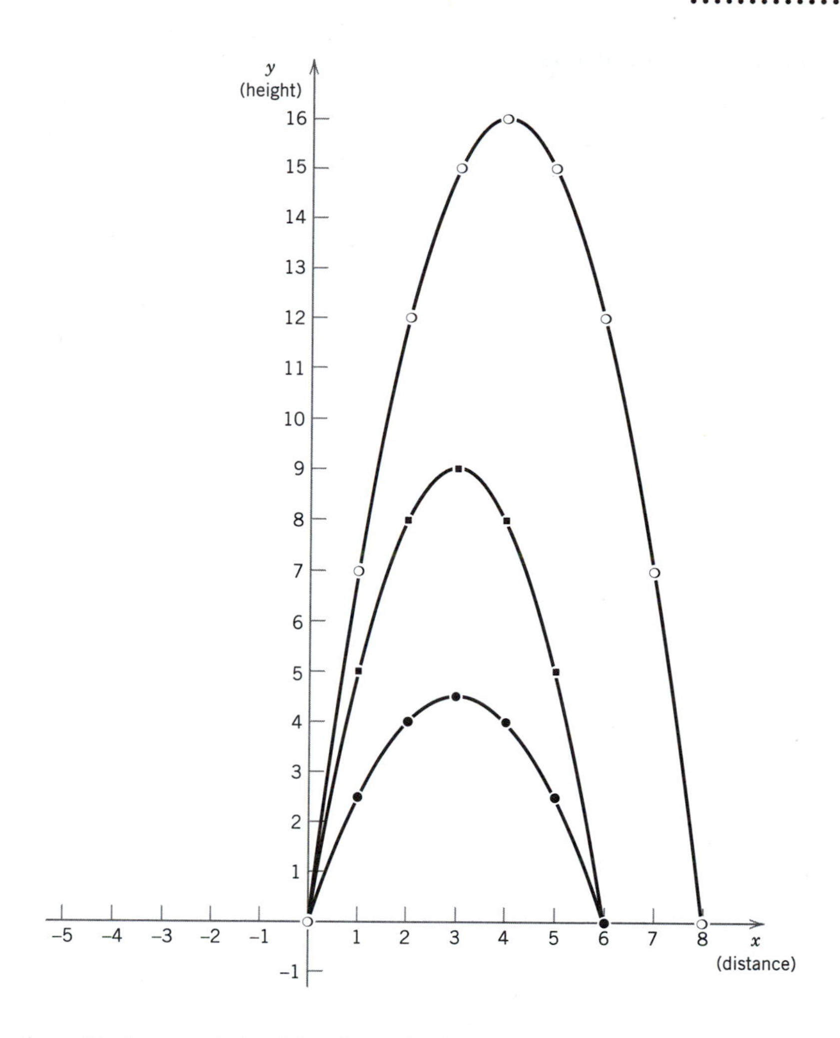

How does this last parabola differ from the first two?

Right! By increasing the $6x$ term to $8x$, we've increased the distance traveled to 8. If you wanted the ball to go a distance of 5, how should you modify the equation?

(Answer: replace $6x$ by $5x$)

We now can see how the distance and height of the parabola $y = -x^2 + 6x$ can be altered.

To increase its height, multiply $-x^2 + 6x$ by a positive number larger than 1. To decrease its height, multiply $-x^2 + 6x$ by a positive number less than 1. To alter its distance, replace $6x$ by Kx, where K is the desired distance.

PRACTICE

3. Write the equation for the trajectory of a golf ball that will go a distance of 10.

4. Write the equation for the trajectory of a golf ball that will go a distance of 10, but only half as high as the one in Practice Exercise 3.

5. Write the equation for the trajectory of a golf ball that will go a distance of 10 and 3 times as high as the one in Practice Exercise 3.

5.8 WHAT DO YOU KNOW?

If you have worked carefully through this chapter, then you are familiar with three kinds of mathematical relationships: linear, exponential, and quadratic. The exercises in this section are designed to test and refine your ability to work with the mathematics of relationships.

1. Draw the graph of a linear relationship with positive slope.
2. Draw the graph of a linear relationship with negative slope.
3. Find the coordinates of the points A–E.

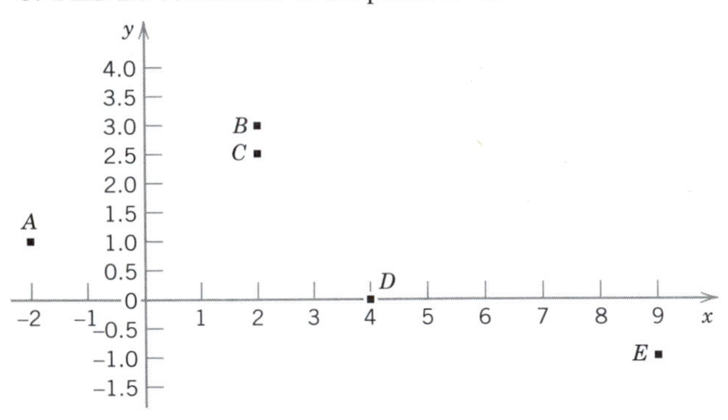

4. Locate the following points on the picture in Exercise 3. $F(1, 1)$, $G(6, 2)$, $H(-1, -1.5)$, $I(0, 3)$
5. Classify each of the following as linear, exponential, quadratic, or none of these:

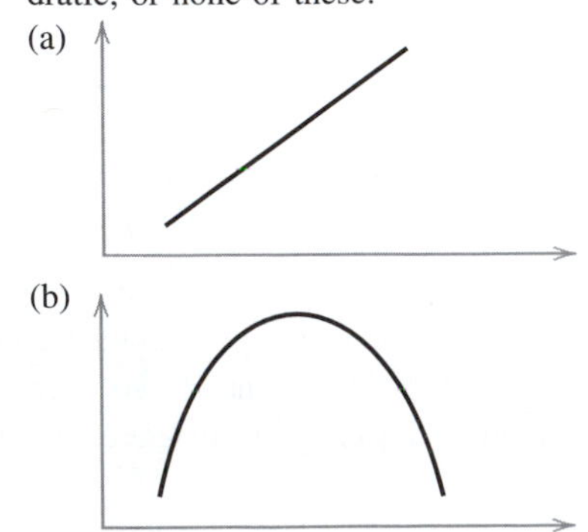

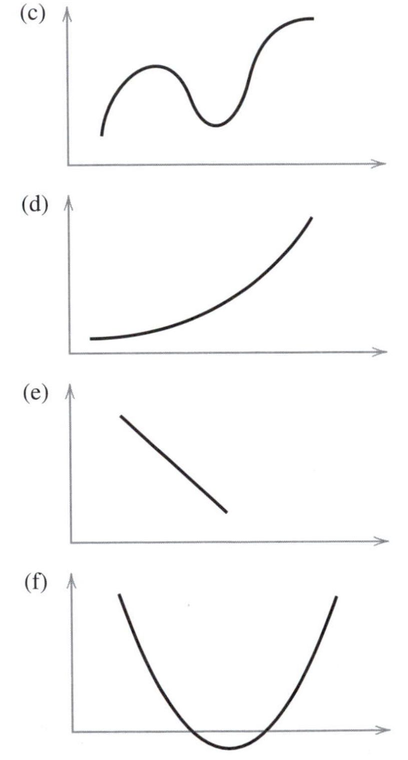

6. If the relationship between a person's annual income and the number of pairs of shoes that he or she owns is linear, is the graph of that relationship more likely to have positive or negative slope? Explain.
7. If the relationship between a person's age and the number of times per week that he or she plays with a teddy bear is linear, is the graph of that relationship more likely to have positive or negative slope? Explain.
8. Sketch the graph of each of the following linear equations:
   (a) $y = 3x + 2$
   (b) $y = 4x - 6$

(c) $y = -2x + 5$
(d) $y = 4x$
(e) $y = 4$
(f) $x = 6$
(g) $y = -2x - 0.5$

9. The cost of producing counted cross-stitch kits is linearly related to the number of kits produced. It costs $250 to produce 10 kits and $282 to produce 18 kits.
(a) Construct a linear model for the relationship between the cost of producing counted cross-stitch kits and the number of kits produced.
(b) Draw the graph of this relationship.
(c) How much will it cost to produce 50 counted cross-stitch kits?
(d) How many counted cross-stitch kits can be produced for $1000?

10. The number of ice cream cones sold by a park vendor is linearly related to the temperature in the park at noon. The vendor sold 130 cones when it was 80° and 150 cones when it was 90°.
(a) Construct a linear model for the relationship between the number of ice cream cones sold and the temperature.
(b) Draw the graph of this relationship.
(c) How many ice cream cones will be sold when the temperature reaches 101°?
(d) If the vendor sold 140 cones, how hot was it?

11. Melba's investments are increasing linearly over time. In 1980, she had investments totalling $12,000. In 1993, her investments totalled $160,000.
(a) Construct a linear model for the relationship between Melba's total investments and time.
(b) Draw the graph of this relationship.
(c) What will Melba's investments total in 1999?
(d) When will Melba have investments totalling $500,000?

12. Find the regression line for each of the following sets of data:
(a) (1, 0), (2, 3), (4, 6)
(b) (2, 3), (3, 4), (4, 5), (5, 6), (7, −2), (8, 9)
(c) (1, 0), (2, 0), (3, −1), (4, −2)

13. Compute and interpret the r-value for each of the data sets in Exercise 12.

14. Mei is working in a chemical laboratory. She is investigating the time it takes for a certain chemical mixture to solidify. She has been varying the percentage of chemical x in the mixture and then recording the solidification time. Here are her data:

Percentage of x (%)	**Solidification Time**
5	2 minutes
10	3 minutes
15	3.5 minutes
20	4.8 minutes

(a) Construct a linear regression model that relates solidification time to percentage of x.
(b) Use your model to predict the time it will take a mixture that is 35% chemical x to solidify.
(c) How good is the correlation between percentage of x and solidification time?

15. Take a survey of 10 people that you know. Find their heights and weights. Investigate these questions.
(a) In your data set are height and weight linearly related?
(b) Is the correlation different for males than females in your data?
(c) Find an eleventh person. Ask them for their height and weight. Predict their weight from their height using the model you've constructed. Compare your prediction with reality. Discuss the comparison.

16. Fern has discovered that the number of species inhabiting a tract of rain forest decreases linearly with the decrease in area of the tract. A 200-hectare tract has 300,000 species. A 10-hectare tract has 60,000 species.
(a) Construct a linear model for the relationship between the number of species and the area of rain forest.
(b) Draw the graph of this relationship.
(c) How many species inhabit a 1000-hectare tract of rain forest?
(d) How many species inhabit a 2-hectare tract of rain forest?

17. The Malthusian population model predicts that population is exponentially related to time by the equation

$$P = P_0K^t$$

where

P = the population
t = the time in years
P_0 = the initial population
K = some constant that varies from species to species

Find the population in 10 years for each of the following:
(a) $P_0 = 1000$, $K = 1.012$
(b) $P_0 = 1000$, $K = 2.012$
(c) $P_0 = 12{,}000$, $K = 0.567$
(d) $P_0 = 12{,}000$, $K = 1.004$

18. Explain how the change in K in Problem 13 affected population growth.

19. Another exponential relationship is between time and the amount of carbon-14 in an object. In living things, the ratio of ordinary carbon to radioactive carbon-14 stays about the same (approximately 1 : 1,000,000,000,000). After an animal or plant dies, the carbon-14 begins to decay, altering the ratio because the ordinary carbon remains the same. It takes about 5700 years (called the **half-life**) for half the carbon-14 to decay. Suppose that a plant has 1,000,000,000 carbon-14 atoms in it when it dies.
(a) How many carbon-14 atoms will it have in 5700 years?
(b) How many carbon-14 atoms will it have in 11,400 years?

(c) The equation that describes this relationship is

$$A = 1{,}000{,}000{,}000 \cdot (0.5)^t$$

where t is the time expressed in intervals of 5700 years. Thus, $t = 1$ means 5700 years, $t = 2$ means 11,400 years, and so on. Sketch the graph of this relationship.

(d) How many carbon-14 atoms will remain after 79,800 years?

20. An amoeba splits into two amoebas every 10 minutes. Thus, if we have one amoeba now, in 10 minutes we will have two, in 20 minutes we will have four, and so forth.

(a) Write an exponential equation that relates the number of amoeba to time.

(b) Draw the graph of this relationship.

(c) How many amoeba will we have in 1 hour?

21. Suppose that you had a regular checkerboard (32 black and 32 red squares). Put 1 penny on the square in the lower-left-hand corner. Put 2 pennies on the square next to it, 4 pennies on the next square, 8 on the next, 16 on the next, and so on until you reach the last square.

(a) How many pennies will be stacked on the 64th square?

(b) How tall would this stack of pennies be?

22. Sketch the graph of each of the following. Identify each curve as linear, exponential, or quadratic.

(a) $y = 5x - 2$

(b) $y = 3 \cdot 2^x$

(c) $y = 3x^2 + x$

(d) $y = -2 \cdot (0.2)^x$

(e) $y = -0.2x + 5$

(f) $y = -0.2x^2 + 5$

23. The equation of a soccer ball kicked into the air at the best angle for distance is given by

$$y = (-32/v^2) \cdot x^2 + x$$

where

x = the distance traveled by the ball
y = the height of the ball at the distance x
v = the velocity with which the ball is kicked

(a) Suppose the ball is kicked with a velocity of 32 feet per second. How high is the ball when it is 5 feet from the point where it was kicked?

(b) Suppose the ball is kicked with a velocity of 32 feet per second. How high is the ball when it is 15 feet from the point where it was kicked?

(c) Suppose the ball is kicked with a velocity of 32 feet per second. How high is the ball when it is 32 feet from the point where it was kicked?

(d) Suppose the ball is kicked with a velocity of 64 feet per second. Experiment with various values of x until you can determine how far the ball will go before it hits the ground ($y = 0$).

24. In Chapter 1 we looked at compounding interest semiannually, quarterly, monthly, and weekly. It is reasonable to ask if we could compound interest daily, hourly, or even every second. In fact, some banks compound interest continuously. Theoretically this means that the interest is compounded at infinitely small time intervals, but in practice continuous compounding is accomplished by the formula

$$A = Pe^{rt}$$

where

P = the principle
r = the rate of interest
t = time in years
A = the total amount in the account accrued
e = a special constant whose value is approximately 2.7183.

Notice that the continuous compounding formula is an exponential equation.

Suppose Elroy deposits $1000 into an account that is continuously compounded at a rate of 3%. How much will Elroy have in his account in 5 years?

ANSWERS TO PRACTICE EXERCISES

Section 5.1

1. Not linear
2. Linear, positive slope
3. Linear, negative slope
4. Not linear
5. Linear, negative slope
6. Linear, positive slope
7. $A(-4, 2)$, $B(0, 0)$, $C(2, 5)$, $D(-2, -3)$
8.

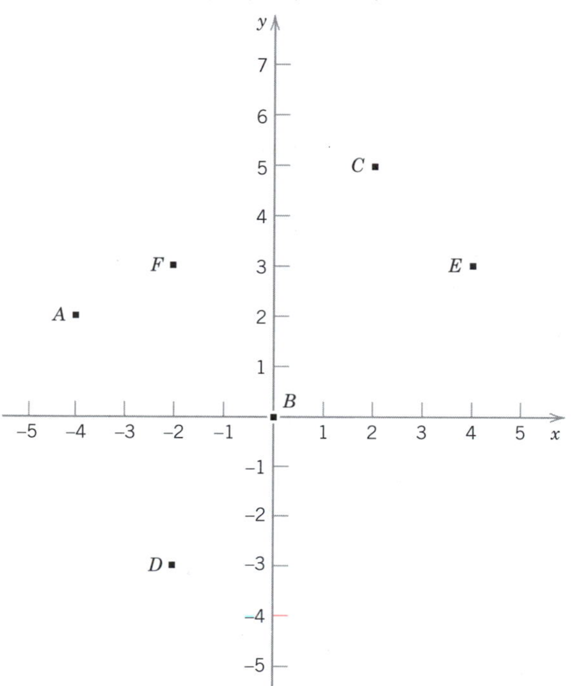

Section 5.2

1. $y = 2x + 1$
   Two solutions: $x = 3, y = 7$ and $x = 0, y = 1$
   These are not the only possible solutions! Yours may be different, but your line should look like mine.

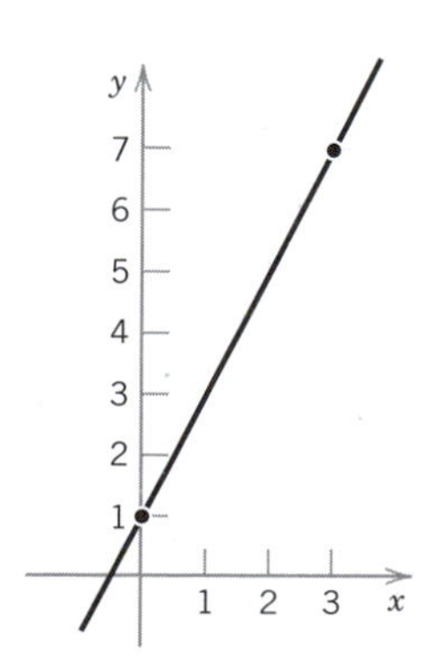

2. $y = -3x + 5$
   Two solutions: $x = 0, y = 5$ and $x = 2, y = -1$
   These are not the only possible solutions! Yours may be different, but your line should look like mine.

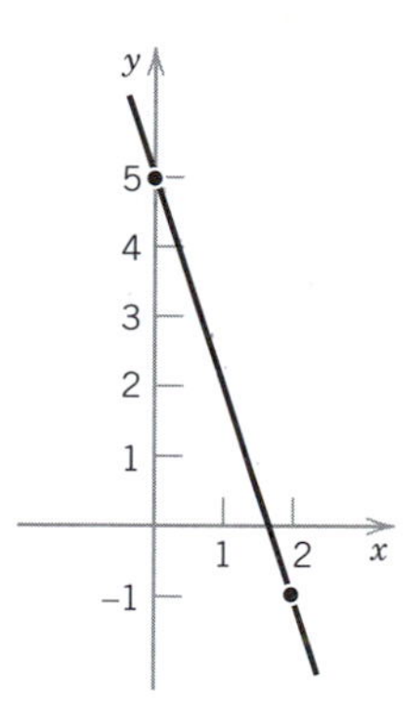

3. $y = 3x - 1$
   Two solutions: $x = 0, y = -1$ and $x = 2, y = 5$
   These are not the only possible solutions! Yours may be different, but your line should look like mine.

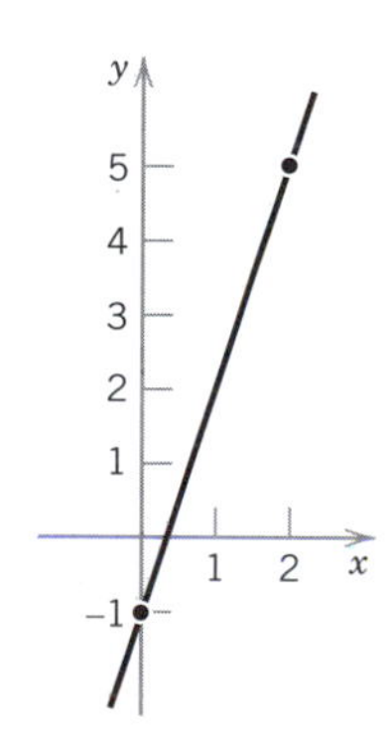

4. $y = -4x - 5$
   Two solutions: $x = -1, y = -1$ and $x = 1, y = -9$
   These are not the only possible solutions! Yours may be different, but your line should look like mine.

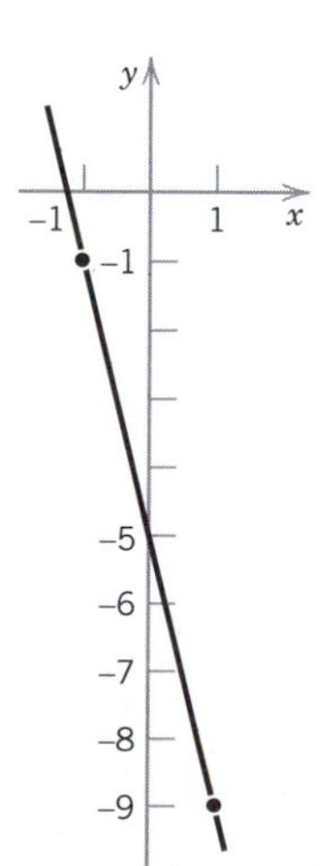

5. Practice Exercises 1 and 3 have positive slope. Practice Exercises 2 and 4 have negative slope.
If the value multiplied by x in the equation is positive, then the slope is positive. If the value multiplied by x in the equation is negative, then the slope is negative.

Section 5.3

1. (90, 1.29) and (93, 1.34)
2. $x_1 = 90, y_1 = 1.29$ and $x_2 = 93, y_2 = 1.34$
3. $m = (y_2 - y_1)/(x_2 - x_1) = (1.34 - 1.29)/(93 - 90)$ $= 0.05/3 = 0.0167$
4. $y = 0.0167x + 1.29 - 0.0167 \cdot 90$
5. $y = 0.0167x - 0.213$
6. $y = -3x + 72$
7. $y = 0.1x + 1$
8. (a) 160 pounds
   (b) 4 feet, 9 inches (57 inches)
9. (a) $1.39 per pound
   (b) 102.57°F (Whew!)

Section 5.4

1. $n = 3, \quad \bar{x} = 2, \quad \bar{y} = 3.67$
2. $n = 6, \quad \bar{x} = 13.83, \quad \bar{y} = 550$
3.

x	y	x^2	xy
1	2	1	2
2	4	4	8
3	5	9	15
6	11	14	25

4.

x	y	x^2	xy
10	200	100	2000
12	400	144	4800
13	500	169	6500
14	700	196	9800
16	700	256	11,200
18	800	324	14,400
83	3300	1189	48,700

5. $$m = \frac{\Sigma xy - \Sigma x \cdot \Sigma y/n}{\Sigma x^2 - \Sigma x \cdot \Sigma x/n} = \frac{25 - 6 \cdot 11/3}{14 - 6 \cdot 6/3} = \frac{3}{2} = 1.5$$

$b = \bar{y} - m \cdot \bar{x} = 3.67 - 1.5 \cdot 2 = 0.67$

$y = 1.5x + 0.67$

6. $$m = \frac{\Sigma xy - \Sigma x \cdot \Sigma y/n}{\Sigma x^2 - \Sigma x \cdot \Sigma x/n} = \frac{48{,}700 - 83 \cdot 3300/6}{1189 - 83 \cdot 83/6} = 74.69$$

$b = \bar{y} - m \cdot \bar{x} = 550 - 74.69 \cdot 13.83 = -482.96$

$y = 74.69x - 482.96$

7. (3, 4), (4, 3), (5, 1), (6, −1)

x	y	x^2	xy
3	4	9	12
4	3	16	12
5	1	25	5
6	−1	36	−6
18	7	86	23

$n = 4, \quad \bar{x} = 18/4 = 4.5, \quad \bar{y} = 7/4 = 1.75$

$$m = \frac{\Sigma xy - \Sigma x \cdot \Sigma y/n}{\Sigma x^2 - \Sigma x \cdot \Sigma x/n} = \frac{23 - 18 \cdot 7/4}{86 - 18 \cdot 18/4} = -1.7$$

$b = \bar{y} - m \cdot \bar{x} = 1.75 + 1.7 \cdot 4.5 = 9.4$

$y = -1.7x + 9.4$

8.

x	y	x^2	y^2	xy
1	2	1	4	2
2	4	4	16	8
3	5	9	25	15
6	11	14	45	25

$$r = \frac{\Sigma xy - \Sigma x \Sigma y/n}{\sqrt{(\Sigma x^2 - \Sigma x \Sigma x/n)(\Sigma y^2 - \Sigma y \Sigma y/n)}}$$
$$= \frac{25 - 6 \cdot 11/3}{\sqrt{(14 - 6 \cdot 6/3)(45 - 11 \cdot 11/3)}}$$
$$= \frac{3}{\sqrt{2 \cdot 4.67}}$$
$$= 0.98$$

9.

x	y	x^2	y^2	xy
10	200	100	40,000	2000
12	400	144	160,000	4800
13	500	169	250,000	6500
14	700	196	490,000	9800
16	700	256	490,000	11,200
18	800	324	640,000	14,400
83	3300	1189	2,070,000	48,700

$$r = \frac{\Sigma xy - \Sigma x \Sigma y/n}{\sqrt{(\Sigma x^2 - \Sigma x \Sigma x/n)(\Sigma y^2 - \Sigma y \Sigma y/n)}}$$
$$= \frac{48{,}700 - 83 \cdot 3300/6}{\sqrt{(1189 - 83 \cdot 83/6)(2{,}070{,}000 - 3300 \cdot 3300/6)}}$$
$$= \frac{3050}{\sqrt{40.83 \cdot 255{,}000}}$$
$$= 0.945$$

10.

x	y	x^2	y^2	xy
3	4	9	16	12
4	3	16	9	12
5	1	25	1	5
6	−1	36	1	−6
18	7	86	27	23

$$r = \frac{\Sigma xy - \Sigma x \Sigma y/n}{\sqrt{(\Sigma x^2 - \Sigma x \Sigma x/n)(\Sigma y^2 - \Sigma y \Sigma y/n)}}$$

$$= \frac{23 - 18 \cdot 7/4}{\sqrt{(86 - 18 \cdot 18/4)(27 - 7 \cdot 7/4)}}$$

$$= \frac{-8.5}{\sqrt{5 \cdot 14.75}}$$

$$= -0.99$$

Section 5.5

1. 19,683 dandelions. April 6 corresponds to $t = 9$.

2. (a) 14.7 pounds per square inch

(b) 11.9 pounds per square inch

(c) 9.85 pounds per square inch

(d) 4.60 pounds per square inch

(e)

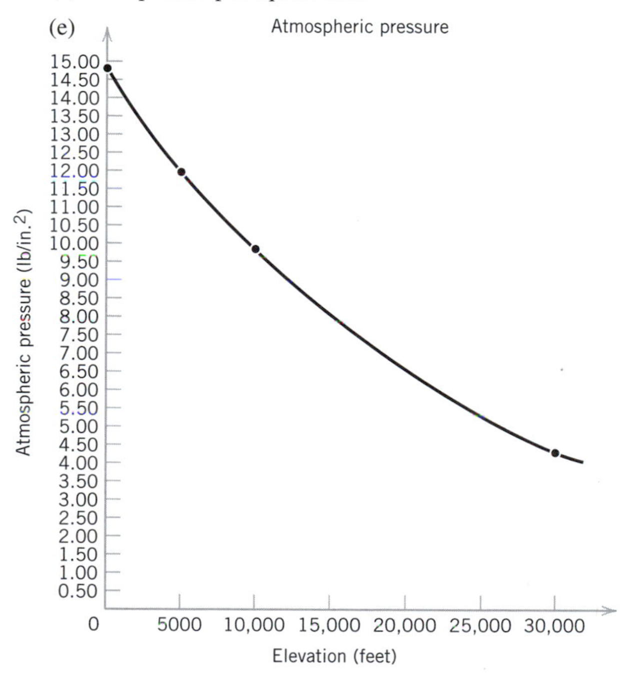

3.

J	K in R_1	K in R_2	K in R_3
0	5	5	5
1	5	10	2.5
2	5	20	1.25
3	5	40	0.625

4. In R_1, as J gets larger, K stays at 5.
In R_2, as J gets larger, K gets larger.
In R_3, as J gets larger, K gets smaller.
The value of the base (1, 2, or 0.5) makes the difference.

5. (a) (increasing)

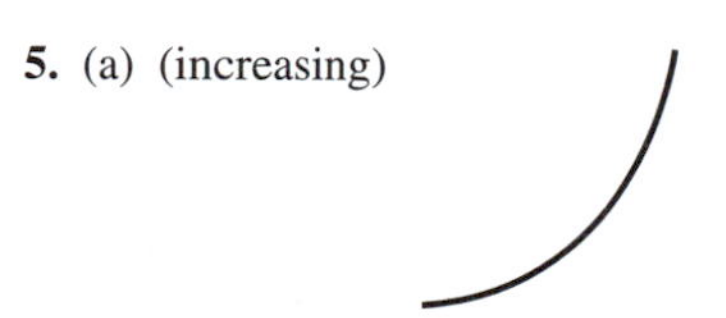

(b) (decreasing)

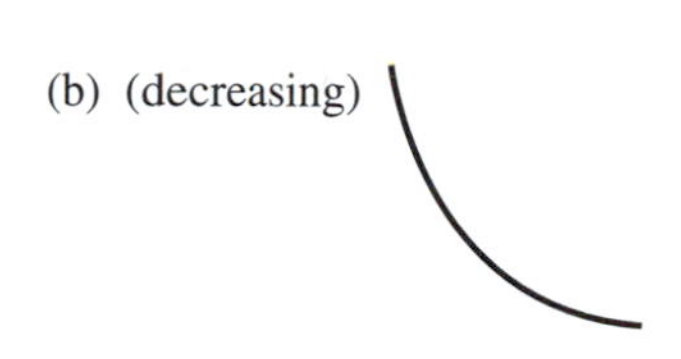

Section 5.6

1.

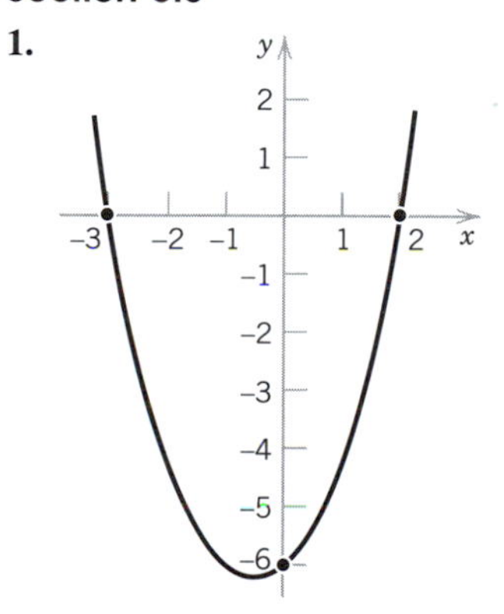

2.

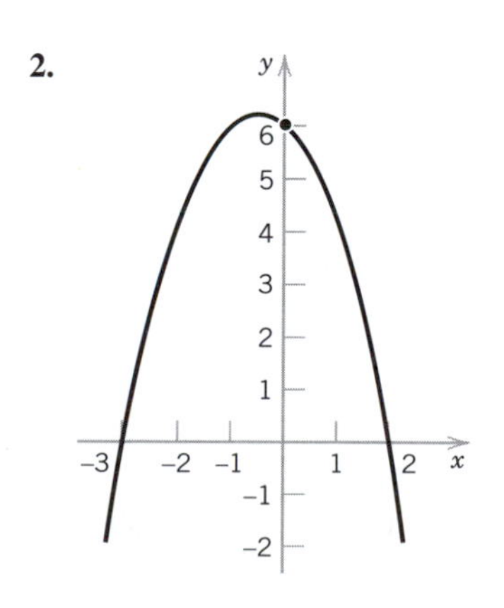

3.

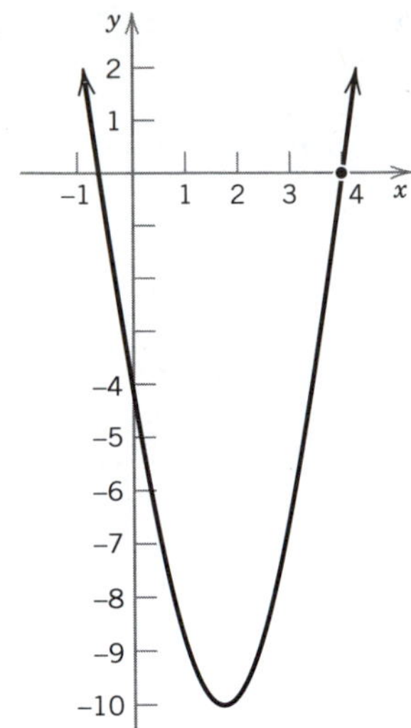

4.

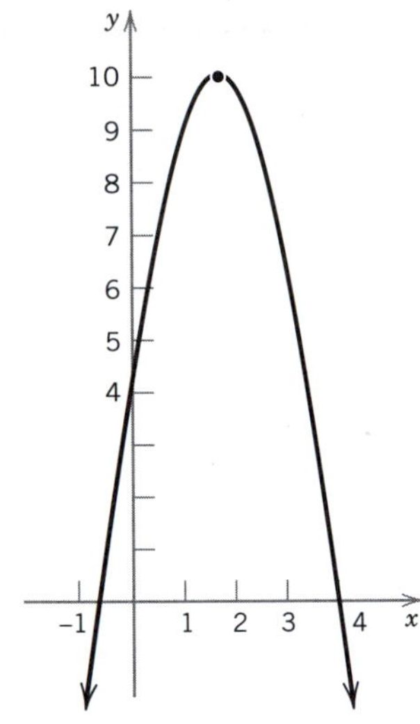

5. Practice Exercises 1 and 3 open up. Practice Exercises 2 and 4 open down. If the coefficient of x^2 in the quadratic equation is positive, then the curve opens up. If that coefficient is negative, then the curve opens down.

Section 5.7

1. The graph $y = -2x^2 + 12x$ goes twice as high as that of $y = -x^2 + 6x$, but still goes only as far as 6.
2. The graph of $y = -0.25x^2 + 1.5x$ goes 1/4 as high as that of $y = -x^2 + 6x$, but still goes only as far as 6.
3. $y = -x^2 + 10x$
4. $y = -0.5x^2 + 5x$
5. $y = -3x^2 + 30x$

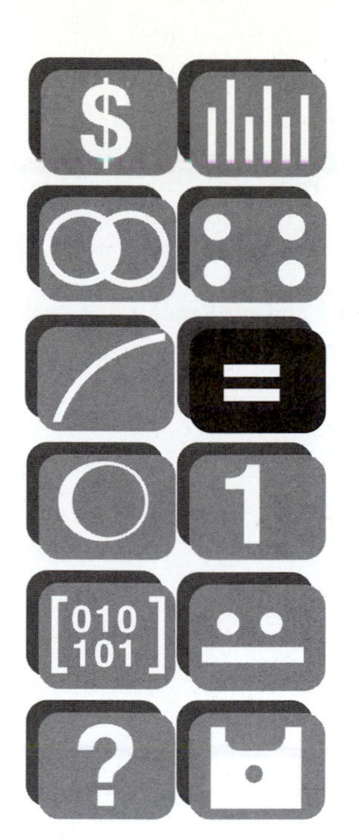

Chapter Six

Mathematics of Optimization

What's the Point? Modern life is complicated. Often, a decision must be based on a multitude of mutually conflicting considerations. Although mathematics cannot provide you with a formula for making decisions, it can, in some limited situations, help you to select the best combination of several variables. The choice suggested by mathematics is called the **optimal** solution. The mathematics that helps us to find optimal solutions is called the mathematics of optimization. This chapter shows you how to find optimal solutions to some small, well-defined problems.

Situation 20 Breaking Even

You are beginning a small manufacturing business, Sun at Night, Inc., that will make and sell solar-powered flashlights. You know that your cost for producing the flashlights is linearly related to the number of flashlights you make by the equation

$$y = 2.5x + 4000$$

where

x = the number of flashlights produced
y = the cost (in dollars) of producing the flashlights

You intend to sell the flashlights for $3.95.

How much will it cost you to make 100 flashlights?

(Answer: $4250)

How much revenue will you make if you sell 100 flashlights?

(Answer: $395)

As you can see, the manufacture and sale of 100 flashlights will leave you at a loss because your costs greatly exceed your revenue. This is typical for a starting business because of certain initial costs that cannot be attached reasonably to the sale price. Thus, for some quantities of flashlights sold, you will not make a profit. How many flashlights will you need to sell in order to make a profit?

The following table will give us some idea of how many flashlights need to be sold:

Number (x) of Flashlights	Cost $2.5x + 4000$	Revenue from Sales ($3.95x$)	Revenue − Cost
100	$4250	$395	−$3855
500	$5250	$1975	−$3275
1000	$6500	$3950	−$2550
5000	$16,500	$19,750	+$3250

Somewhere between 1000 and 5000 flashlights, we begin to make a profit. In the next two sections, we will see how to find the exact number of flashlights that are needed to make a profit.

6.1 SYSTEMS OF LINEAR EQUATIONS

The linear equation relating cost of production to number of flashlights in Situation 20 is

$$y = 2.5x + 4000$$

The revenue earned from selling x flashlights can be described by the equation

$$y = 3.95x$$

where y is the revenue earned in dollars.

Below are the graphs of both these equations. The ❍'s mark the graph of $y = 2.5x + 4000$, and the ● the graph of $y = 3.95x$.

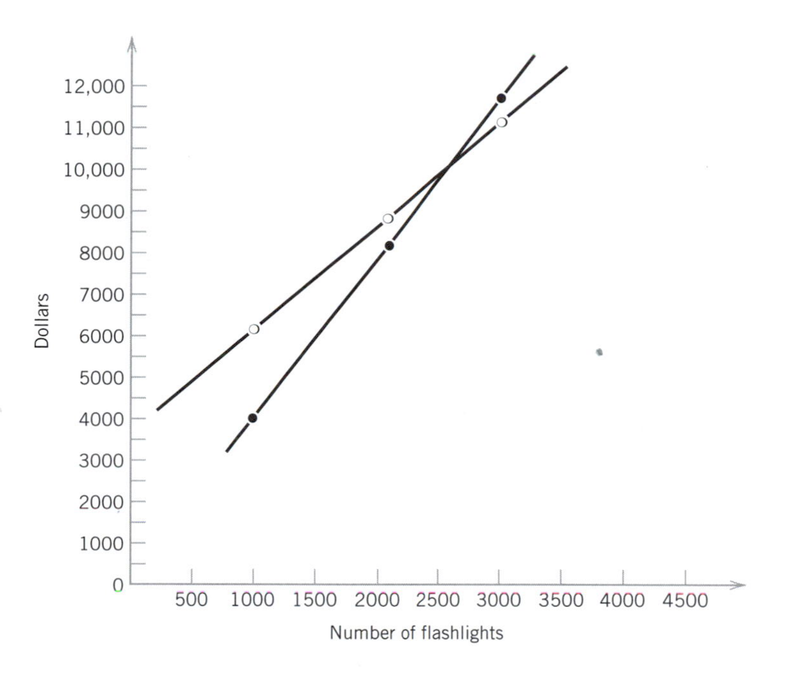

For the x values for which the revenue (●) line is below the cost (❍) line, we are operating at a loss. For the x values for which the revenue (●) line is above the cost (❍) line, we are operating at a profit. The point where the two lines cross is called the **break-even point,** because it is at this point that the revenue and cost are exactly the same.

What are the coordinates of the break-even point in the graph above?

Yes, it is hard to tell what the coordinates of the break-even point are. Our graphs can give us some *ballpark* estimates, but these are not really very precise. Therefore, we need an algebraic method for finding the exact coordinates of the point where two lines cross.

Let's look at the two equations for cost and revenue again.

$$y = 2.5x + 4000$$
$$y = 3.95x$$

When two linear equations are written together like this, they become a **system of linear equations.** A solution to a system of linear equations is a pair of values (x, y) that is a solution to *both* the equations.

Where is a solution to a system of two linear equations located on their graphs?

Right! The solution to the system is the point where the two graphs cross. Thus, to find the break-even point, we need only to find the solution to the system of equations.

6.2 SOLVING SYSTEMS OF LINEAR EQUATIONS

The solution to a system of linear equations can be found using some elementary algebra. First, note that since $y = 2.5x + 4000$ and $y = 3.95x$, then

$$2.5x + 4000 = 3.95x$$

Why?

We have eliminated y from the equations because since $2.5x + 4000$ equals y and $3.95x$ equals y, the two expressions involving x must equal each other.

Next, we solve the equation

$$2.5x + 4000 = 3.95x$$

for x. (In case you may have forgotten some of your high school algebra, I've explained each step along the way.)

$2.5x + 4000 = 3.95x$	(Start)
$2.5x - 2.5x + 4000 = 3.95x - 2.5x$	(Subtract $2.5x$ from both sides to get together all the x terms)
$4000 = 1.45x$	(Simplify)
$4000/1.45 = 1.45x/1.45$	(Divide both sides by 1.45 to isolate x)
$2758.62 = x$	(Simplify)

We can check this answer by substituting it into each of the original equations. If it is correct, then the two y values we obtain should be the same.

Cost equation: $y = 2.5 \cdot 2758.62 + 4000 = \$10{,}896.55$

Revenue equation: $y = 3.95 \cdot 2758.62 = \$10{,}896.55$

The break-even point is reached when we make and sell 2758.62 flashlights. Anything above this number of flashlights will earn us a profit. What is the minimum number of flashlights that must be sold before we earn a profit?

(Answer: 2759)

How much profit will be made when we sell 2759 flashlights? Well, we know that our cost will be $2.5 \cdot 2759 + 4000 = \$10{,}897.50$. And our revenue will be $3.95 \cdot 2759 = \$10{,}898.05$. Thus, our profit will be $\$10898.05 - \$10{,}897.50 = \$0.55$.

PRACTICE

1. Complete the solution to the following problem:

Myrna manufactures and sells Raggedy Ann dolls. The cost of producing the dolls y is linearly related to the number of dolls made x by the equation

$$y = 10x + 500$$

Myrna can sell each doll for $32.95.

(a) Write the system of equations whose solution is the break-even point for Myrna's business.

(b) Eliminate y from the system of equations in part (a).

(c) Solve the equation in part (b) for x.

(d) How many dolls must Myrna make and sell in order to make a profit?

(e) What will her profit be when she sells these dolls?

2. Paula is an independent manufacturer of commemorative T-shirts. Paula's cost of production y is linearly related to the number of T-shirts she produces x by the equation $y = 4x + 300$. Paula can sell each T-shirt that she produces for \$11.50. How many T-shirts must she produce and sell in order to make a profit?

Not every system of linear equations has a solution. Two lines can intersect in a single point, or they can not intersect at all (i.e., they are parallel), or they can coincide (i.e., they are the same line).

Draw the two lines of each system of equations on the same picture. Do the lines intersect in one point, no points, or do they coincide?

3. $y = 2x + 4$
$y = 2x + 7$

4. $y = 3x + 1$
$y = 3x$

5. $y = 4x + 2$
$y = 4x + 2$

6. $y = 2x + 3$
$y = -2x + 5$

7. $y = -5x - 4$
$y = -5x - 4$

Let's see what happens when we try to solve a system of linear equations in which the two lines do not intersect. Take as an example Practice Exercise 3.

$$y = 2x + 4$$
$$y = 2x + 7$$

First, we eliminate the y.

$$2x + 4 = 2x + 7$$

Now solve for x.

$$2x - 2x + 4 = 2x - 2x + 7$$
$$4 = 7$$

The result $4 = 7$ is obviously not true! Thus, if a system of linear equations consists of two nonintersecting lines, we will get a false equation when we try to solve the system. These systems are called **inconsistent.**

Now let's look at what happens when we try to solve a system of linear equations in which the two lines coincide. We'll take the system in Practice Exercise 5.

$$y = 4x + 2$$
$$y = 4x + 2$$

Eliminate y.

$$4x + 2 = 4x + 2$$

Solve for x.

$$4x - 4x + 2 = 4x - 4x + 2$$
$$2 = 2$$

Now, $2 = 2$ is true no matter what the values of x and y are. If we get an equation that is true for *all* x and y, then the system of linear equations consists of two coincident lines. These systems are called **dependent.**

Here is another example. Solve the system of linear equations

$$y = 4x + 2$$
$$2y = 6x + 7$$

The second equation here presents a small problem in that the $6x + 7$ is set equal to $2y$ instead of y. In order to eliminate the y from the system, we can multiply the first equation by 2 to get

$$2y = 8x + 4$$

Now the two expressions involving x are set equal to the same term $2y$, and we can eliminate the y to get

$$8x + 4 = 6x + 7$$

Next solve for x.

$$8x - 6x + 4 = 6x - 6x + 7$$
$$2x + 4 = 7$$
$$2x + 4 - 4 = 7 - 4$$
$$2x = 3$$
$$2x/2 = 3/2$$
$$x = 3/2$$

The y value when $x = 3/2$ can be found from either of the equations in the system. Using the first one, we get $y = 4 \cdot (3/2) + 2 = 8$. Thus, the solution is (3/2, 8).

Here is one last example. Solve the system of linear equations

$$4x + \quad y = 2$$
$$2y = 6x + 7$$

The first equation does not appear in the form we are used to, so we fix it by isolating the y.

$$4x + y = 2$$
$$4x - 4x + y = -4x + 2$$
$$y = -4x + 2$$

The second equation also presents a small problem in that the $6x + 7$ is set equal to $2y$ instead of y. In order to eliminate the y from the system, we can multiply the first equation by 2 to get

$$2y = -8x + 4$$

Now the two expressions involving x are set equal to the same term $2y$, and we can eliminate the y to get

$$-8x + 4 = 6x + 7$$

Next solve for x.

$$-8x + 8x + 4 = 6x + 8x + 7$$
$$4 = 14x + 7$$
$$4 - 7 = 14x + 7 - 7$$
$$-3 = 14x$$
$$-3/14 = 14x/14$$
$$-0.214 = x$$

The y value when $x = -0.214$ can be found from either of the equations in the system. Using the first one, we get $y = -4 \cdot (-0.214) + 2 = 2.856$. Thus, the solution is (−0.214, 2.856).

PRACTICE

Solve each of the following systems. Identify any inconsistent or dependent systems.

8. $y = 4x + 3$
$y = 2x + 5$

9. $y = 2x - 5$
$3y = x + 3$

10. $y = 4x + 3$
$2y = 8x + 6$

11. $y = 2x + 5$
$y = 2x + 7$

12. $x + 3y = 4$
$2x + y = 9$

Situation 21

The Hottest Chili in the West

The two fundamental ingredients of chili are beans and chili peppers. The *hotness* of a serving of chili can be given a numerical value using the Rawlins hot equation:

$$H = 8 \cdot C - 2 \cdot B$$

where

H = the hotness
C = the number of chili peppers per serving
B = the number of beans per serving

Oziel wants to make the hottest chili possible to enter in the Albany County Fair. He cannot just use an unlimited number of chili peppers because of the Texas flavor rule.

> *Texas Flavor Rule:* In any serving of chili, five times the number of chili peppers minus four times the number of beans cannot exceed 6. If this rule is violated, the chili loses its flavor balance.

Also, Oziel wants to make sure that the total number of beans and chili peppers in each serving is at least three. Otherwise, he knows that the judges at the fair will find his chili too *soupy*. Finally, Oziel wants to use only the finest ingredients in his chili. The Sarawakan minichili peppers cost 1¢ each and the Andean beans 8¢ each. Oziel cannot afford to spend more than 32¢ per serving (Oziel's aunt will buy the ingredients other than chili peppers and beans for him).

How many chili peppers and beans should Oziel use per serving to make the hottest chili possible?

Oziel's chili problem is an example of an optimization problem. He wants to maximize hotness but is limited by cost, quantity, and the Texas flavor rule. We say that hotness is Oziel's **objective function,** and his objective is subject to the **constraints** of cost, quantity, and flavor.

Our solution to Oziel's chili problem begins with an investigation of **linear inequalities.**

6.3 LINEAR INEQUALITIES

Let's look again at the Texas flavor rule.

> *Texas Flavor Rule:* In any serving of chili, five times the number of chili peppers minus four times the number of beans cannot exceed 6. If this rule is violated, the chili loses its flavor balance.

We can express this rule mathematically in terms of chili peppers and beans this way:

$$5C - 4B \leq 6$$

where

C = the number of chili peppers per serving
B = the number of beans per serving

The symbol $\leq$ is read *less than or equal to*. It is one of four **inequality symbols** used in mathematics. Here is a list of these inequality symbols, their meaning, and some of the English expressions that are equivalent to them.

Symbol	Literal Meaning	English Equivalents
$\leq$	Less than or equal to	Cannot exceed, is no more than, is at most
$\geq$	Greater than or equal to	Is at least, is no less than
$>$	Greater than	Is more than, exceeds
$<$	Less than	Is fewer than, is below

PRACTICE

Fill in the blanks.

English	Symbols
Twice x plus y is no more than 3.	$2x + y \leq 3$
The sum of x and y exceeds 6.	$x + y > 6$
1. Three times x minus twice y is at least 5.	$3x - 2y$ ____ 5
2. Four times x plus y is at most 4.	________ ≤ 4
3. Five times Z minus three times Q is less than 18.	____________
4. Forty times A plus half of B is more than C.	$40A + 0.5B$ ____

A **solution to a linear inequality** is any pair of values that makes the inequality true. For example, for the Texas flavor rule

$$5C - 4B \leq 6$$

$C = 2$ and $B = 3$ is a solution because $5 \cdot 2 - 4 \cdot 3 = 10 - 12 = -2 \leq 6$. On the other hand, $C = 5$ and $B = 3$ is not a solution because $5 \cdot 5 - 4 \cdot 3 = 25 - 12 = 13$, which is not less than or equal to 6.

PRACTICE

Decide whether or not each of the following pairs of values is a solution to the Texas flavor rule linear inequality $5C - 4B \leq 6$:

5. $C = 0, \quad B = 0$

6. $C = 7, \quad B = 1$

7. $C = 2, \quad B = 2$

8. $C = 6, \quad B = 7$

9. $C = 4, \quad B = 4$

10. $C = 2, \quad B = 6$

11. $C = 5, \quad B = 1$

12. $C = 8, \quad B = 3$

The solutions to the Texas flavor rule inequality represent the number of chili peppers and beans per serving that will produce a proper flavor balance. There are many solutions and many pairs that are not solutions. So, we would like to have a way of seeing easily which pairs are solutions and which are not. One way we can do this is with a graph.

If the $\leq$ in $5C - 4B \leq 6$ is replaced by an $=$, we get the equation $5C - 4B = 6$. We can rewrite this equation like this

$$5C - 4B = 6$$
$$-4B = 6 - 5C$$
$$4B = -6 + 5C$$
$$4B = 5C - 6$$
$$B = 1.25C - 1.5$$

The last equation is a linear equation. Let's graph it using C as the horizontal axis and B as the vertical axis. I've used the points $(0, -1.5)$ and $(2, 1)$.

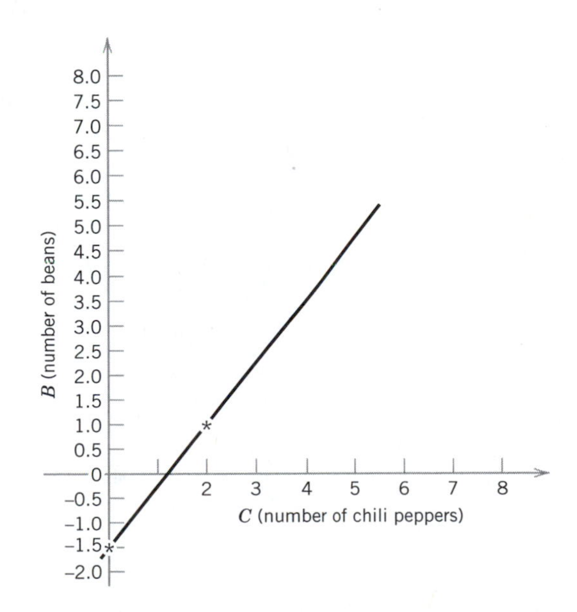

Now let's locate the points from Practice Exercises 5–12 on this picture. I'll use ● to mark the points that are solutions to the inequality and ○ to mark those that are not.

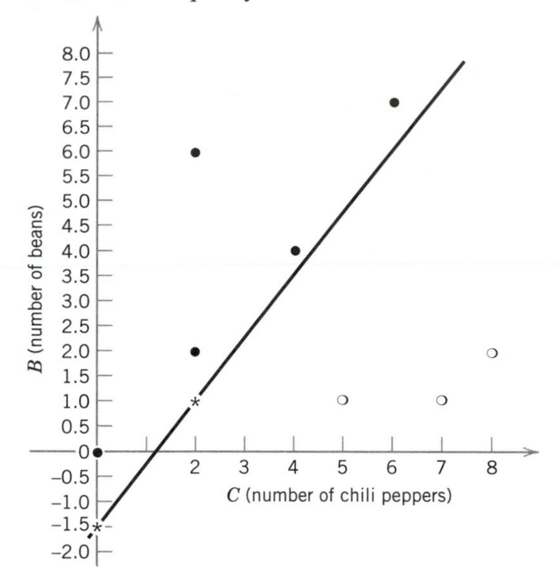

Look carefully at this last graph. What do you notice about the positions of the points that are solutions (●) and those that are not (○) relative to the line?

____

____

____

What you observed is true of all linear inequalities. All the points that are solutions lie on one side of the line that corresponds to the equality associated with the inequality, and all the points that are not solutions lie on the other side of the line. Furthermore, all the points on the same side of the line with any solution are themselves solutions. We therefore will shade the entire side of the line containing the solutions to indicate that all these points are solutions. Like this:

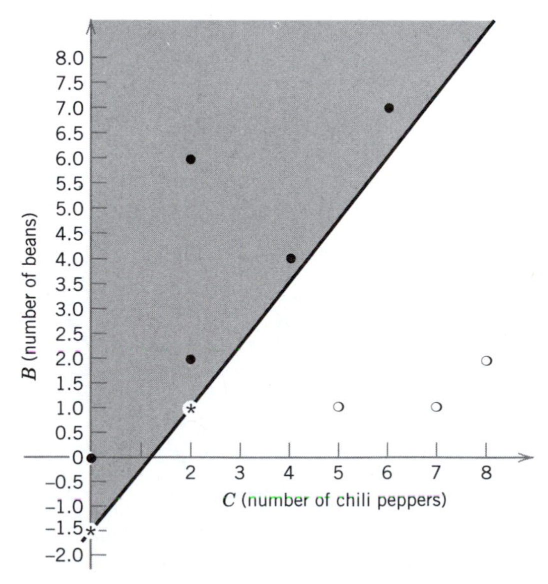

Finally, to complete the picture, if the line itself is not a solution to the inequality (it won't be if the inequality is $<$ or $>$), then we make it a broken line. Otherwise, it remains a solid line.

The procedure for drawing the solutions of a linear inequality then consists of these four steps:

Step 1: Draw the graph of the corresponding linear equality.

Step 2: Select any point not on the line and see if it is a solution to the inequality.

Step 3: If the test point is a solution, then shade the side of the line containing it. Otherwise, shade the other side.

Step 4: Make your line a dotted line if the inequality is $<$ or $>$.

Here is another example. Draw the solutions of $2x + 3y > 6$.

Step 1: First draw the line $2x + 3y = 6$. I'll use x for the horizontal axis and y for the vertical axis. My points are (3, 0) and (0, 2).

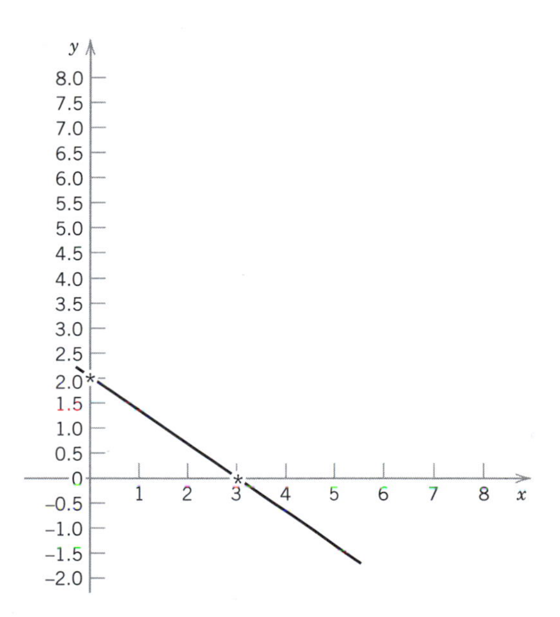

Step 2: I'll take (5, 5) as a test point because it is well away from the line. I find that (5, 5) is a solution because $2 \cdot 5 + 3 \cdot 5 = 10 + 15 = 25$ is greater than 6.

Step 3: Because (5, 5) is a solution, I'll shade the side of the line containing (5, 5). I've marked (5, 5) with a ● so you can see clearly where it is.

Step 4: I make the line a dotted line because my inequality is >.

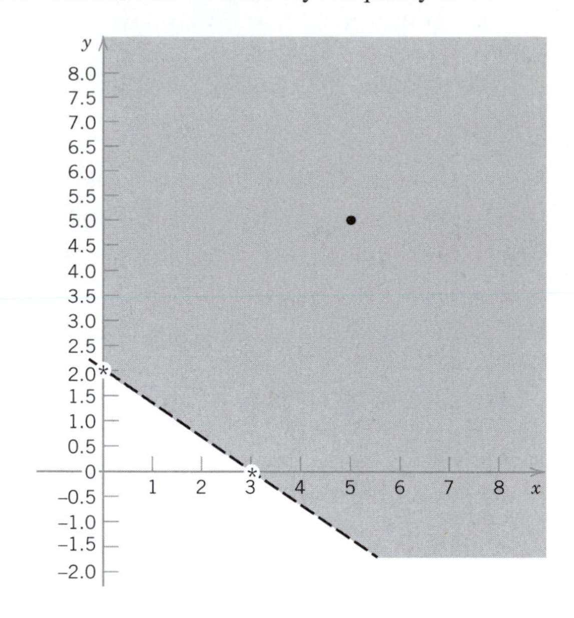

PRACTICE

Draw the solutions of each of the following linear inequalities. Use x as the horizontal axis and y as the vertical axis and (2, 4) as the test point in each case.

13. $3x + y \leq 6$

14. $4x - y \geq 8$

15. $-2x + y < 4$

16. $x + y > 3$

6.4 SYSTEMS OF LINEAR INEQUALITIES

In the last section, we drew a picture of all the solutions to the Texas flavor rule inequality. However, Oziel must also consider cost and quantity as he makes his chili. Concerning the quantity of chili peppers and beans, we were told that

> *Oziel wants to make sure that the total number of beans and chili peppers in each serving is at least three. Otherwise, he knows that the judges at the fair will call his chili too soupy.*

Using C for the number of chili peppers and B for the number of beans per serving, we can write this quantity constraint as

$$C + B \geq 3$$

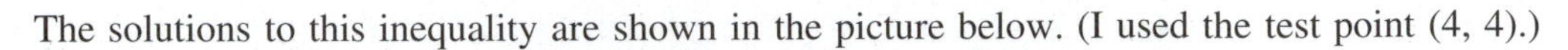

The solutions to this inequality are shown in the picture below. (I used the test point (4, 4).)

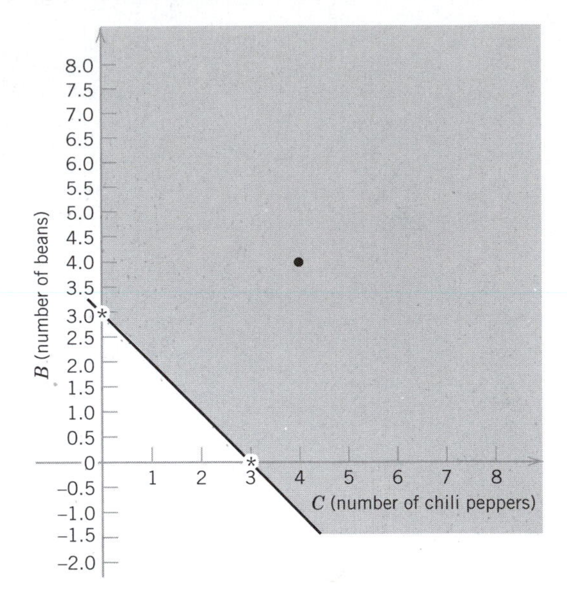

The number of chili peppers and beans represented by the points in this shaded region satisfy Oziel's quantity condition. Some of them, like (5, 5), also satisfy the Texas flavor rule. Others, like (5, 1), do not satisfy the Texas flavor rule.

Oziel's chili needs to satisfy *both* the Texas flavor rule and the quantity condition. These solutions can be found by drawing the solutions to both inequalities in the same picture and shading the area where the two separate shadings overlap. In the picture below, the points marked with ● satisfy only the Texas flavor rule, the points marked with ❍ satisfy only the condition $C + B \geq 3$, and the points marked with ✾ satisfy both conditions.

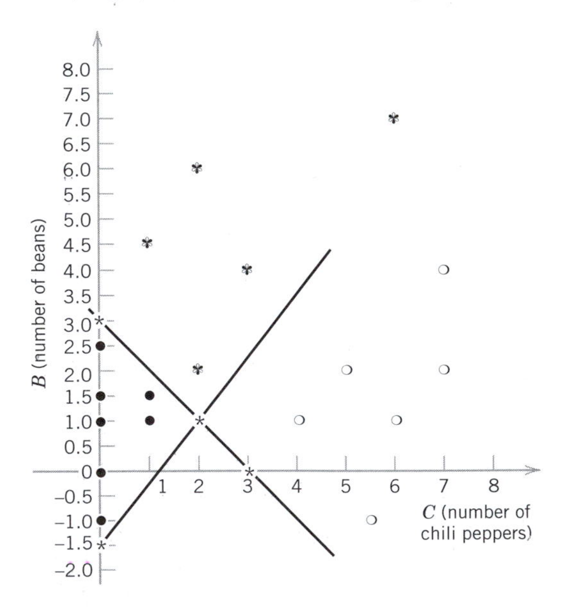

If we now shade the region marked by the ✾ points, we get the points that satisfy both $5C - 4B \leq 6$ and $C + B \geq 3$. These points comprise the **solution to the system of linear inequalities:**

$$5C - 4B \leq 6$$

$$C + B \geq 3$$

The solution looks like this:

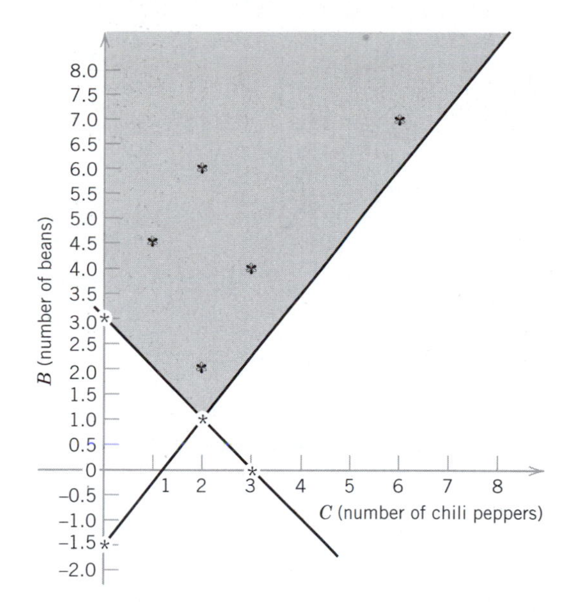

Finally, recall that Oziel wants to use only the finest ingredients in his chili. The Sarawakan minichili peppers cost 1¢ each and the Andean beans 8¢ each. Oziel cannot afford to spend more than 32¢ per serving. This constraint on money can be expressed as the linear inequality

$$C + 8B \leq 32$$

The solution to this money constraint is shown below. I used (5, 1) as my test point.

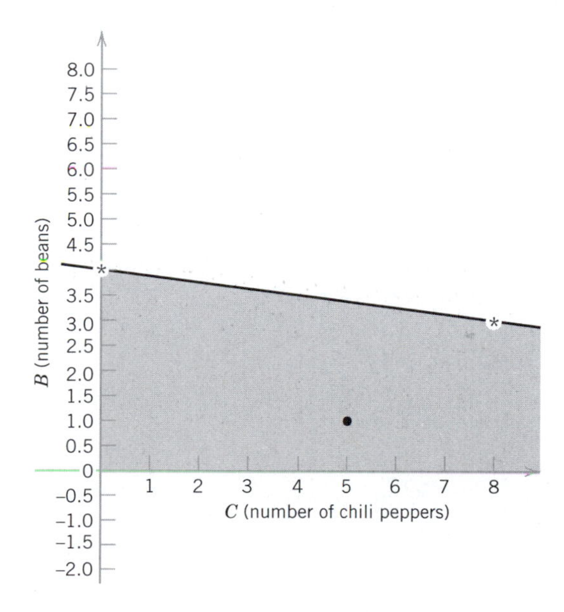

To find the points that satisfy all three of our constraints (Texas flavor rule, quantity, and money), we include this last solution in our picture. I also have used two additional constraints imposed on the problem by the real world. Because neither the number of chili peppers nor the number of beans can be negative, the points that satisfy all the inequalities must lie within the area where C and B are both positive.

Solution to the System of Inequalities

$$5C - 4B \leq 6 \quad \text{(Flavor rule)}$$

$$C + B \geq 3 \quad \text{(Quantity)}$$

$$C + 8B \leq 32 \quad \text{(Money)}$$

$$C \geq 0$$

$$B \geq 0$$

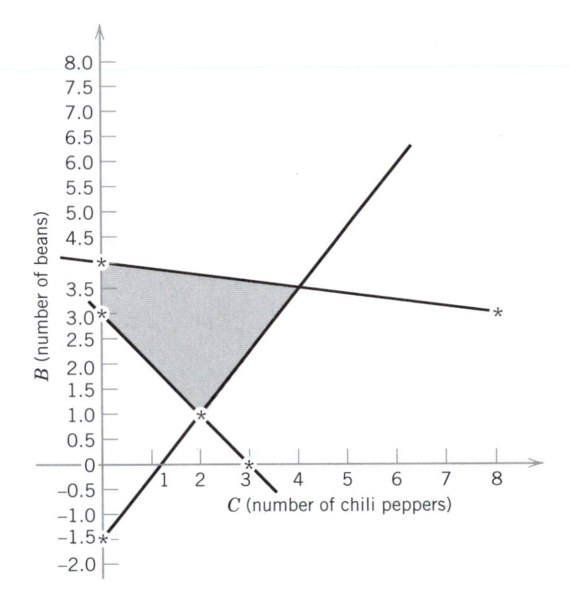

The shaded region of this last graph represents all the points that satisfy all three of the inequalities important to Oziel, as well as the two reality constraints, $B \geq 0$ and $C \geq 0$.

This set of points is called the **feasible region** for the system of inequalities. It is the region of points that are solutions to every inequality in the system.

To find the feasible region for a system of inequalities, we draw the solution of each inequality on the same graph. The overlap of all the shadings is the feasible region.

PRACTICE

Draw the feasible region for each of the following systems of inequalities. Let x be the horizontal axis and y the vertical axis.

1. $x + y \geq 4$

$x + y \leq 6$

$x \geq 0$

$y \geq 0$

2.
$$\begin{aligned} x + y &\geq 4 \\ 5x + 3y &\leq 15 \\ x &\geq 0 \\ y &\geq 0 \end{aligned}$$

3.
$$\begin{aligned} x + y &\leq 4 \\ x + 3y &\geq 6 \\ 3x - 2y &\leq 6 \\ x &\geq 0 \\ y &\geq 0 \end{aligned}$$

4.
$$\begin{aligned} 0.5x + y &\geq 1 \\ -x + y &\leq 1 \\ 5x + 8y &\leq 40 \\ x &\geq 0 \\ y &\geq 0 \end{aligned}$$

We have now reduced the possible combinations of chili peppers and beans for Oziel's chili to the feasible region of the system of linear inequalities:

$$5C - 4B \leq 6 \quad \text{(Flavor rule)}$$
$$C + B \geq 3 \quad \text{(Quantity)}$$
$$C + 8B \leq 32 \quad \text{(Money)}$$
$$C \geq 0$$
$$B \geq 0$$

Now every combination of ingredients taken from this region will satisfy the constraints placed on Oziel's cooking, but which combination will provide the hottest chili? To answer this last question, we must consider the **corner points.**

6.5 CORNER POINTS

The corner points of a feasible region are exactly what they sound like, the points at the corners of the region. The corner points of the feasible region for our chili problem are labeled P–S in the picture below.

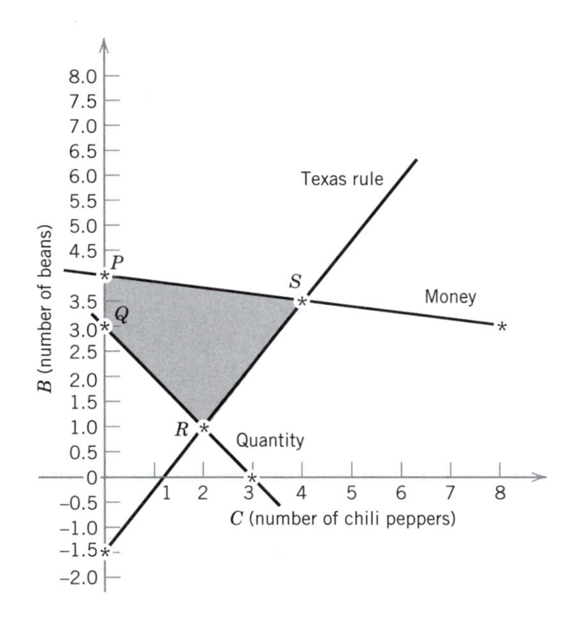

Notice that the corner points lie at the intersections of the lines that form the boundaries of the feasible region. For example, corner point R is the intersection of the line determined by the quantity constraint and that determined by the Texas flavor rule.

To find the coordinates of the corner points then, we only need to find the point of intersection of the appropriate lines.

For corner point R, we need to solve the system of linear equations

$$5C - 4B = 6$$
$$C + B = 3$$

To solve this system, we rewrite the equations.

$$5C = 4B + 6$$
$$C = -B + 3$$

Then multiply the second equation by 5.

$$5C = 4B + 6$$
$$5C = -5B + 15$$

Now eliminate C and solve for B.

$$4B + 6 = -5B + 15$$
$$4B + 5B + 6 = -5B + 5B + 15$$
$$9B + 6 = 15$$
$$9B + 6 - 6 = 15 - 6$$
$$9B = 9$$
$$9B/9 = 9/9$$
$$B = 1$$

If we now substitute 1 for B in the rewritten second original equation, we get

$$C = -B + 3$$
$$= -1 + 3$$
$$= 2$$

Thus, the coordinates of corner point R are (2, 1).

Similarly, corner point S occurs at the intersection of the money constraint line and the Texas flavor rule line. So, the coordinates of corner point S is the solution to the system

$$5C - 4B = 6$$
$$C + 8B = 32$$

We can solve this system by first rewriting the equations.

$$5C = 4B + 6$$
$$C = -8B + 32$$

Then multiplying the second equation by 5 yields

$$5C = 4B + 6$$
$$5C = -40B + 160$$

Then eliminating C and solving for B, we get

$$4B + 6 = -40B + 160$$
$$4B + 40B + 6 = -40B + 40B + 160$$

$$44B + 6 = 160$$
$$44B = 154$$
$$44B/44 = 154/44$$
$$B = 3.5$$

The value of C is then

$$C = -8B + 32$$
$$= -8 \cdot 3.5 + 32$$
$$= 4$$

The coordinates of corner point S are (4, 3.5).

Corner point P is at the intersection of the money constraint line and the y-axis. Thus, its C coordinate is 0. Substitute this value into the money line equation and solve for B.

$$C + 8B = 32$$
$$0 + 8B = 32$$
$$8B = 32$$
$$8B/8 = 32/8$$
$$B = 4$$

The coordinates of point P are (0, 4).

Corner point Q occurs at the intersection of the quantity constraint line and the y-axis. Thus, its C coordinate is 0. Substitute this value into the quantity line equation and solve for B.

$$C + B = 3$$
$$0 + B = 3$$
$$B = 3$$

The coordinates of point Q are (0, 3).

The table below summarizes our work in finding the corner points.

CORNER POINTS FOR FEASIBLE REGION OF HOT CHILI PROBLEM

Corner Point	Coordinates
P	(0, 4)
Q	(0, 3)
R	(2, 1)
S	(4, 3.5)

It is among the corner points that we will find the optimal balance of ingredients for our chili.

PRACTICE

1. Sketch the feasible region and find the coordinates of the corner points for the following system of linear inequalities:

$$x + y \leq 4 \quad \text{(Constraint I)}$$

$$x + 3y \geq 6 \quad \text{(Constraint II)}$$

$$3x - 2y \leq 6 \quad \text{(Constraint III)}$$

$$x \geq 0$$

$$y \geq 0$$

6.6 LINEAR PROGRAMMING

In Situation 21, Oziel's primary goal was to produce the hottest chili possible. To measure hotness, he used the Rawlins hot equation:

$$H = 8 \cdot C - 2 \cdot B$$

where

H = the hotness

C = the number of chili peppers per serving

B = the number of beans per serving

Oziel wants to make the value of H as large as possible. In other words, he wants to *maximize* H. The values for C and B that Oziel can use in computing H were limited or **constrained** by considerations of flavor, quantity, and money. All the constraints were linear relationships.

We can restate the problem posed in Situation 21 mathematically, like this.

$$\text{Maximize } H = 8C - 2B$$

subject to the constraints

$$5C - 4B \leq 6 \quad \text{(Texas flavor rule)}$$
$$C + B \geq 3 \quad \text{(Quantity constraint)}$$
$$C + 8B \leq 32 \quad \text{(Money constraint)}$$
$$C \geq 0 \quad \text{and} \quad B \geq 0 \quad \text{(Reality constraints)}$$

This is an example of a **linear programming problem.** A linear programming problem asks that you optimize (i.e., maximize or minimize) a particular quantity while being restricted by one or more constraints. In addition, the **objective** (the quantity you wish to optimize) and all the constraints are expressed as linear relationships.

A solution to a linear programming problem can be found using the following optimal value rule.

Optimal Value Rule: The optimal value of the objective in a linear programming problem will be found at one of the corner points of the feasible region of the system of linear inequalities given by the constraints of the linear programming problem.

The assertions of the optimal value rule come intuitively from the observation that the corner points represent the extreme points of the feasible region and, as a result, are where we would expect to find the extreme values of the objective function. All of the feasible regions in this book are **bounded,** which means that they are completely enclosed by the constraint lines. In some cases linear programming problems produce **unbounded** feasible regions. When a feasible region is unbounded the objective function may not have a minimum or maximum value.

We can now solve Oziel's problem! We have the coordinates of the corner points for the feasible region of the chili problem. So, all we need do now is see which of these corner points produces the highest value of H. The following table shows the necessary calculations.

CORNER POINTS FOR FEASIBLE REGION OF HOT CHILI PROBLEM

Corner Point	Coordinates (C, B)	Hotness Value $H = 8C - 2B$
P	(0, 4)	−8
Q	(0, 3)	−6
R	(2, 1)	14
S	(4, 3.5)	25

Notice that the hottest chili is produced when Oziel uses 4 chili peppers and 3.5 beans per serving. Remember that every point in the feasible region meets the conditions of the constraints, but none of them will give a higher H value than (4, 3.5).

It has taken us quite a while to solve the chili problem, so it is probably a good idea to take a look back and summarize the steps we took to solve this linear programming problem.

Solving a Linear Programming Problem
Step 1: Identify the objective and write the constraints.
Step 2: Draw the feasible region for the system of linear inequalities given by the constraints.
Step 3. Find the coordinates of the corner points of the feasible region.
Step 4: Find the value of the objective at each of the corner points and select the optimal solution.

Here is another example of a linear programming problem.

Problem

Elmira manufactures little rubber duckies. Each Ninja Ducky sells for $5. Each Jurassic Ducky sells for $4. Each Ninja Ducky requires 6 ounces of plastic and 4 ounces of rubber. Each Jurassic Ducky requires 8 ounces of plastic and 2 ounces of rubber. Elmira can obtain no more than 2000 ounces of rubber and 4000 ounces of plastic each week. How many of each kind of Ducky should Elmira make each week in order to maximize her revenue?

Solution

Step 1: Identify the objective and write the constraints.

Elmira's objective is to maximize her revenue. Her revenue is dependent on the number of each kind of Ducky she sells. If we let N represent the number of Ninja Duckies sold per week and J the number of Jurassic Duckies sold per week, then Elmira's revenue per week R is calculated by

$$R = 5N + 4J$$

Why is it $5N + 4J$?

____

____

____

There are two constraints in this problem. Both the constraints have to do with limited resources. Elmira is limited to 2000 ounces of rubber per week. Since each Ninja Ducky will use 4 ounces of rubber and each Jurassic Ducky 2 ounces of rubber, we have the constraint

$$4N + 2J \leq 2000$$

What is the constraint that expresses the limited amount of plastic that Elmira can obtain per week?

____

____

____

(Answer: $6N + 8J \leq 4000$)

Finally, because rubber duckies can only come in nonnegative quantities, we have the two real-life constraints

$$N \geq 0$$

$$J \geq 0$$

The complete linear programming problem looks like this.

$$\text{Maximize } R = 5N + 4J$$

subject to the constraints

$$4N + 2J \leq 2000$$

$$6N + 8J \leq 4000$$

$$N \geq 0$$

$$J \geq 0$$

Step 2: Draw the feasible region for the system of linear inequalities given by the constraints.

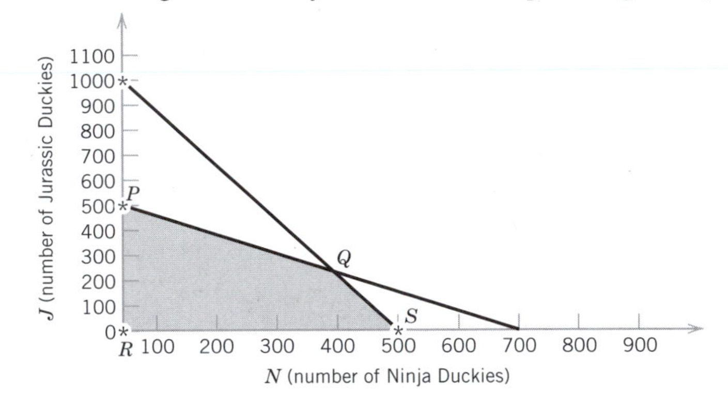

Step 3: Find the coordinates of the corner points of the feasible region.

Corner Point	Coordinates
P	(0, 500)
Q	(400, 200) (Check this!!!!)
R	(0, 0)
S	(500, 0)

Step 4: Find the value of the objective at each of the corner points and select the optimal solution.

Corner Point	Coordinates (N, J)	Value of R $R = 5N + 4J$
P	(0, 500)	\$2000
Q	(400, 200)	\$2800
R	(0, 0)	\$0
S	(500, 0)	\$2500

The maximum revenue of \$2800 per week is reached when Elmira makes and sells 400 Ninja Duckies and 200 Jurassic Duckies per week. ■

PRACTICE

1. Go back to Situation 21. Read it again. Now solve it using the four-step method without looking at my solution.

6.7 WHAT DO YOU KNOW?

If you have worked carefully through this chapter, then you can solve simple systems of linear equations and also elementary linear programming problems. The exercises in this section are designed to test and refine your skills in the mathematics of optimization.

Solve the following systems of linear equations:

1. $y = 3x + 5$
   $y = 2x - 8$
2. $y = -2x + 7$
   $y = 4x + 1$
3. $y = x - 9$
   $2y = x + 3$
4. $3y = x - 5$
   $y = 2x + 7$
5. $2y = 4x + 7$
   $3y = x - 9$
6. $x + 3y = 6$
   $2x - y = -2$
7. $x + 4y = 7$
   $x + 4y = 8$
8. $2x + 7y = 3$
   $4x + 14y = 6$
9. $x = 7$
   $x = y - 9$
10. $2x - 3y = 18$
    $y = 2x - 5$
11. $3x + y = 0$
    $x - 4y = 13$
12. $2x - 4y = 18$
    $-x + 2y = -9$

Sketch the feasible region for each of the following systems of linear inequalities.

13. $x + y \geq 4$
    $x + y \leq 7$
    $x \geq 0$
    $y \geq 0$
14. $2x + y \geq 4$
    $3x + 5y \leq 15$
    $x \geq 0$
    $y \geq 0$
15. $x + 2y \geq 4$
    $x + y \leq 7$
    $-x + y \geq 1$
    $x \geq 0$
    $y \geq 0$
16. $2x + y \geq 4$
    $x + 5y \leq 10$
    $x - y \leq 8$
    $x \geq 0$
    $y \geq 0$
17. $2x + y \leq 2$
    $x + 2y \leq 2$
    $x \geq 0$
    $y \geq 0$
18. $x + y \leq 5$
    $-x + y \leq 1$
    $x - y \leq 1$
    $x \geq 0$
    $y \geq 0$
19. Find the coordinates of the corner points for the feasible region in Exercise 13.
20. Find the coordinates of the corner points for the feasible region in Exercise 14.
21. Find the coordinates of the corner points for the feasible region in Exercise 15.
22. Find the coordinates of the corner points for the feasible region in Exercise 16.
23. Find the coordinates of the corner points for the feasible region in Exercise 17.
24. Find the coordinates of the corner points for the feasible region in Exercise 18.
25. The cost of producing waterproof baskets is linearly related to the number of baskets produced by the equation

$$y = 20x + 280$$

where

y = the cost in dollars
x = the number of baskets produced

(a) If each basket can be sold for \$30, find the break-even point.
(b) How many baskets must be made and sold in order to make a profit?
(c) What is the minimum profit that can be made?

26. The cost of producing water slides is linearly related to the number of slides produced by the equation

$$y = 3000x + 15{,}000$$

where

y = the cost in dollars
x = the number of water slides produced

(a) If each slide can be sold for \$5000, find the break-even point.
(b) How many slides must be made and sold in order to make a profit?
(c) What is the minimum profit that can be made?

27. Melena has discovered that it costs her \$450 to make two neural networks and \$975 to make five neural networks.

She can sell every neural network she produces for \$300.

(a) Write a linear equation that relates Melena's cost of production y to the number of neural networks x that she makes.

(b) Find Melena's break-even point.

(c) How many neural networks must she manufacture in order to make a profit?

(d) What is the minimum profit that can be made?

28. Francisco weaves luxury bread box covers. It costs him \$547 to weave 5 covers and \$632 to weave 10 covers. He can sell every cover that he makes for \$33.

(a) Write a linear equation that relates Francisco's cost of production y to the number of bread box covers x that he makes.

(b) Find Francisco's break-even point.

(c) How many bread box covers must Francisco produce in order to make a profit?

(d) What is the minimum profit that he can make?

(e) Francisco says that he can retire from bread box cover weaving after he has made a total profit of \$100,000. How many bread box covers must he sell before he can retire?

29. Melvin manufactures two types of plastic fasteners for solar-powered flashlights. He must make at least three fasteners every hour or the plastic will harden and clog his machinery. Each small fastener requires 4 ounces of a special bonding additive. Each large fastener requires 6 ounces of the special bonding additive. The bonding additive is supplied by a machine that can produce no more than 24 ounces of the additive each hour. Melvin makes a profit of \$7 on each small fastener and \$9 on each large fastener.

(a) How many of each fastener should Melvin produce each hour to maximize his profit?

(b) What will his maximum profit be?

30. Flaubert is on a special diet. He is required to have at least 3300 units of Vitamin C per day and at least 4200 units of Vitamin B per day. Dietary supplement I is a pill that contains 500 units of Vitamin C and 400 units of Vitamin A. Dietary supplement II is a pill that contains 300 units of Vitamin C and 500 units of Vitamin A. Flaubert's physician does not want him to take more than 15 pills per day.

(a) If each dietary supplement I pill costs 30¢ and each dietary supplement II pill 25¢, how many of each pill should Flaubert take each day to meet his vitamin requirements and minimize his cost?

(b) What will his minimum cost be?

31. Sam and Janet Evening operate a small musical instrument business. They make dulcimers and sitars. They use only high-quality imported Bulgarian steel strings and aged Laotian teak for their instruments. Each dulcimer requires 10 feet of steel string and 16 ounces of teak. Each sitar requires 20 feet of steel string and 5 pounds of wood. Sam, who handles all imports, is able to get no more than 90 feet of steel string and no more than 18 pounds of teak each month. It takes Sam 5 hours to make a dulcimer. It takes Janet 20 hours to make a sitar. The Evenings can work no more than a total of 400 hours each month. They can sell each dulcimer they make for \$80 and each sitar for \$375.

(a) How many of each instrument should Sam and Janet Evening make to maximize their revenue?

(b) What will their maximum revenue be?

32. A farmer has 60 acres of land on which to grow beans and corn. An acre of beans will cost her \$70 to cultivate, but at harvest will return a profit of \$350. An acre of corn will cost \$40 to cultivate and at harvest will return a profit of \$200. The farmer has only \$2800 available for cultivation.

(a) How many acres should the farmer devote to each crop to maximize her profit?

(b) What will her maximum profit be?

33. A dairy company is offering two new kinds of ice cream for sale: Triple Hitter and Major League. Both are vanilla ice cream with flavored chips. Each gallon of Triple Hitter contains 10 ounces of chocolate chips, 10 ounces of mint chips, and 10 ounces of butterscotch chips. Each gallon of Major League contains 20 ounces of chocolate chips, 10 ounces of mint chips, and 5 ounces of butterscotch chips. Each week the company can obtain 5000 ounces of chocolate chips, 2000 ounces of mint chips, and 1000 ounces of butterscotch chips. The company can make a profit of \$1 on each gallon of Triple Hitter that it sells and \$1.50 on each gallon of Major League. Suppose that the company can sell every gallon of ice cream that it makes.

(a) How many gallons of each type of ice cream should the company make to obtain a maximum weekly profit?

(b) What is that maximum weekly profit?

ANSWERS TO PRACTICE EXERCISES

Section 6.2

1. (a) $y = 10x + 500$
$y = 32.95x$

(b) $10x + 500 = 32.95x$

(c)
$$\begin{aligned} 10x + 500 &= 32.95x \\ 10x - 10x + 500 &= 32.95x - 10x \\ 500 &= 22.95x \\ 500/22.95 &= 22.95x/22.95 \\ 21.79 &\approx x \end{aligned}$$

(d) 22 dolls

(e) \$4.90

2. 41 T-shirts (40 T-shirts is the break-even point!)

3. Do not intersect.

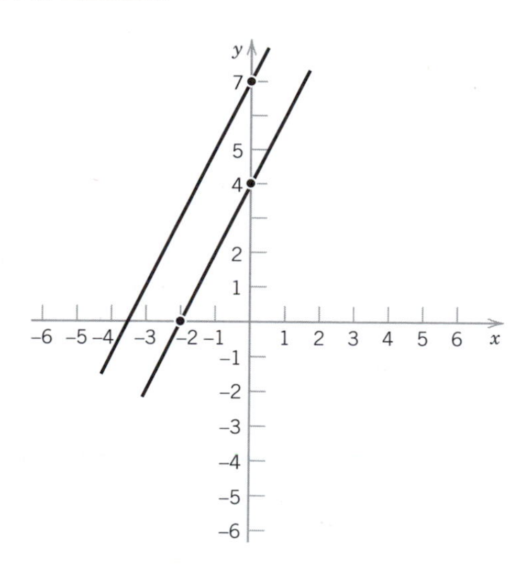

4. Do not intersect.

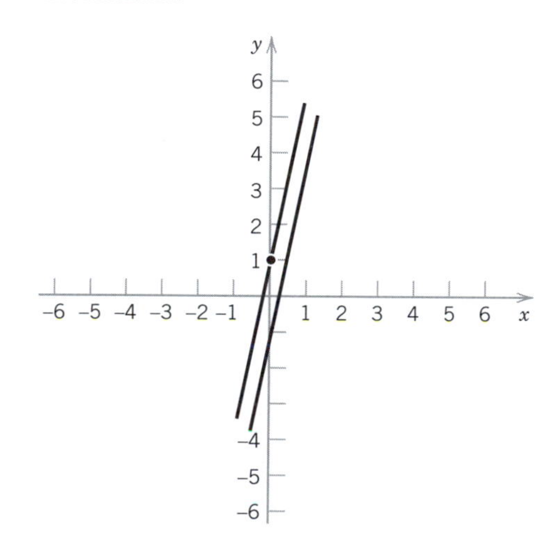

5. Coincide.

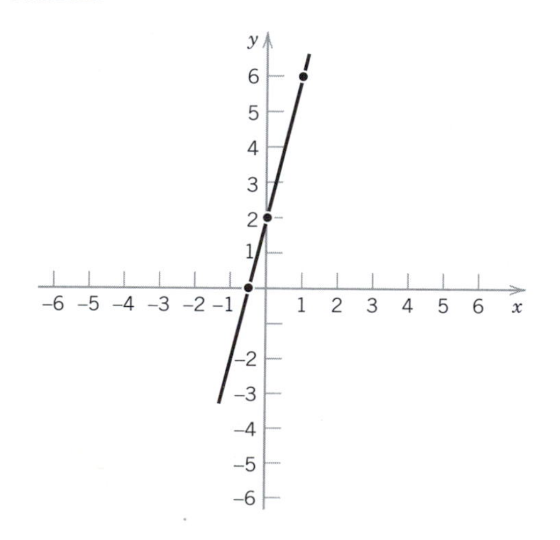

6. Intersect in one point.

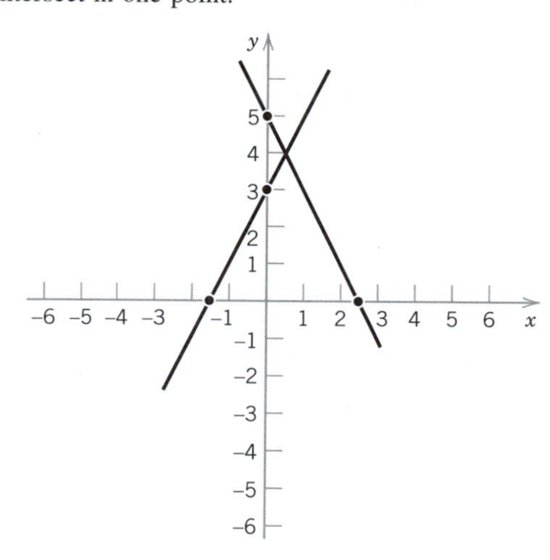

7. Coincide.

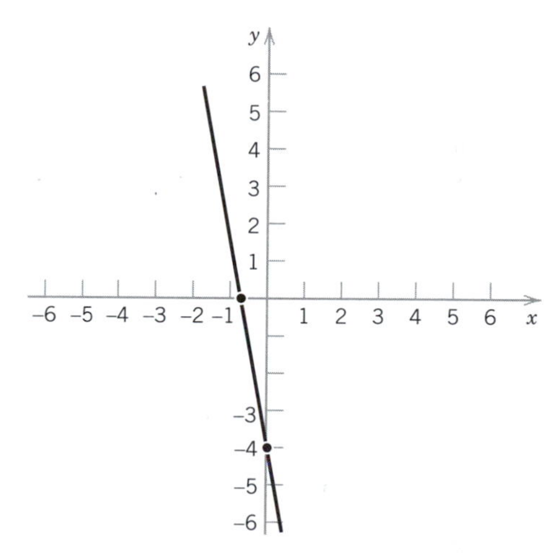

8. (1, 7)

9. $y = 2x - 5$
$3y = x + 3$

Multiply the first equation by 3: $3y = 6x - 15$

Eliminate y: $6x - 15 = x + 3$

Solve for x: $6x - x - 15 = x - x + 3$

$$5x - 15 = 3$$
$$5x + 15 - 15 = 3 + 15$$
$$5x = 18$$
$$5x/5 = 18/5$$
$$x = 3.6$$

Substitute into first equation: $y = 2 \cdot 3.6 - 5$

$$y = 2.2$$

Solution is (3.6, 2.2).

10. Dependent

11. Inconsistent

12. $x + 3y = 4$
$2x + y = 9$

Isolate y in both equations: $3y = -x + 4$

$$y = -2x + 9$$

Multiply the second equation by 3: $3y = -x + 4$

$$3y = -6x + 27$$

Eliminate y: $-x + 4 = -6x + 27$

Solve for x: $-x + 6x + 4 = -6x + 6x + 27$

$$5x + 4 = 27$$
$$5x + 4 - 4 = 27 - 4$$
$$5x = 23$$
$$5x/5 = 23/5$$
$$x = 4.6$$

Substitute into the second equation: $y = -2 \cdot 4.6 + 9 = -0.2$

The solution is $(4.6, -0.2)$.

Section 6.3

1. $\geq$
2. $4x + y$
3. $5Z - 3Q < 18$
4. $> C$
5. Is a solution
6. Is not a solution
7. Is a solution
8. Is a solution
9. Is a solution
10. Is a solution
11. Is not a solution
12. Is not a solution

13.

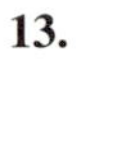

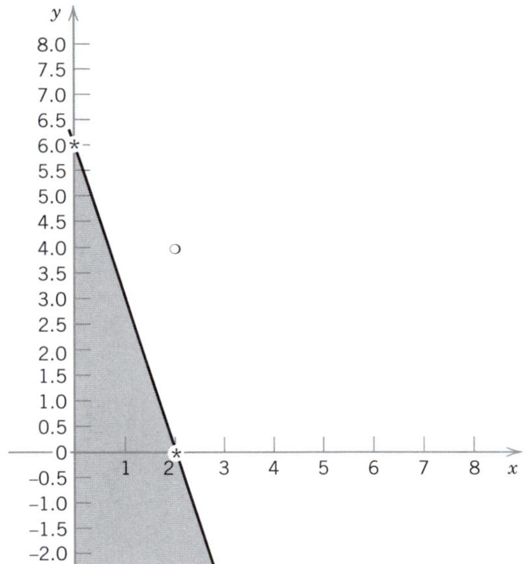

14.

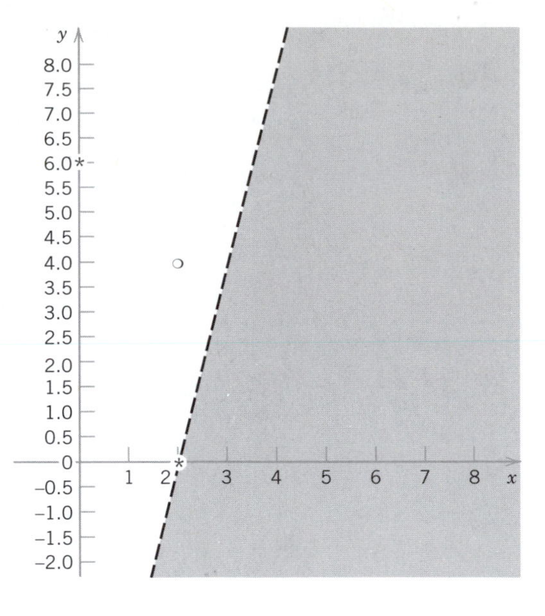

15.

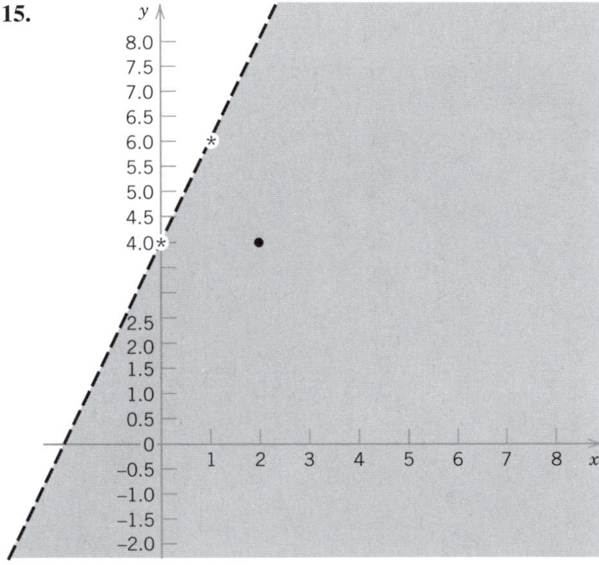

16.

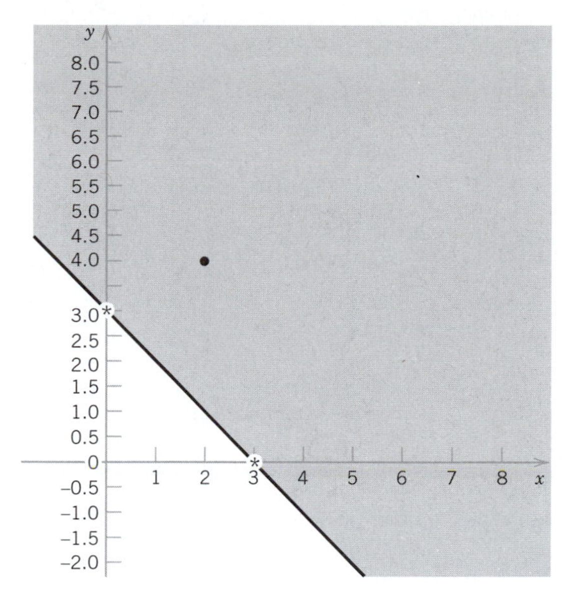

Section 6.4

1. $x + y \geq 4$
$x + y \leq 6$
$x \geq 0$
$y \geq 0$

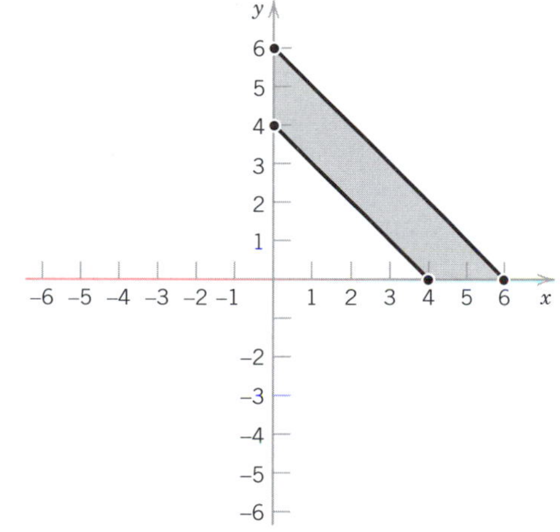

2. $x + y \geq 4$
$5x + 3y \leq 15$
$x \geq 0$
$y \geq 0$

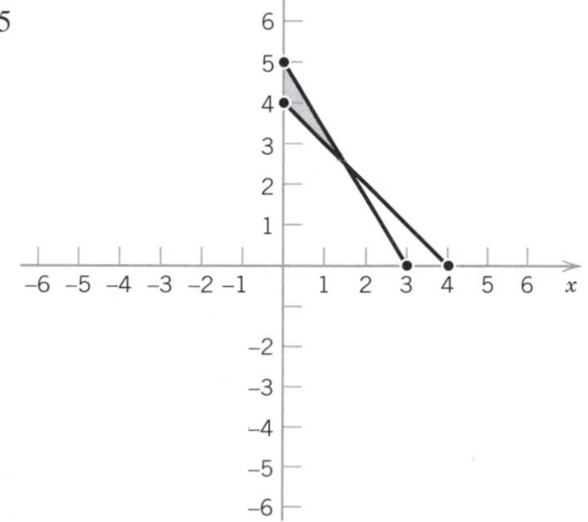

3. $x + y \leq 4$
$x + 3y \geq 6$
$3x - 2y \leq 6$
$x \geq 0$
$y \geq 0$

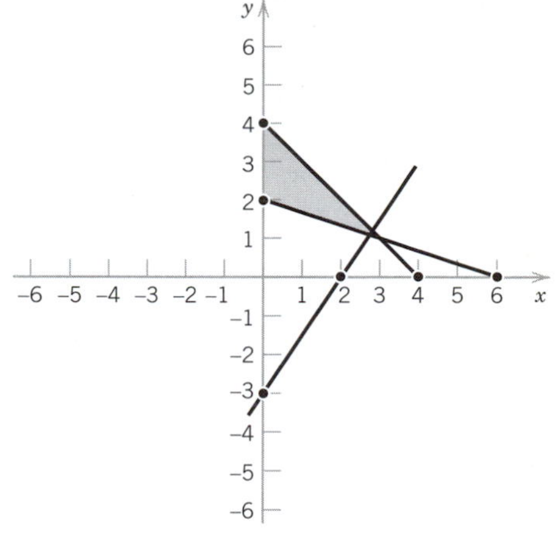

4. $0.5x + y \geq 1$
$-x + y \leq 1$
$5x + 8y \leq 40$
$x \geq 0$
$y \geq 0$

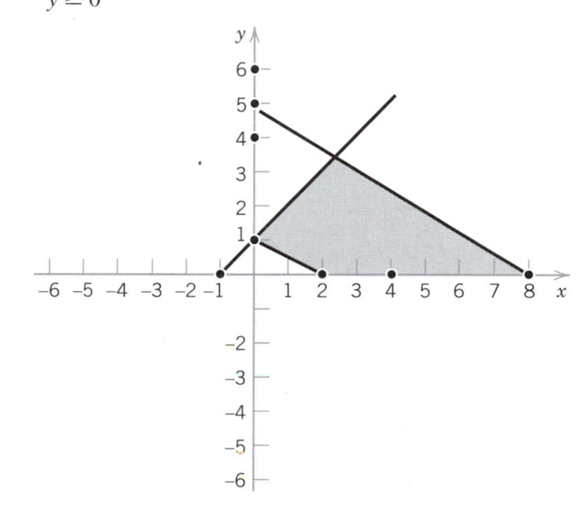

Section 6.5

1. The feasible region.

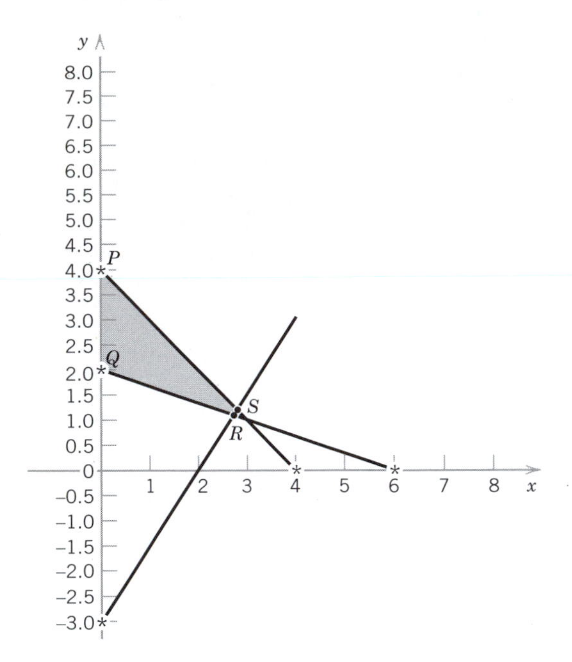

Corner point P: (0, 4)
Corner point Q: (0, 2)
Corner point R is at the intersection of

$$x + 3y = 6 \quad \text{and}$$
$$3x - 2y = 6$$

Rewrite the equations, multiply the first by 3, eliminate x, and solve for y.

$$x = -3y + 6$$
$$3x = 2y + 6$$

$$3x = -9y + 18$$
$$3x = 2y + 6$$

$$-9y + 18 = 2y + 6$$
$$18 = 11y + 6$$
$$12 = 11y$$
$$y = 12/11 = 1.09$$

Then $x = -3y + 6 = -3 \cdot 1.09 + 6 = 2.73$.
The coordinates of corner point R are (2.73, 1.09).
The coordinates of corner point S are (2.8, 1.2).

Section 6.6

1. Maximum H is 25. This maximum is reached using 4 chili peppers and 3.5 beans per serving. (Reread Sections 6.5 and 6.6 for details.)

Chapter Seven

Mathematics of Space

What's the Point? The measurement of space is an everyday occurrence in many professions and a common occurrence in almost everybody's life. How much paint is needed to cover the walls in this room? How much fencing do we need to enclose the garden? What is the best way to connect the telephone system? This chapter will help you to understand some of the basic concepts of **geometry,** the mathematics of space.

Situation 22 Painting Your Apartment

It's time to repaint the living room of your luxury apartment. You still have the brochure from the rental company that gives the floor plan of your living room. Here it is. (The diagrams in this chapter with labeled measurements are not drawn to scale unless I say so. This is done intentionally so that you will focus on the mathematical techniques instead of the pictures. If this bothers you, you are encouraged to make your own scale drawings.)

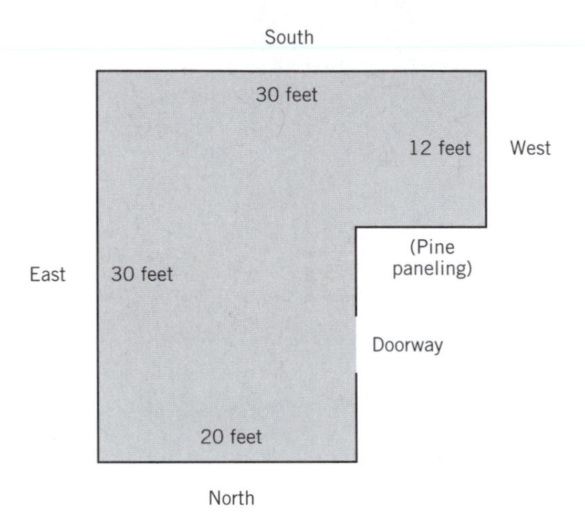

You measure the height of your ceiling and find that it is 11 feet high. You also measure the position of your windows and make the following drawings. There aren't any windows on the north wall. The section of wall marked *pine paneling* is covered with natural white pine panels and will not be painted.

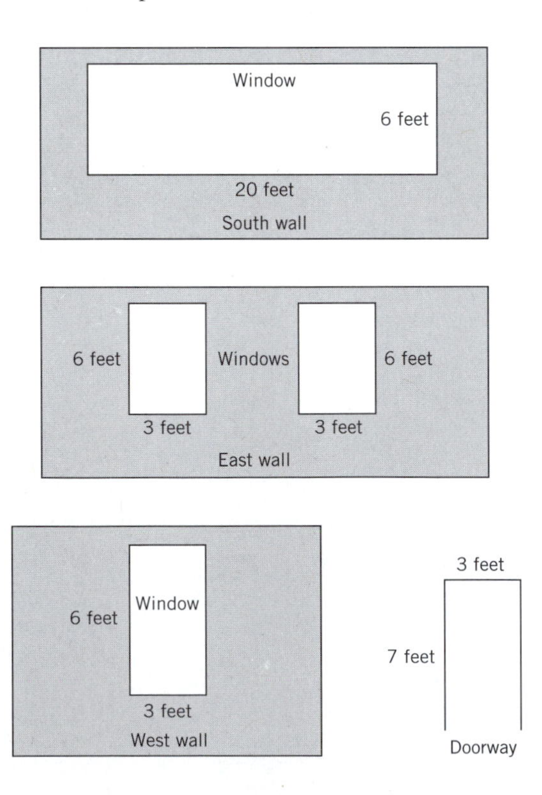

The paint that you've selected costs $16.95 per gallon, and it says on the paint can that 1 gallon will cover 400 square feet of area. How many gallons of paint should you buy to paint your apartment?

____

____

____

The key to resolving the problem is to find the number of square feet of wall in your apartment's living room. From that you should be able to determine the number of gallons of paint needed by using proportions.

7.1 AREA

The **area** of a geometric figure is the amount of surface enclosed by the figure. In the picture below, the figure on the left has a larger area than the figure on the right because the figure on the left encloses a larger surface.

Area is measured in squares of particular sizes. Here is a **square inch.**

Get out a ruler and measure each side of the square inch. How long is each side? _____

Similarly, a **square foot** is a square, each of whose sides is 1 foot in length; a **square yard** is a square, each of whose sides is 1 yard (3 feet) in length; and a **square mile** is a square, each of whose sides is 1 mile in length.

If we say that the area of a figure is *12 square inches,* this means that we can fit 12 of our square-inch squares inside the figure. The figure below has an area of 12 square inches. Notice that there are 12 square-inch squares inside of it.

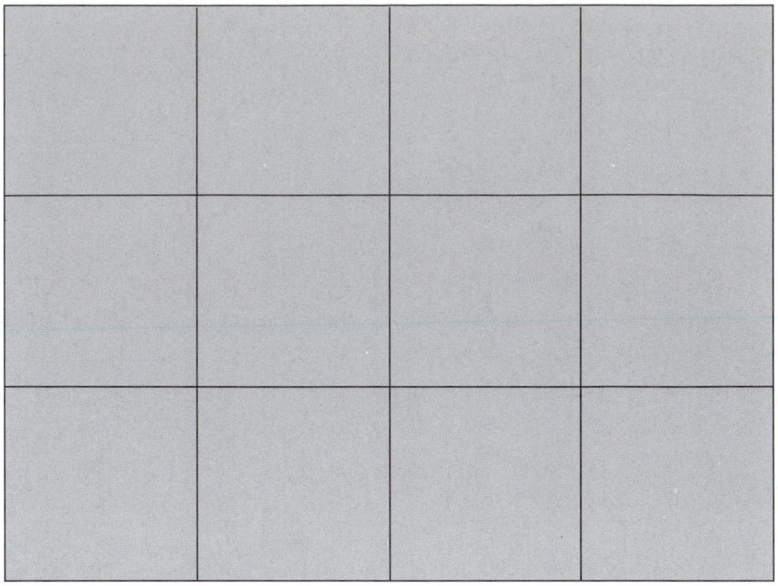

How long is the figure above? __________

How wide is the figure above? __________

Can you see any relationship between the length and width of the figure and its area?

____

____

Right! The area is the product of the length and width, $12 = 4 \cdot 3$. This is a special case of one of the classical formulas for area that you probably first saw in school when you were a child. Below is a table that includes this formula for the area of a rectangle and several others.

To use this table, you need to be able to identify the kind of figure whose area you want and then determine the values of the variables needed to calculate its area.

For example, if your figure is a rectangle, then you must know its length L and width W before you can find its area. The area of a figure will always be expressed in square units. If you measure the figure in feet, then the area will be in square feet. If you measure the figure in inches, then the area will be in square inches. It is important that you use the same measurement units for all variables involved in the area calculation. Don't mix feet with inches, inches with miles, yards with feet, and so forth!

Classical Area Formulas

Figure	Name of Figure	Area of Figure
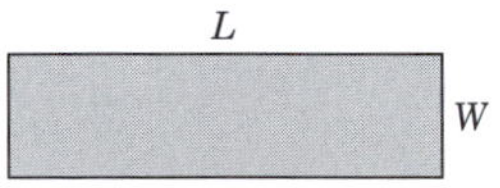	Rectangle	*Length* · *Width*
	Square	*Side* · *Side*
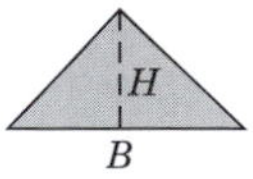	Triangle	0.5 · *Base* · *Height*
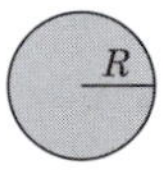	Circle	π · *Radius* · *Radius*

The calculation of the area of the circle requires the use of a special number called π (pi). The approximate value of π is 3.14159. I'll have more to say about π in Chapter 8.

PRACTICE

Find the area of each of the following figures:

1.

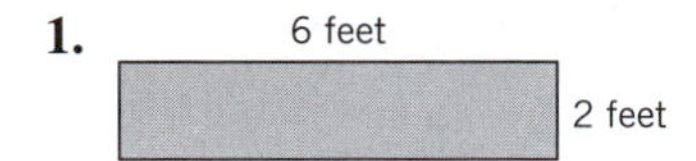

2.

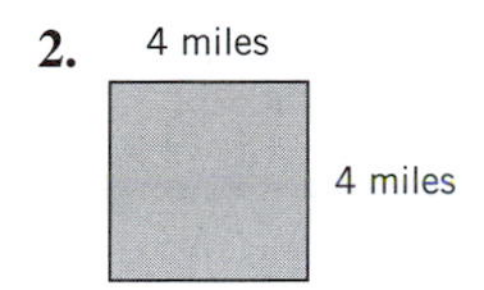

3.

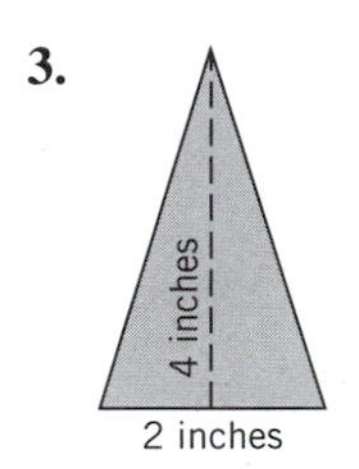

4.

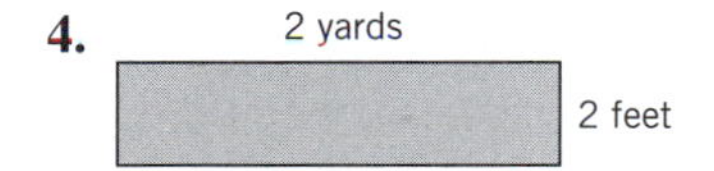

5.

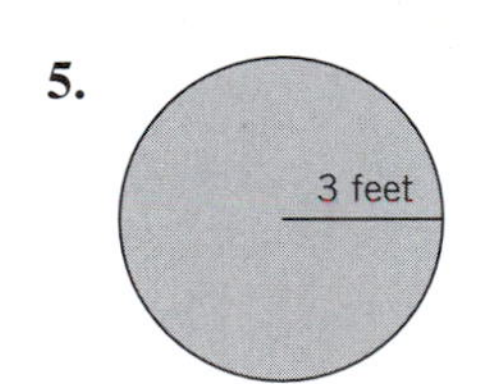

6.

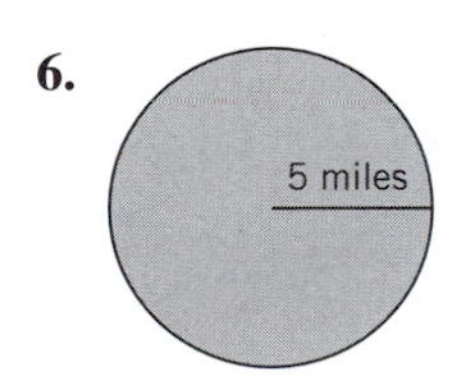

In Situation 22, the walls that we want to paint are all rectangular. So, we can use the formula for the area of a rectangle to compute their areas.

The north wall is 20 feet long and 11 feet high. The area of the north wall is $20 \cdot 11 = 220$ square feet.

The south wall is 30 feet long and 11 feet high. The area of the south wall is $30 \cdot 11 = 330$ square feet. But we aren't going to paint all of the south wall because we certainly don't want to paint the big picture window. The area of the window on the south wall is $20 \cdot 6 = 120$ square feet. We can subtract this area from the total area of the south wall to obtain the area of the part of the wall that we will actually paint: $330 - 120 = 210$ square feet.

The west wall has an area of $12 \cdot 11 = 132$ square feet. The window in the west wall has an area of $3 \cdot 6 = 18$ square feet. Thus, we will need to paint $132 - 18 = 114$ square feet of the west wall.

How many square feet of the east wall will we need to paint?

(Answer: $30 \cdot 11 - (3 \cdot 6 + 3 \cdot 6) = 330 - 36 = 294$ square feet)

Finally, the area around the doorway needs to be painted. The wall containing the doorway has an area of $18 \cdot 11 = 198$ square feet. Why?

If we subtract the size of the doorway from the total area of the wall containing the doorway, we get the area that needs to be painted: $198 - 3 \cdot 7 = 198 - 21 = 177$ square feet.

The total area that needs painting is

$$220 \text{ (north wall)} + 210 \text{ (south wall)} + 114 \text{ (west wall)}$$

$$+ 294 \text{ (east wall)} + 177 \text{ (doorway wall)} = 1015 \text{ square feet}$$

How many gallons of paint do we need if each gallon will cover 400 square feet?

Right! We will need 2.54 gallons. To find this, we set up the proportion

$$1 \text{ gallon} : 400 \text{ square feet} = x \text{ gallons} : 1015 \text{ square feet}$$

$$1/400 = x/1015$$

$$1015 = 400x$$

$$1015/400 = x$$

$$2.54 = x$$

The paint store won't sell you 0.54 gallons of paint. So, you will need to buy 3 gallons of paint to paint the living room of the apartment.

How will the answer change if the paint you are using comes in quarts?

(Answer: You'll need to buy 2 gallons and 3 quarts)

Finally, you may notice that area measure is given in different units, depending on the product that is being used to cover the area. Paint coverage is expressed in square feet, carpeting is measured in square yards, and land is measured in acres (4840 square yards) or square miles. It is therefore useful to have some correspondence values handy.

One square foot is a square, each of whose sides is 1 foot in length. But a foot is 12 inches. Thus, a square foot is a square, each of whose sides is 12 inches in length. The area of a square with sides 12 inches long is $12 \cdot 12 = 144$ square inches. Thus, 1 square foot = 144 square inches. The table below gives some other correspondences of area measure. You should be able to check the accuracy of the table yourself!

AREA MEASURE CORRESPONDENCES

Square Inches	Square Feet	Square Yards	Acre	Square Miles
1	0.00694	0.000716	0.000000148	0.000000000249
144	1	0.111	0.0000230	0.0000000359
1296	9	1	0.000207	0.000000323
6,272,640	43,560	4840	1	0.00156
4,014,489,600	27,878,400	3,097,600	641.03	1

PRACTICE

7. Suppose you wish to carpet the living room of the apartment in Situation 22. Carpet is measured in square yards. How many square yards of carpet will you need?

7.2 PERIMETER

Another basic measurement of space and shape is perimeter. The **perimeter** of an enclosed figure is the distance you would travel if you walked completely around its boundary (the boundary itself is sometimes also referred to as the perimeter). Perimeters are measured in length units like inches, feet, yards, and miles.

If you wanted to put a decorative floral strip of wallpaper just below the ceiling all around the apartment living room in Situation 22, then you would need to know the perimeter of the room before you bought the right length of strip. Here's the floor plan again. I've labeled the walls connecting the west wall to the north wall A and B, respectively.

To find the perimeter, add up the lengths of all the boundary lines:

$$30 \text{ (south wall)} + 12 \text{ (west wall)} + 10\ (A)$$
$$+ 18\ (B) + 20 \text{ (north wall)} + 30 \text{ (east wall)} = 120 \text{ feet}$$

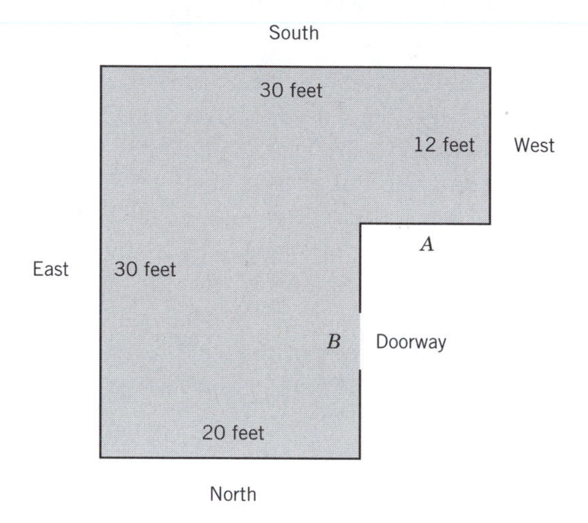

You can find the perimeter of any shape bounded by straight lines by just adding up the lengths of all the lines that bound the shape. However, circles have a special formula for perimeter.

The perimeter of a circle, called the **circumference,** is calculated by the formula $C = 2 \cdot \pi \cdot R$, where R, the **radius,** is the distance from the center of the circle to its boundary.

PRACTICE

Calculate the perimeter of each of the following figures:

1.

2 yards

4 yards

2.

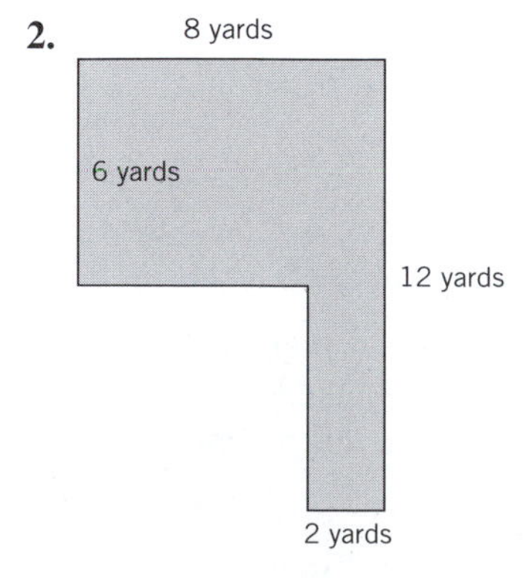

3.

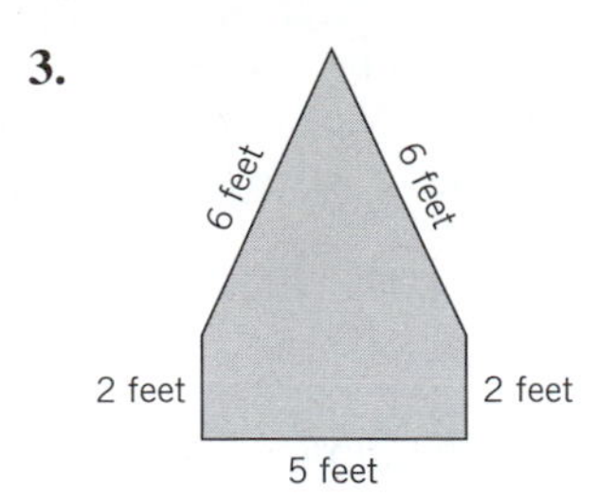

4.

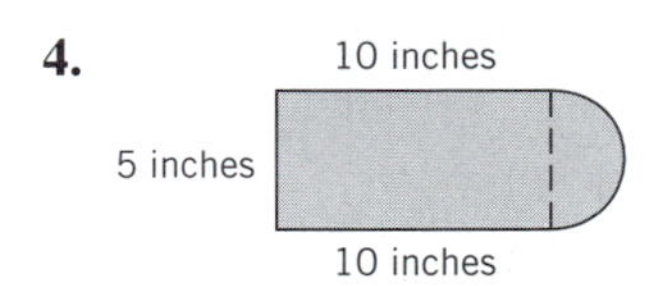

5.

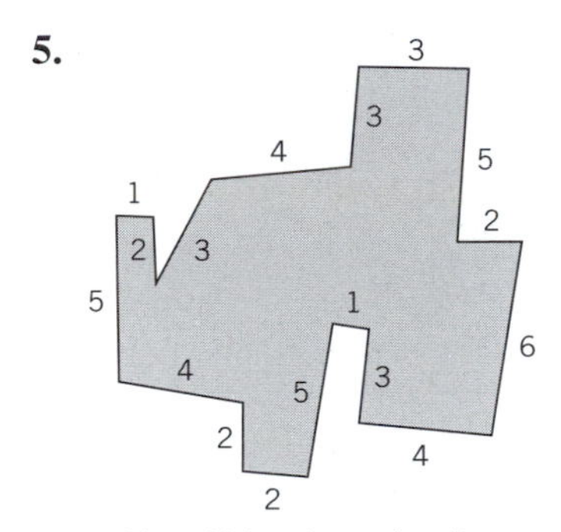

7.3 VOLUME

The last basic measurement of space and shape that we will consider is volume. The **volume** of a solid object is a measure of the amount of space contained within that object. Volumes are measured in terms of cubic lengths like cubic inches, cubic feet, cubic yards, cubic miles, and so on.

A **cubic inch** is a cube (like a die or sugar cube), each of whose sides is 1 inch in length. Here is a picture of a cubic inch.

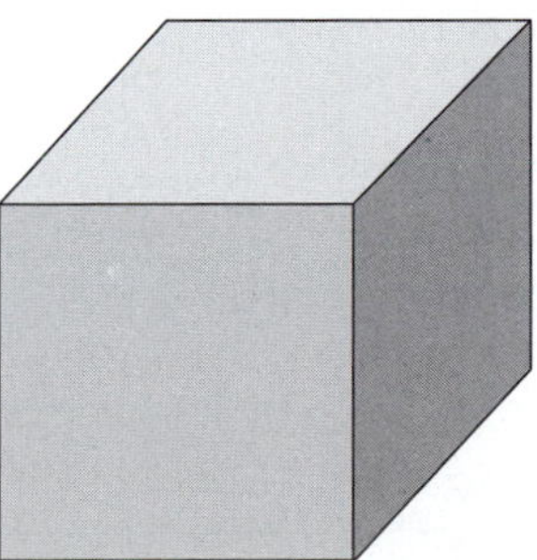

If we could stack 30 of these cubic inches into a box in such a way that there were no gaps between the little cubes and the box was completely filled, then we would say that the box had a volume of 30 cubic inches.

The volume of containers that hold liquid is also measured in units of liquid measure like pints, quarts, and gallons. It is not unusual to refer to the volume of the liquid itself. Thus, you might say, "I drank a pint of orange juice" instead of saying, "I drank all the orange juice in the pint container."

Below is a summary table of the relationships between the various common liquid measures and cubic measures.

Measures of Volume

1 gallon = 4 quarts = 231 cubic inches	1 cubic foot = 1728 cubic inches
1 quart = 2 pints = 57.75 cubic inches	1 cubic yard = 27 cubic feet
1 pint = 28.875 cubic inches	1 cubic mile = 5,451,776,000 cubic yards

PRACTICE

Fill in the blanks

1. 2 gallons = _______ pints

2. 5 pints = _______ quarts

3. 200 cubic inches = _______ gallons

4. 20 cubic feet = _______ cubic yards

The volume of solid objects with vertical sides (called right solids) can be calculated by multiplying the area of the base of the solid by its height. To see why this is so, consider a box with a rectangular base that is 5 inches long and 3 inches wide. In addition, the sides of the box are vertical and 4 inches tall.

Here's the box.

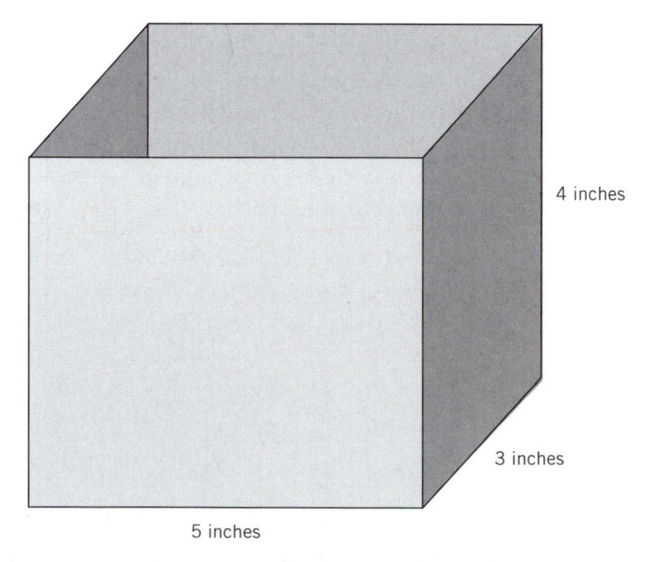

Scale: 1/2 inch = 1 inch.

In order to find the volume of this box, we must determine how many cubic-inch cubes can be packed into the box. Let's first make a single layer of cubes in the bottom of the box, like this.

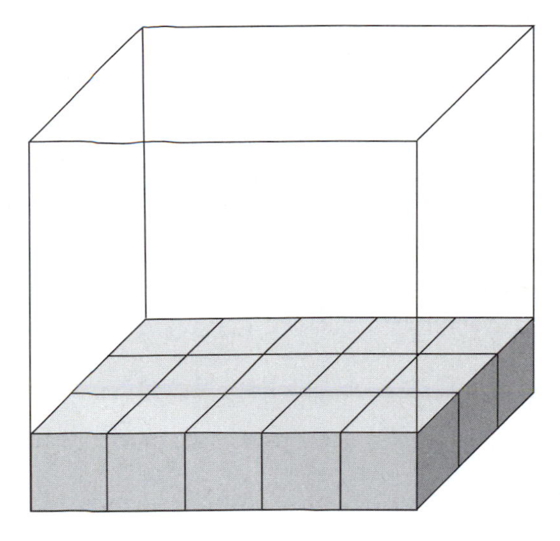

Scale: 1/2 inch = 1 inch.

How many cubes were needed to cover the bottom of the box?

So far, then, we have 15 cubic inches of volume. How many cubes will it take to make the next layer?

Right! So now, we have two layers of 15 cubes each, yielding a total volume of 30 cubic inches. Now each layer that we make will take 15 cubes. A cube is an inch high, so to reach the top of the 4-inch-tall box, we will need a total of four layers of cubes.

Hence, the total volume will be $15 \cdot 4 = 60$ cubic inches. The 15 represents the area of the base of the box and the 4 the height of the box.

PRACTICE

Find the volume of each of the following objects. Each object has vertical sides with the indicated height and a base as described.

5. Base: square with side 3 inches
Height: 5 inches

6. Base: rectangle with length 8 feet and width 2 feet
Height: 7 feet

7. Base: circle with radius 5 yards
Height: 15 yards

8. Base: triangle with base 4 feet and height 2 feet
Height: 10 feet

The volume of a sphere (a shape like a ball) is calculated using the formula

$$V = 4\pi R^3/3$$

where R is the radius.

9. Find the volume of a sphere with radius 3 inches.

10. Find the volume of a sphere with radius 10 miles.

Suppose that you have just finished painting your apartment and have decided to celebrate by filling your bathtub with champagne. Your bathtub has a rectangular base and vertical sides. It is 7 feet long, 3 feet wide, and 4 feet deep. How many quarts of champagne will you need to fill your bathtub to a foot below the top?

(Answer: The volume of champagne you need is $(7 \cdot 3) \cdot 3 = 63$ cubic feet. You know that 1 quart = 57.75 cubic inches. So, you need to convert 63 cubic feet into cubic inches. One cubic foot is 1728 cubic inches, so 63 cubic feet is 108,864 cubic inches. 108,864 cubic inches is 1,885.09 quarts. Hence, you need 1,885.09 quarts of champagne (but you would probably settle for 1885 quarts).)

Situation 23

Driving in England

Fern and Zork are finally taking that driving tour of England they have been planning for years. They obtain their rental car and set off on the highway. They are careful to drive on the left and soon become used to it. However, they soon become very aware that England, like the rest of Europe and most of the world, uses the **metric system** of measurement. Distances are expressed in **kilometers,** not miles, gasoline is measured in **liters,** not gallons, and low bridges are measured in **meters,** not feet.

What is this metric system? How does it work? Is 100 kilometers a long drive? Is 40 kilometers per liter good gas mileage?

7.4 THE METRIC SYSTEM

The system of measurement used in the United States can be confusing to someone learning it for the first time. Look at these relationships:

1 foot = 12 inches

1 yard = 3 feet

1 mile = 5280 feet

1 gallon = 4 quarts

1 quart = 2 pints

1 pound = 16 ounces

1 bushel = 4 pecks

1 peck = 8 quarts

What a jumble of numbers! There doesn't seem to be any system to the numbers at all. Why not 12 feet to a yard? Why not 2 quarts to a pint? And while we're talking about quarts, there are two kinds, dry and liquid. A quart of strawberries (dry) is about 67.20 cubic inches, but a quart of milk (liquid) is 57.75 cubic inches.

The **metric system** is a system of measurement that measures weight, volume, and length in a systematic fashion. Devised by the French Academy of Sciences in 1790, the metric system has been adopted by every country in the world except for a dozen (including the United States of America, Liberia, Myanmar, Nauru, and Ghana).

The idea behind the metric system is that each step up a scale of measurement should be the same number, namely, 10. Take length as an example. One of the length units of the metric system is the centimeter.

———

This is a centimeter.

A centimeter is 10 times the length of the unit below it, the millimeter, and a tenth of the unit above it, the decimeter.

Furthermore, the metric system employs a series of prefixes that mean the same thing for all types of measure: length, weight, area, and volume. Here are the prefixes, their symbols, and their meanings.

METRIC PREFIXES

Prefix	Symbol	Meaning
Kilo	k	1000
Hecto	h	100
Deka	da	10
Deci	d	0.1
Centi	c	0.01
Milli	m	0.001

Notice that there is no prefix that means 1. The basic units of each kind of measurement fill that role. Here they are.

BASIC METRIC UNITS

Unit (symbol)	What measured	Approximate U.S. System Equivalent
Meter (m)	length	39.37 inches (a little longer than a yardstick)
Liter (L)	volume	1.06 quarts (a little more than a quart)
Gram (g)	weight	0.035 ounces (a nickel weighs 5 grams)

You can combine a prefix with a basic unit to get measurements bigger or smaller than the basic unit. For example, *hecto* plus *gram* gives *hectogram,* which means *100 grams* and is symbolized by *hg*. Here are some more examples. You should fill in the blanks.

Kilo + liter = kiloliter = 1000 liters = kL

Centi + meter = centimeter = 0.01 meters = cm

Deci + gram = __________ = 0.1 ________ = dg

Hecto + ________ = hectoliter = ________ liters = ________

Milli + gram = __________ = ________ ________ = mg

________ + ________ = __________ = 10 grams = _____

(Answers: decigram, grams, liter, 100, hL, milligram, 0.001 grams, deka, gram, dekagram, dag)

Conversion from one measure to the next is a simple matter of multiplying or dividing by 10. If you are going to a smaller measure, then you multiply by 10 at each step. If you are going to a higher measure, then you divide by 10 at each step.

Here is the progression of prefixes again from smallest to largest. Multiply by 10 as you move left. Divide by 10 as you move right.

——— divide by 10 ———→

milli- centi- deci- base unit deka- hecto- kilo-

←——— multiply by 10 ———

For example, to convert 45 decimeters to hectometers, note that the move from deci- to hecto- takes three steps to the right. Thus, I'll divide 45 by 10 three times in succession, yielding 0.045. So, 45 decimeters is 0.045 hectometers.

Here are some more. To convert 30 grams to milligrams: base unit to milli- is three steps to the left. So, I multiply 30 by 10 three times: $30 \cdot 10 \cdot 10 \cdot 10 = 30{,}000$ milligrams. To convert 4.8 centimeters to kiloliters: centi- to kilo- is five steps to the right. So, I divide 4.8 by 10 five times to get 0.000048 kiloliters.

PRACTICE

Fill in the blanks.

1. 34 centimeters = _____ meters

2. 2.09 liters = _____ hectoliters

3. 0.008 milliliters = _____ dekaliters

4. 1030 grams = _____ kilograms

5. 12.6 kg = _____ hg

6. 140 dm = _____ cm

7. 1,000,000 m = _____ mm

As you were doing the practice exercises, did you notice how the decimal points seemed to move around? Take the number 1.098 and divide it by 10 three times. What do you get? _______

What is the position of the decimal point in 0.001098 as compared to 1.098?

____

____

____

Make up a general statement about how many places to the left or right the decimal point of a number seems to move after you've multiplied or divided the number by 10 several times.

____

____

____

Just like there is a relationship between cubic inches and gallons in the U.S. system of measurement, a relationship exists between cubic length and volume in the metric system. In the metric system, 1 cubic centimeter = 1 milliliter.

PRACTICE

8. A bathtub with vertical walls and a rectangular base is 2 meters long, 1 meter wide, and 0.5 meters deep. How many liters of water will the bathtub hold?

7.5 CONVERSION BETWEEN METRIC SYSTEM AND U.S. SYSTEM

Although working within the metric system is straightforward, converting between the U.S. system and metric system is sometimes necessary. Fortunately, all the conversions can be accomplished using proportions once you know the basic relationships.

The table below gives some of the relationships between the two systems. If you want to know more, consult an encyclopedia or a metric conversion handbook.

SHORT TABLE OF METRIC–U.S. RELATIONSHIPS

Metric	U.S.
1 meter	39.37 inches
1 meter	3.28 feet
1 kilometer	0.62 miles
1 liter	2.12 pints
1 liter	1.06 quarts
1 liter	0.265 gallons
1 gram	0.035 ounces
1 kilogram	2.2 pounds
1 square kilometer	0.386 square miles

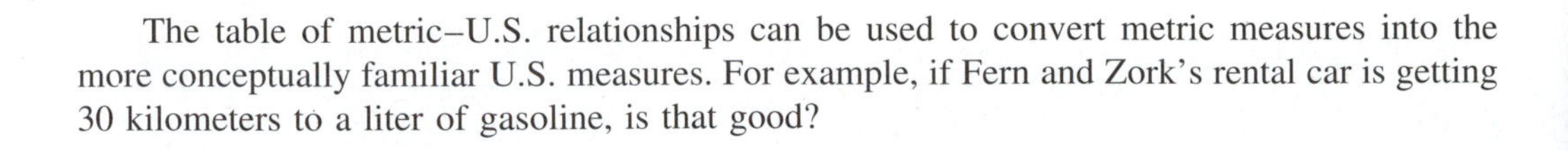

The table of metric–U.S. relationships can be used to convert metric measures into the more conceptually familiar U.S. measures. For example, if Fern and Zork's rental car is getting 30 kilometers to a liter of gasoline, is that good?

Here's one way to decide. First, convert 30 kilometers to miles.

$$1 \text{ kilometer}/0.62 \text{ miles} = 30 \text{ kilometers}/x \text{ miles}$$

$$x = 0.62 \cdot 30 = 18.6 \text{ miles}$$

Next, convert 1 liter to gallons. That conversion can be read from the table: 1 liter = 0.265 gallons.

So, in the U.S. system, the gas mileage of the rental car is 18.6 miles per 0.265 gallons. You can now use a proportion to determine how many miles per gallon that is.

$$18.6/0.265 = x/1$$

$$70.19 = x$$

Wow! The rental car is getting 70.19 miles per gallon! That *is* good mileage.

Here is another conversion situation. A road sign says that London is 450 kilometers away. Is that a long drive?

Since 1 kilometer is 0.62 miles, 450 kilometers is 279 miles. That is about the same as the distance from Los Angeles to Las Vegas. It is a long drive, but one that can easily be made in a day.

PRACTICE

1. A package of ground beef weighs 1.2 grams. Will that be enough to make hamburgers for six people?

2. You are visiting friends in Belgium and planning a party for them. Your party punch recipe calls for 2.5 pints of orange juice. The convenience store near your hostel sells orange juice in 300-milliliter bottles. How many bottles will you need for your punch?

3. You are driving through Guatemala in your father's Winnebago, which is 9 feet tall. Up ahead you see a road bridge; a sign warns that the clearance is 3.5 meters. Will the Winnebago fit under the bridge?

7.6 SCIENTIFIC NOTATION

Although the metric system is convenient to work with because of its regularity, it sometimes leads to rather cumbersome numbers. For example,

$$0.0004 \text{ milliliters} = 0.0000000004 \text{ kilometers}$$

and

$$1{,}000{,}000 \text{ kilometers} = 1{,}000{,}000{,}000{,}000 \text{ millimeters}$$

Scientists, who customarily work with very large and very small numbers, use a convenient shorthand called **scientific notation** to represent these numbers.

The scientific notation shorthand is a compact way of indicating the presence of strings of zeros in a number. The notation is based on a peculiar property of the number 10. Look at these values of powers of 10.

$$10^1 = 10$$

$$10^2 = 100$$

$$10^3 = 1000$$

$$10^4 = 10{,}000$$

$$10^5 = 100{,}000$$

$$10^6 = 1{,}000{,}000$$

Look at the number of zeros in the expanded number and compare that to the exponent on 10. What can you conclude?

____

____

____

How many zeros do you think 10^9 will have when you multiply it out? ________ Do it to check your conjecture.

Now look at this pattern.

$$5 \cdot 10^1 = 50$$

$$5 \cdot 10^2 = 500$$

$$5 \cdot 10^3 = 5000$$

$$5 \cdot 10^4 = 50{,}000$$

$$5 \cdot 10^5 = 500{,}000$$

$$5 \cdot 10^6 = 5{,}000{,}000$$

Notice that the power of 10 on the left side of each equal sign tells you how many zeros are after the 5 on the right side.

Furthermore,

$$5.234 \cdot 10^1 = 52.34$$

$$5.234 \cdot 10^2 = 523.4$$

$$5.234 \cdot 10^3 = 5234.0$$

$$5.234 \cdot 10^4 = 52{,}340.0$$

$$5.234 \cdot 10^5 = 523{,}400.0$$

$$5.234 \cdot 10^6 = 5{,}234{,}000.0$$

Notice that the power of 10 on the left side of each equation indicates the number of places the decimal point moves to the right to obtain the number on the right side.

Now look at these.

$$1/10^1 = 0.1$$

$$1/10^2 = 0.01$$

$$1/10^3 = 0.001$$

$$1/10^4 = 0.0001$$

$$1/10^5 = 0.00001$$

$$1/10^6 = 0.000001$$

Count the number of decimal places to the right of the decimal point on the right side of each of the equations above and compare that to the power of 10. What do you see?

____

For convenience, we will write 10^{-n} for $1/10^n$. Thus, 10^{-3} means $1/10^3$. So,

$$5 \cdot 10^{-1} = 0.5$$
$$5 \cdot 10^{-2} = 0.05$$
$$5 \cdot 10^{-3} = 0.005$$
$$5 \cdot 10^{-4} = 0.0005$$
$$5 \cdot 10^{-5} = 0.00005$$
$$5 \cdot 10^{-6} = 0.000005$$

and

$$5.2 \cdot 10^{-1} = 0.52$$
$$5.2 \cdot 10^{-2} = 0.052$$
$$5.2 \cdot 10^{-3} = 0.0052$$
$$5.2 \cdot 10^{-4} = 0.00052$$
$$5.2 \cdot 10^{-5} = 0.000052$$
$$5.2 \cdot 10^{-6} = 0.0000052$$

Describe the pattern you see here.

A number appears in **scientific notation** when it is written in the form

$$A.B \cdot 10^k$$

where

A = any nonzero integer between -9 and 9
B = any string of positive integers
k = an integer

(An integer is a positive or negative whole number. You will learn more about integers in Chapter 8.)

Here are some numbers written in scientific notation.

$$2.345 \cdot 10^4$$
$$-8.00967 \cdot 10^3$$
$$8.067 \cdot 10^{-3}$$

On your calculator, you might see $8.09 \cdot 10^{18}$ written like this

8.09000000E18

or

8.09^{18}

You can convert a scientific notation number

$$A.B \cdot 10^k$$

into a regular decimal by moving its decimal point k places to the right if k is positive and k places to the left if k is negative.

Here are some more numbers in scientific notation with their decimal equivalents.

Scientific Notation	Decimal Equivalent
$1.456 \cdot 10^2$	145.6
$4.5 \cdot 10^5$	450,000.0
$2.3489 \cdot 10^2$	234.89
$-4.0987 \cdot 10^4$	−40,987
$1.23 \cdot 10^{-3}$	0.00123
$-1.089 \cdot 10^{-7}$	−0.0000001089

PRACTICE

Rewrite each of the following as regular decimal numbers:

1. $3.45 \cdot 10^4$

2. $-2.356 \cdot 10^{-3}$

3. $1.2345 \cdot 10^{10}$

4. $6.89 \cdot 10^{-4}$

To write a regular decimal in scientific notation, reverse the process. Move the decimal point until there is a single digit to its left. Count the number of spaces you moved. If you moved k spaces to the right, then the power on 10 is $-k$. If you moved k space to the left, then the power on 10 is k. Ignore trailing or leading zeros when you write the number in scientific notation.

For example, I'll convert 1,245,600,000 to scientific notation. First, I locate the decimal point.

1245600000.
↑ Here it is!

Now move the decimal point until only a single digit is to its left.

1.245600000
↑
Here it is!

I moved nine spaces to the left. So, the scientific notation version is

$$1.2456 \cdot 10^9$$

Here's another example; I'll convert 0.0000000897 to scientific notation. First, I locate the decimal point.

0.0000000897
↑
Here it is!

Now move the decimal point until only a single digit is to its left.

000000008.97
↑ Here it is!

I moved eight spaces to the right. So, the scientific notation version is

$$8.97 \cdot 10^{-8}$$

PRACTICE

Write each of the following in scientific notation:

5. 0.000000789

6. 2345.90

7. −67,000,000,000,000,000,000,000,000

8. 0.0000000000000000000000067

The problems and techniques that we encountered in painting apartments and driving in England were all mostly numerical. In studying the mathematics of space, we needed to measure and compute. The mathematics of space also includes problems that are not quite so numerical. It is to these types of problems that we now turn.

Situation 24 Shoveling Snow

Melena operates a snow-shoveling service. She offers street shoveling for large corporate complexes. Below is a map of the Itsy-Bitsy Microprocessor Corporation.

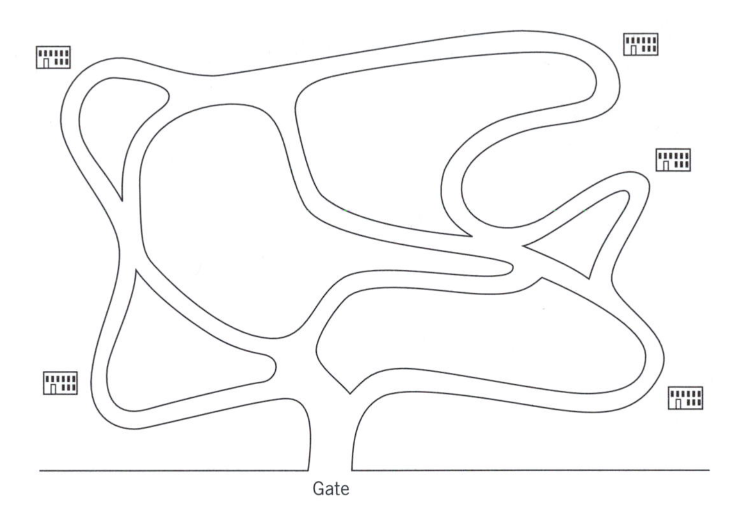

Melena would like to clear the streets of this complex in the most efficient way possible. Specifically, she would like to begin and end at the entrance gate and not drive her snow-shoveling equipment over any road that she has already shoveled (she can, of course, go through an intersection more than once). Draw a path for Melena to follow that will start and end at the entrance gate and go over each road only once.

7.7 GRAPH THEORY

Unlike most of the other situations in this text, you were probably able to provide an answer to Melena's snow-shoveling problem without reading on. Melena's problem is one instance of a wide variety of problems that arise within an area of mathematics called **graph theory.** Graph theory is important in many different contemporary activities including the design of computers, construction of computer programs, design of telecommunications networks, planning of the transportation of goods and materials, and study of the interactions of plants, animals, and the environment.

The graphs that are the subject of graph theory can be thought of as pictures of lines and dots like these.

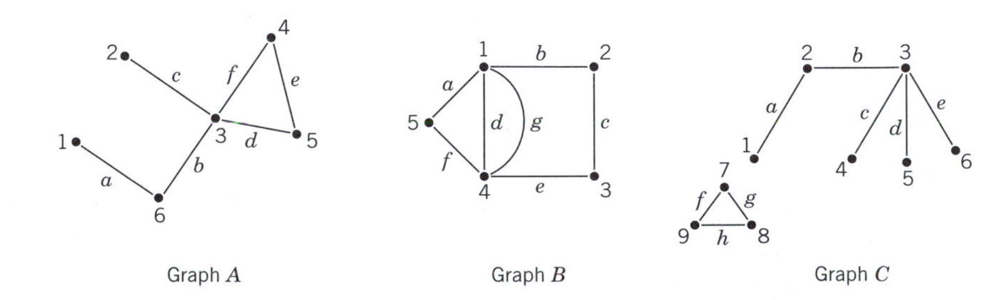

The lines of a graph are called its **edges** and the dots its **vertices** (singular is **vertex**). I will name the vertices with numbers and the edges with letters.

A **path** in a graph is a sequence of edges and vertices that starts and ends with a vertex. Here is a path in graph *A*.

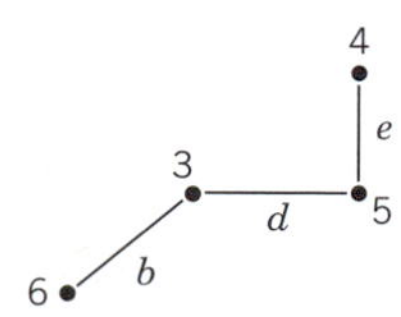

We can describe the path by listing the edges and vertices it uses, namely, (6, *b*, 3, *d*, 5, *e*, 4).

A path in graph *B* is (5, *f*, 4, *d*, 1, *b*, 2, *c*, 3, *e*, 4). Trace this path on the picture of graph *B* shown below.

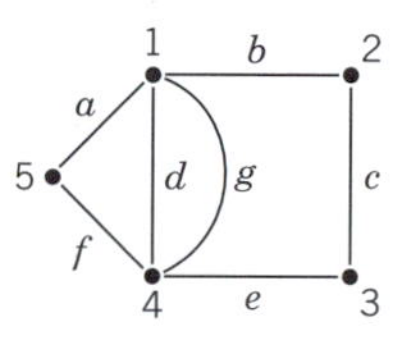

A graph is **connected** if there is a path from any vertex to any other vertex in the graph. Graph *A* is connected.

Is graph *B* connected? ________

Is graph *C* connected? ________

Explain why graph *C* isn't connected.

____

____

____

Graph *C* isn't connected because there isn't a path from vertex 1 to vertex 7.

A **cycle** is a path that begins and ends at the same vertex and doesn't repeat any edge. The path (5, *f*, 4, *d*, 1, *a*, 5) is a cycle in graph *B*.

Find a cycle in graph *C*. ________________________________

An **Euler cycle** is a cycle that includes all the edges of the graph. Here is an Euler cycle for graph *B*.

(5, *f*, 4, *e*, 3, *c*, 2, *b*, 1, *g*, 4, *d*, 1, *a*, 5)

You should trace this cycle on graph *B* to see that it does include all the edges. Do it here.

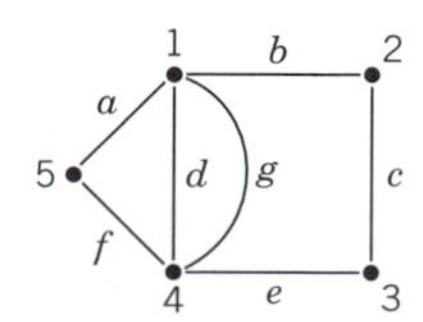

Finding an Euler cycle is equivalent to tracing the graph without going over any line twice and ending where you started. Melena's snow-removal problem was equivalent to finding an Euler cycle through the graph formed by putting vertices at the intersections of the roads. Here is the graph of the Itsy-Bitsy Microprocessor Corporation's road system.

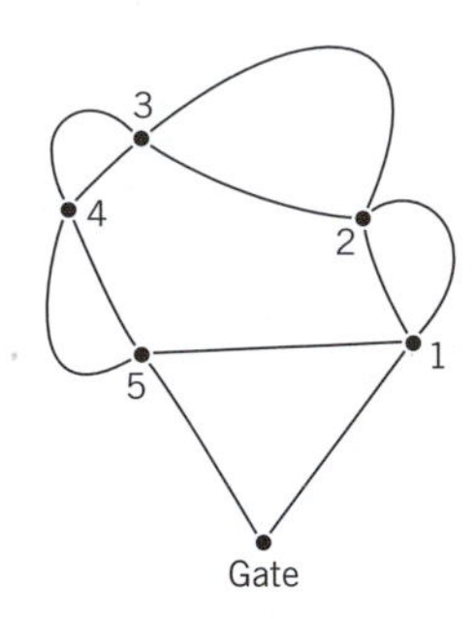

PRACTICE

Find Euler cycles for each of the following graphs:

1.

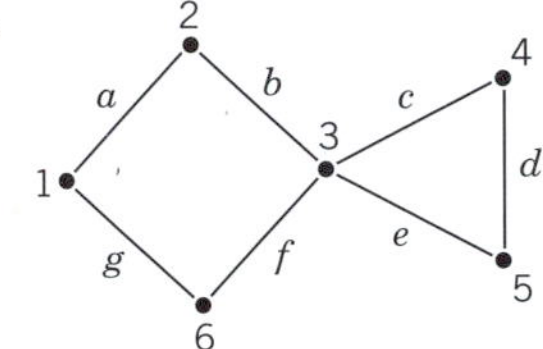

2.

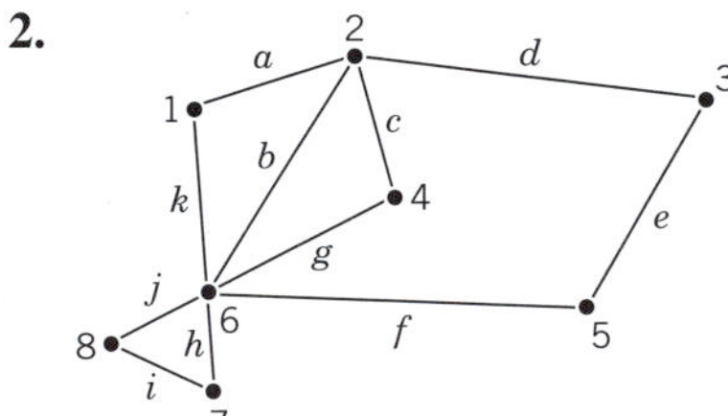

Does the following graph have an Euler cycle? ______

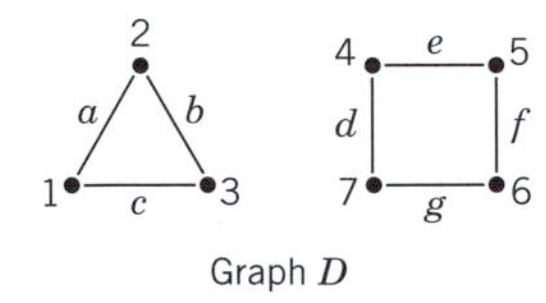

Graph D

Why not?

______

______

______

So, we can see that a graph must be connected before it has an Euler cycle. Does this graph have an Euler cycle?

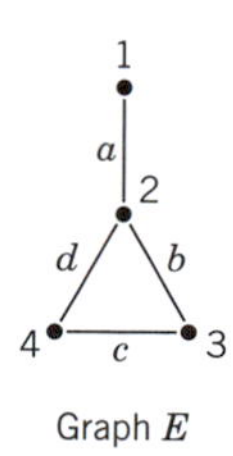

Graph E

Why not?

____

____

____

Right! If we start a path in graph E at vertex 1, there is no way to get back to vertex 1 without going over edge a a second time. Also, if we start anywhere else, once we come out to vertex 1, we can't get back without going over edge a a second time.

PRACTICE

Explain why it is impossible to construct an Euler cycle for each of the following graphs:

3.

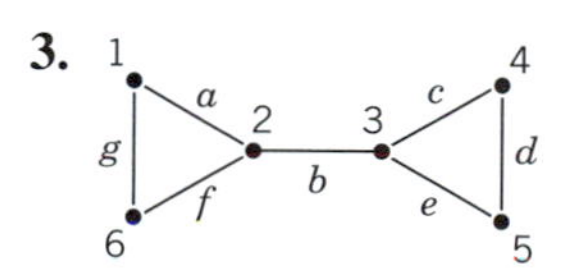

4.

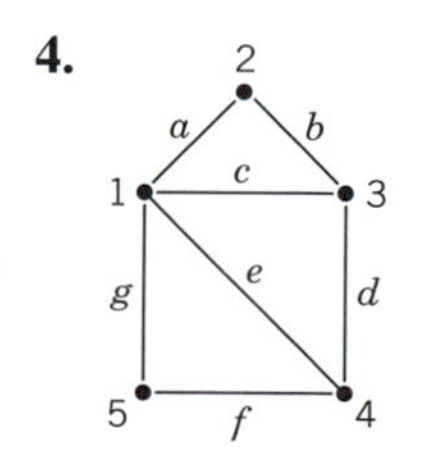

5.

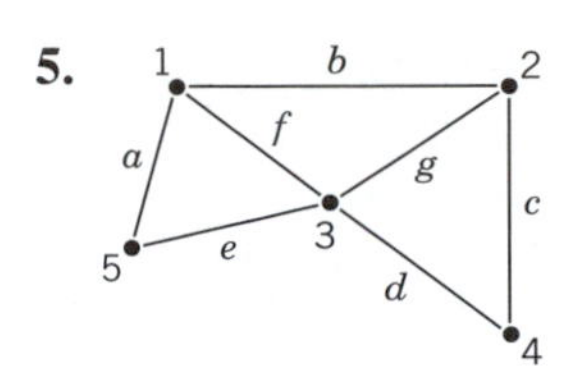

Graph E and the graphs in Practice Exercises 3–5 all were prevented from having Euler cycles by a **problem** vertex. Each of the problem vertices has something in common with the others.

How many edges run into vertex 1 in graph E? ______

How many edges run into vertex 3 in the graph in Practice Exercise 4? ______

How many edges run into vertex 2 in the graph in Practice Exercise 5? ______

Do you see that all the problem vertices have an odd number of edges running into them? The number of edges running into a vertex is called the **degree** of that vertex. Look at graph F below.

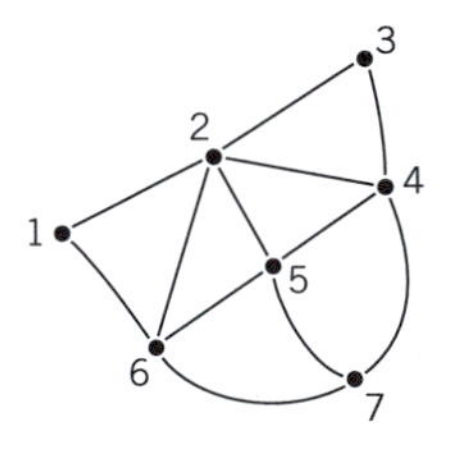

Graph F

Does graph F have an Euler cycle? ______

After a bit of exploring, you should conclude that graph F doesn't have an Euler cycle and the problem lies with vertices 2 and 7. In both cases, any attempt to construct an Euler cycle will force you to go over one of the edges running into vertex 2 or vertex 7 twice. Notice that both vertices 2 and 7 have odd degrees (vertex 2 has degree 5, and vertex 7 degree 3).

The results of these observations on Euler cycles can be summarized in a true proposition about graphs.

A graph has an Euler cycle only if it is connected and no vertex has odd degree.

Melena can use this result to plan her shoveling ahead of time and know whether or not an Euler shoveling is even possible for a road system. Look at the graph of the Itsy-Bitsy Microprocessor Corporation's road system (p. 274) and you will see that Melena can do an Euler shoveling because all the vertices are of even degree.

PRACTICE

Find an Euler cycle or explain why the graph can't have one.

6.

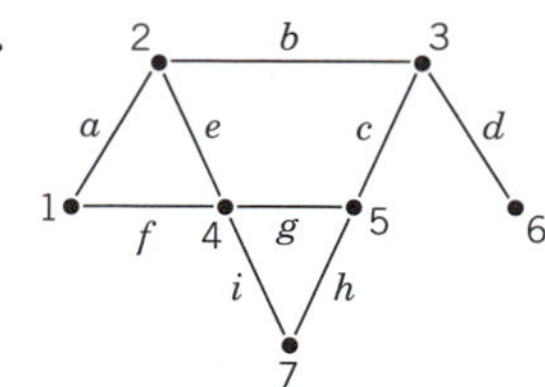

7.

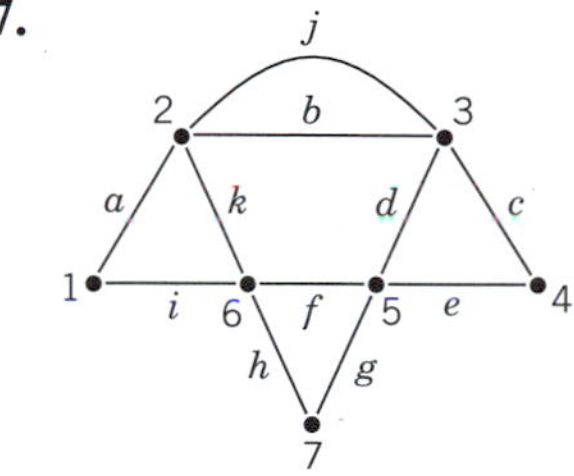

8.

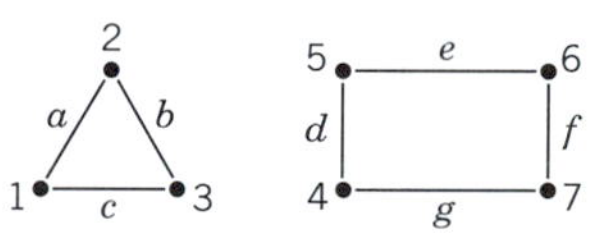

9.

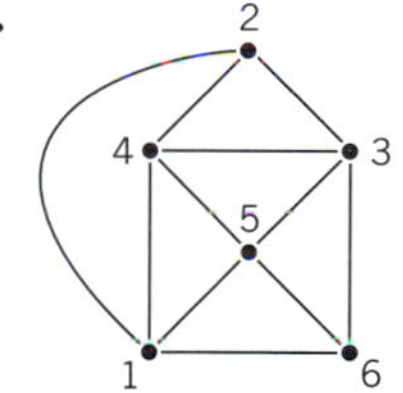

There is much more to graph theory than Euler cycles. The next section of exercises includes some other ideas and terms of graph theory. You should take a course in finite mathematics or consult a graph theory text for more details on these topics.

7.8 WHAT DO YOU KNOW?

If you have worked carefully through this chapter, you can calculate the area, perimeter, and volume of geometric shapes, have an understanding of the metric system, and can decide whether or not a graph has an Euler cycle. The exercises in this section are designed to test and refine your knowledge of the mathematics of space.

1. Compute the areas of each of the following:

(a) A rectangle 25 miles long and 6 miles wide

(b) A square with a side 14 inches long

(c) A rectangle 10 feet long and 6 inches wide

(d) A triangle with a base of 50 feet and height of 20 feet

(e) A circle with radius 12 inches

(f) A triangle with a height of 4 feet and base of 2 yards

2. Compute the areas of each of the following:

(a) A rectangle 25 meters long and 3 meters wide

(b) A square with a side 12 centimeters long

(c) A rectangle 10 meters long and 60 centimeters wide

(d) A triangle with a base of 50 kilometers and height of 20 meters

(e) A circle with radius 30 millimeters

(f) A triangle with a height of 4 meters and base of 2 millimeters

3. Carpeting sells for $5.95 a square yard. How much will it cost to carpet each of the following floors?

(a)

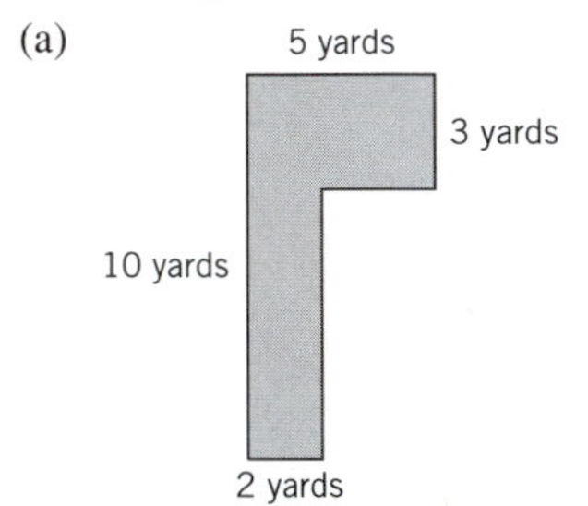

(b)

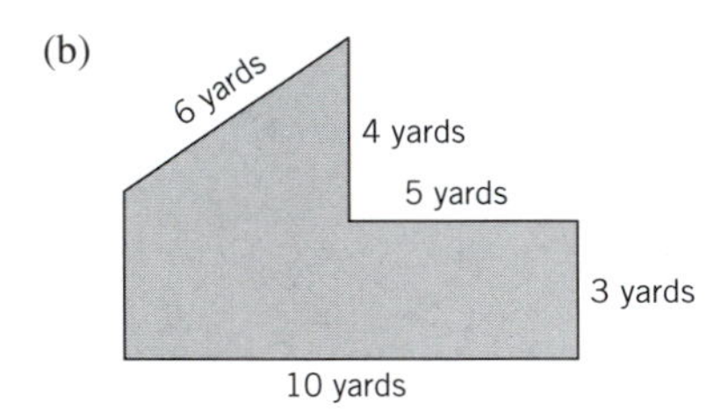

(c)

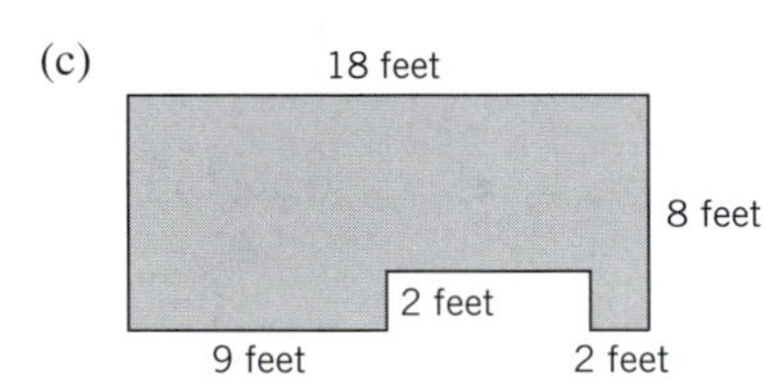

4. Suppose that you want to put new baseboards in the rooms shown in Exercise 3. How many feet of baseboard would you need for each room?

5. The Boody Bombers are repainting their basketball court. They will paint the *key* area a bright red. The key is in the shape of a rectangle with a semicircle on one end as shown in the drawing below.

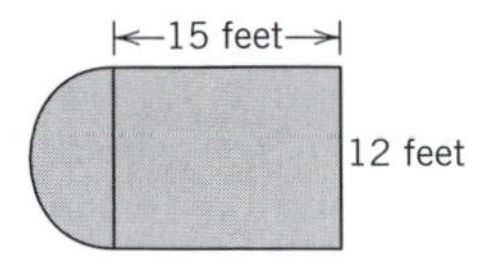

There are two keys on the court. The Bombers have 1 gallon of paint that will cover 200 square feet. Do they have enough to paint their keys? Explain.

6. Find the volume of each of the following containers. Each container has vertical sides and a base as described. Express the volume in cubic feet.

(a) Base: rectangle, 13 feet long, 2 feet wide; height: 5 feet

(b) Base: triangle, height 2 feet, base 5 feet; height: 4 feet

(c) Base: rectangle, 2 yards long, 6 inches wide; height: 5 feet

(d) Base: circle, radius 1.5 feet; height: 7 feet

7. Find the volume of each of the following containers. Each container has vertical sides and a base as described. Express the volume in liters.

(a) Base: rectangle, 13 centimeters long, 2 centimeters wide; height: 5 centimeters

(b) Base: triangle, height 2 meters, base 5 meters; height: 4 centimeters

(c) Base: rectangle, 20 meters long, 6 centimeters wide; height: 5 kilometers

(d) Base: circle, radius 1.5 meters; height: 7 centimeters

8. A soft drink can is 5 inches tall and has a circular base with a radius of 1.25 inches. What is the volume of the can in cubic inches?

9. A common fund-raising game is the *guess-the-number-of-marbles game.* A large jar is filled with marbles and you must guess the number of marbles in the jar. You make a donation to guess, and if you get it right, you will win a prize.

Suppose that the marbles are all spheres of radius 5 millimeters. Suppose that the jar has a circular base and vertical sides. The jar is 40 centimeters tall and the base of the jar has a radius of 10 centimeters.

(a) What is the maximum number of marbles that could fit into the jar?

(b) Is it possible to get that maximum number of marbles into the jar? Explain.

10. A nickel is about 2 millimeters thick.

(a) How many nickels would be in a stack of nickels that was 3 meters high?

(b) How many miles high would a stack of nickels worth $2,000,000 be?

11. Determine which is larger.

(a) 5 gallons or 5 liters

(b) 450 miles or 450 kilometers

(c) 300 milliliters or 2 quarts

(d) 400 cubic centimeters or 100 cubic inches

12. A milk tank truck carries a cylinder that is 8 meters long and has a hemispheric cap on either end. The radius of each cap is 1 meter. How many quarts of milk can this cylinder hold?

13. Write each of the following in scientific notation:

(a) 0.0000000345

(b) 123.567

(c) 14,000,000,000,000,000,000,000,000,000

(d) −0.000000000000000000000000000000000000088

14. The earth is about 93,000,000 miles from the sun. How many inches is the earth from the sun?

15. A trillion dollars ($1,000,000,000,000) is a lot of money. To get an idea of how much, suppose you tried to count from 1 to a trillion. Suppose you counted fast. Suppose you could count four numbers every second. If you counted continuously, how long would it take you to count to a trillion? Express your answer in years.

16. Find an Euler cycle for each of the following graphs or explain why no Euler cycle is possible:

(a)

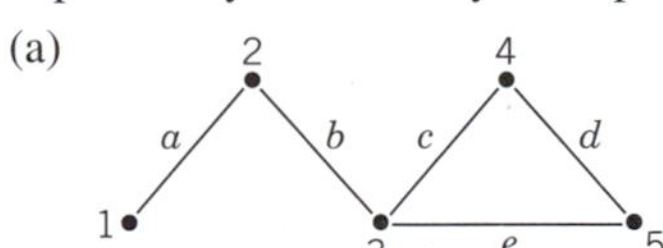

(b)

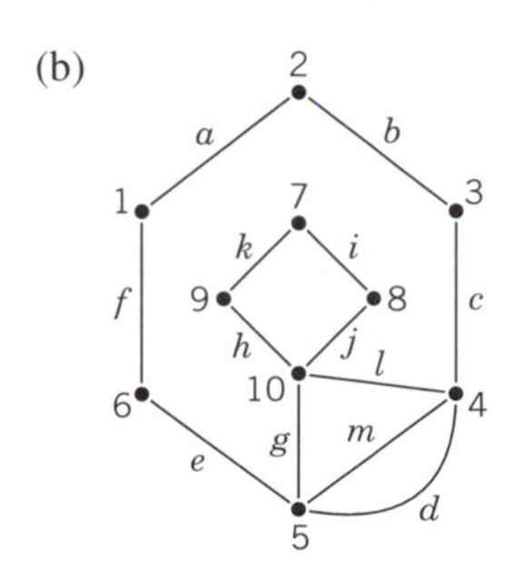

(c)

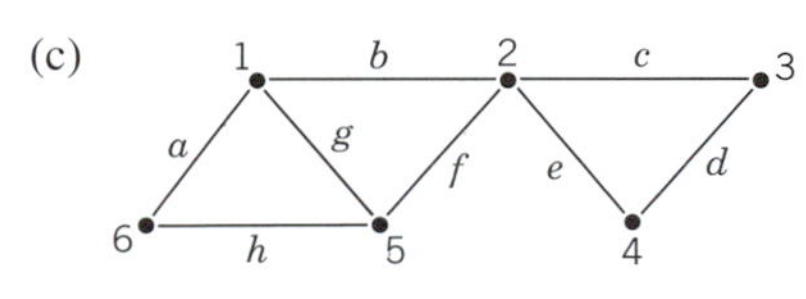

(d)

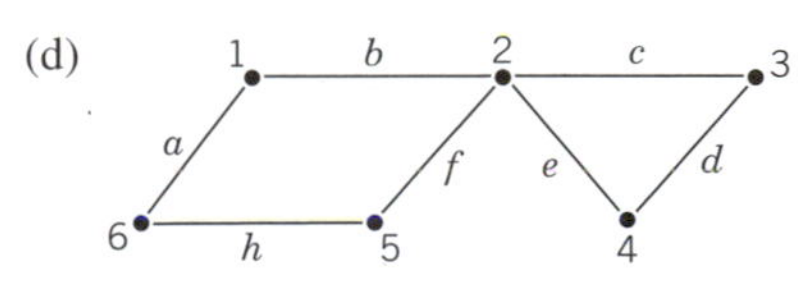

17. A telecommunications company wants to send a test message through all its communications lines. The graph below shows the lines as edges and the customers as vertices.

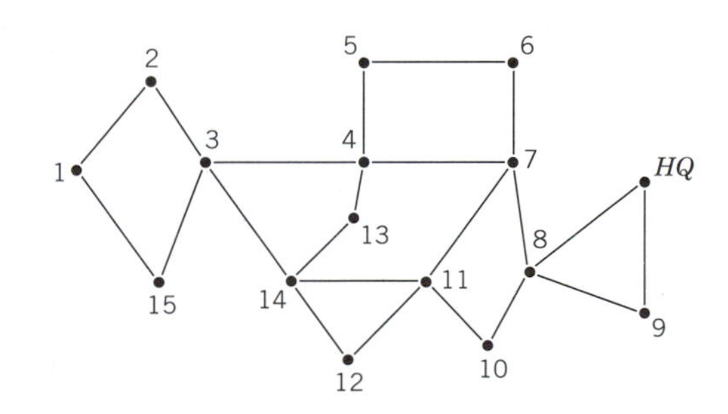

Find a path through the communications lines that begins and ends at headquarters and passes through every line exactly once.

18. A graph is a **tree** if it is connected and has no cycles. Which of the following are trees?

(a)

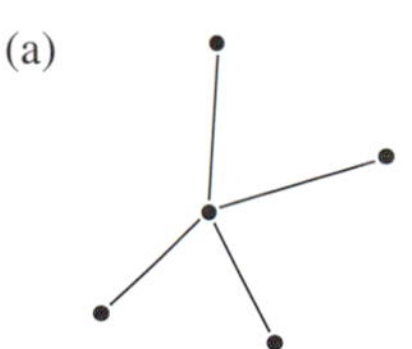

(b)

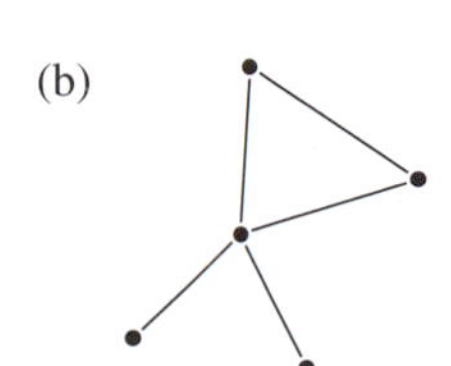

(c)

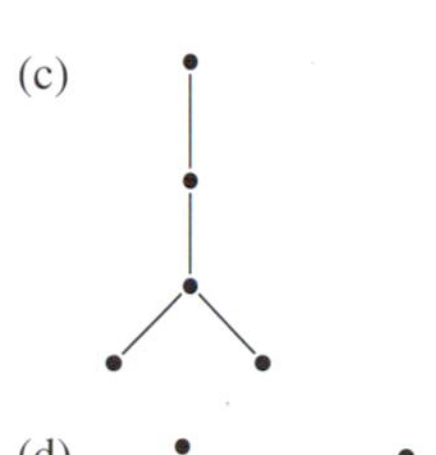

(d)

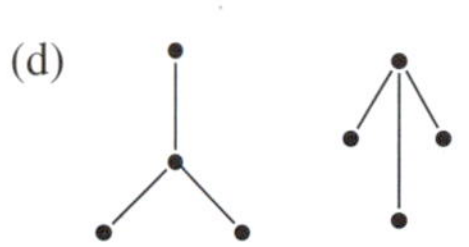

19. There is a colorful collection of terminology associated with trees. If a particular vertex of a tree is designated as the **root** of the tree, then we can view the tree as *growing* downward from the root. From this point of reference, tree vertices may have a **parent, children,** and **siblings.** For example, in the following tree, the root is marked with an R. The parent of vertex X is vertex A. The children of vertex X are vertices C, D, and E. Vertex Y is a sibling of vertex X.

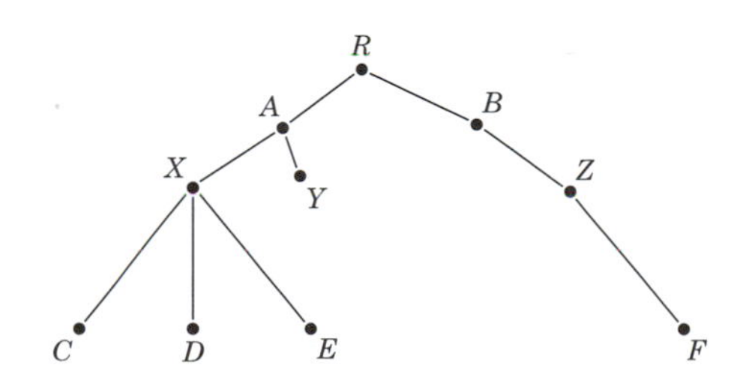

For each of the following trees with root R, find the parent, children, and siblings of vertex X, if they exist.

(a)

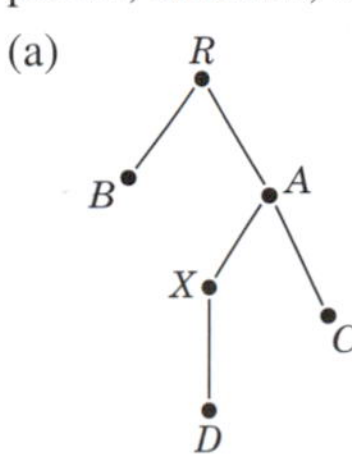

(b)

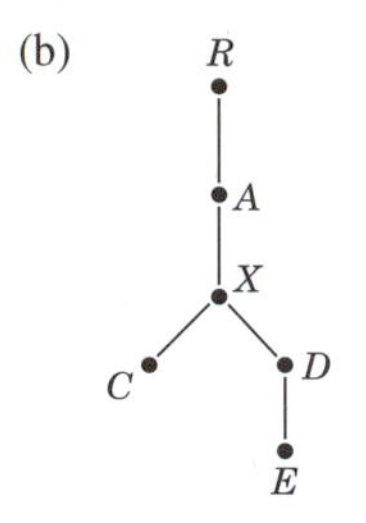

(c)

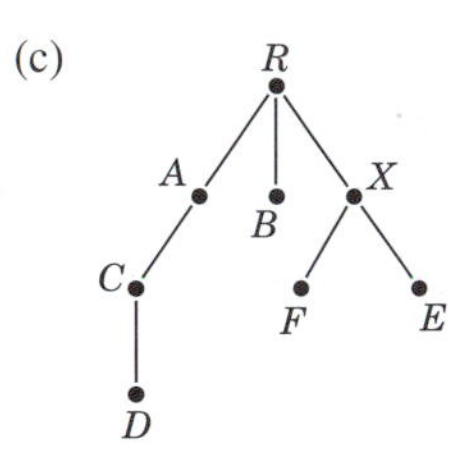

(d)

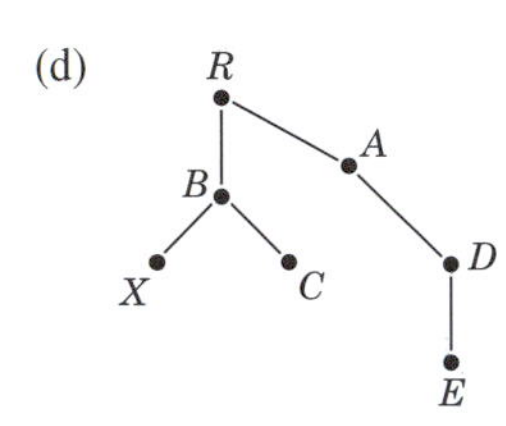

20. Explain why in a tree with root R in which every vertex besides R has a parent, no vertex can have two parents.

21. A graph is **planar** if it can be drawn without any of the edges crossing. For example, this graph

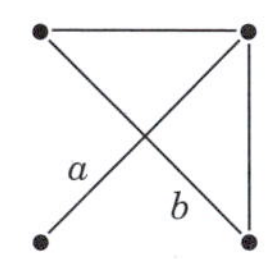

is planar because I can redraw edge a so that it doesn't cross edge b. Like this.

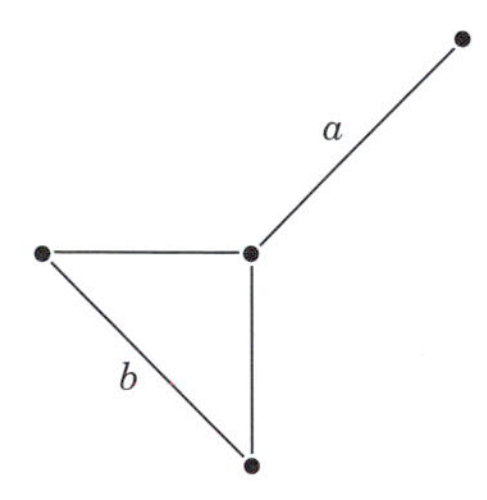

Show that each of the following graphs is planar by redrawing them without crossing edges:

(a)

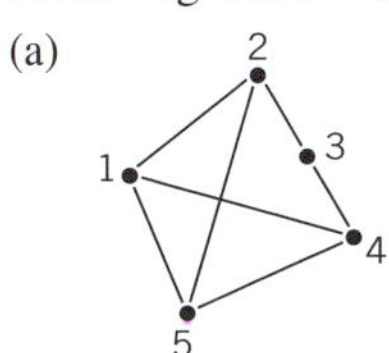

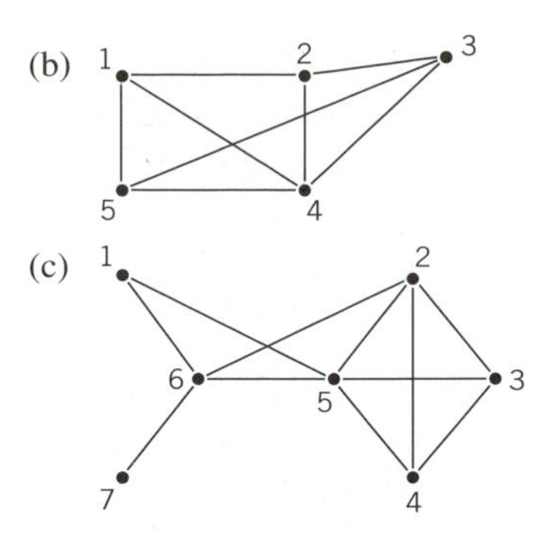

22. Are all trees planar? Explain.

23. Sometimes, the edges of graphs are assigned **weights.** These are numbers that express the cost or difficulty of moving along that edge. For example, in the snow-shoveling problem of Situation 24, if a roadway went steeply uphill, it would have a greater weight than a flat roadway because it would take more energy to plow the snow uphill. I'll give the weight of an edge in parentheses next to its name.

The **weight of a path** is the sum of the weights of the edges in that path. For example, the weight of the path (1, a, 2, b, 3) in the graph below is 8. The weight of the path (1, c, 4, d, 3) is 6.

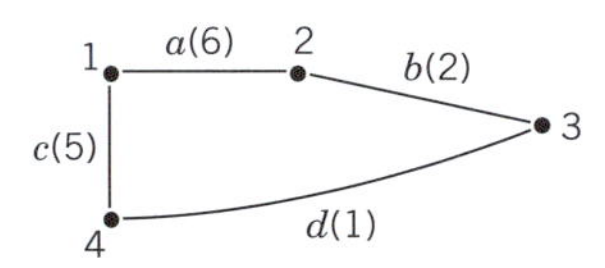

Find the path with the least weight from vertex 1 to 6 in each of the following graphs:

(a)

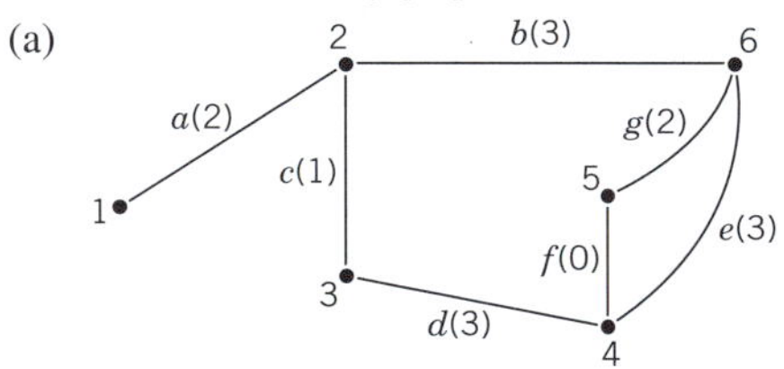

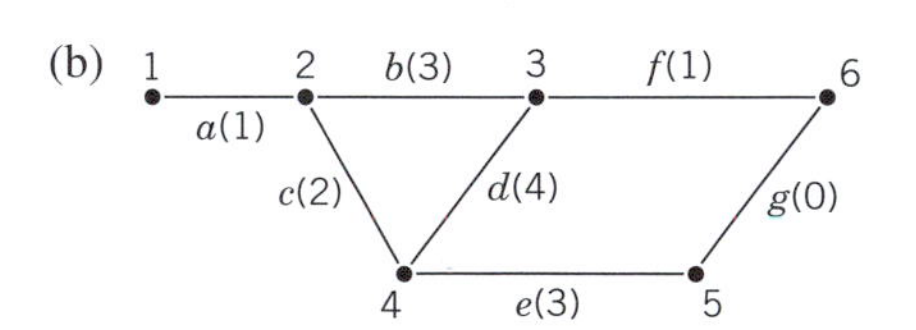

(c)

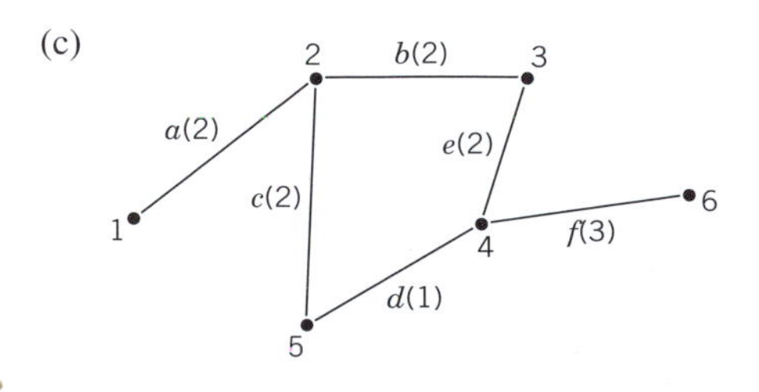

24. A connected graph has a **cut vertex** A if the removal of A and all edges touching A results in a graph that is disconnected. The vertex A is a cut point for the graph shown in (a). Find a cut point for each of the other graphs.

(a)

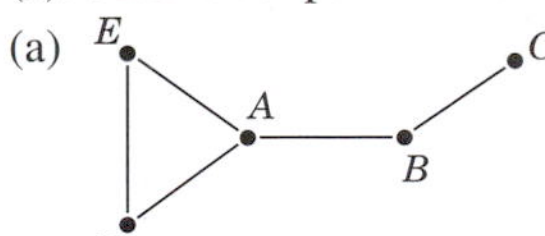

(b)

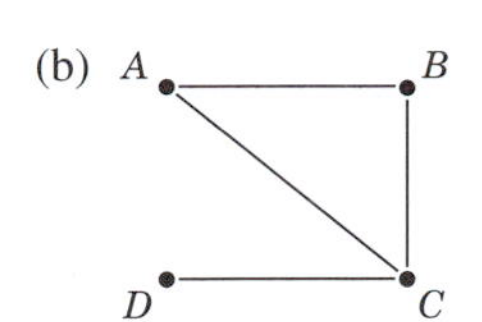

(c)

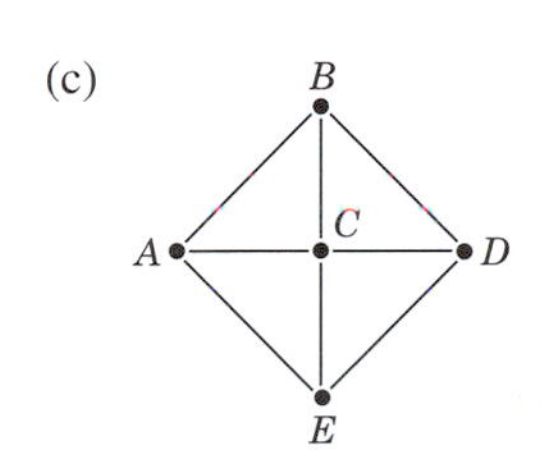

(d)

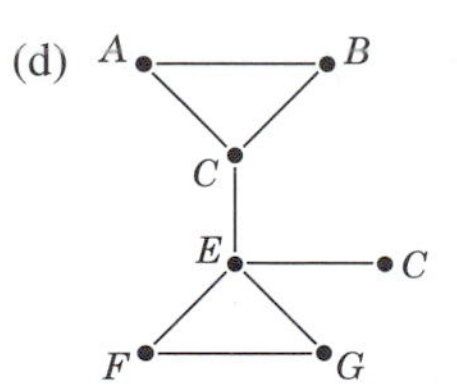

(e)

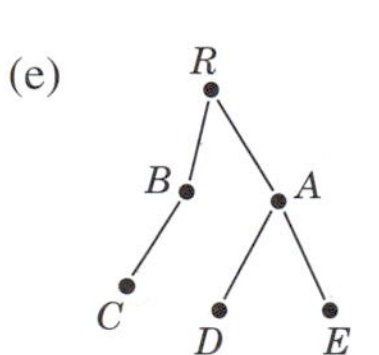

(f)

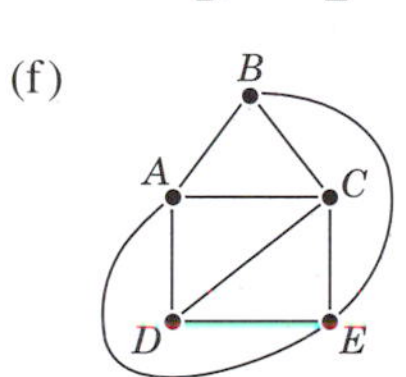

ANSWERS TO PRACTICE EXERCISES

Section 7.1

1. Rectangle. $A = L \cdot W = 6 \cdot 2 = 12$ square feet
2. Square. $A = S \cdot S = 4 \cdot 4 = 16$ square miles
3. Triangle. $A = 0.5 \cdot B \cdot H = 0.5 \cdot 2 \cdot 4 = 4$ square inches
4. Rectangle. The measurement units are not the same. So, I'll change the 2 yards into feet. Two yards = 6 feet. Then, $A = 6 \cdot 2 = 12$ square feet.
5. Circle. $A = \pi \cdot R \cdot R = 3.14159 \cdot 3 \cdot 3 = 28.27431$ square feet
6. Circle. $A = \pi \cdot R \cdot R = 3.14159 \cdot 5 \cdot 5 = 78.53975$ square miles
7. Here's the floorplan of the living room again. I've added a dotted line that divides the floor into two rectangles.

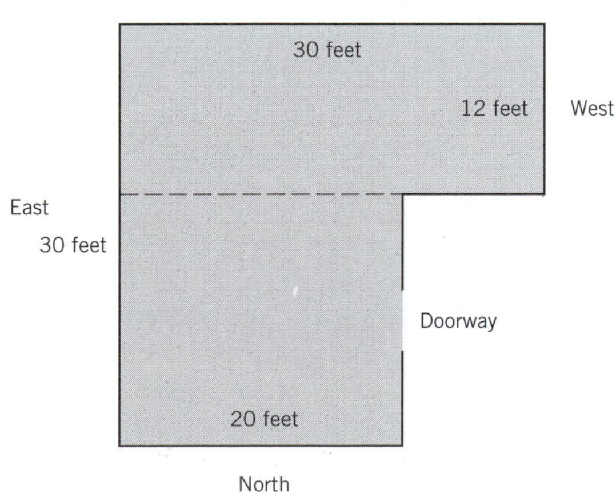

The area of the upper rectangle is $30 \cdot 12 = 360$ square feet. The area of the lower rectangle is $20 \cdot 18 = 360$ square feet. The total floor area is 720 square feet. 720 square feet = 80 square yards. You will need 80 square yards of carpeting.

Section 7.2

1. Perimeter = 4 + 2 + 4 + 2 = 12 yards
2. Perimeter = 6 + 8 + 12 + 2 + 6 + 6 = 40 yards
3. Perimeter = 2 + 2 + 6 + 5 + 6 = 21 feet
4. Perimeter = 5 + 10 + curve + 10 = 25 + curve
   The curve is half the circumference of a circle. The radius of the circle is 2.5 inches. So, the whole circumference would be $2 \cdot \pi \cdot 2.5 \approx 15.71$ inches. Half of this is 7.86 inches. Thus, the perimeter of the figure is 25 + 7.86 = 32.86 inches.
5. Perimeter = 1 + 2 + 3 + 4 + 3 + 3 + 5 + 2 + 6 + 4 + 3 + 1 + 5 + 2 + 2 + 4 + 5 = 55 miles

Section 7.3

1. 2 gallons = 8 quarts = 16 pints
2. 1 quart : 2 pints = x quarts : 5 pints
   $$1/2 = x/5$$
   $$5 = 2x$$
   $$2.5 \text{ quarts} = x$$
3. 1 gallon : 231 cubic inches = x gallons : 200 cubic inches
   $$1/231 = x/200$$
   $$200 = 231x$$
   $$0.866 \text{ gallons} = x$$

4. 1 cubic yard : 27 cubic feet = x cubic yards : 20 cubic feet

$$1/27 = x/20$$
$$20 = 27x$$
$$0.741 \text{ cubic yards} = x$$

5. Volume = $(3 \cdot 3) \cdot 5 = 45$ cubic inches
6. Volume = $(8 \cdot 2) \cdot 7 = 112$ cubic feet
7. Volume = $(\pi \cdot 5 \cdot 5) \cdot 15 = 1178.10$ cubic yards
8. Volume = $(0.5 \cdot 4 \cdot 2) \cdot 10 = 40$ cubic feet
9. Volume = $4 \cdot \pi \cdot 3^3/3 = 113.10$ cubic inches
10. Volume = $4 \cdot \pi \cdot 10^3/3 = 4188.79$ cubic miles

Section 7.4

milli- centi- deci- base unit deka- hecto- kilo-

← Multiply by 10

1. 34 divided by 10 twice: 0.34 meters
2. 2.09 divided by 10 twice: 0.0209 hectoliters
3. 0.008 divided by 10 four times: 0.0000008 dekaliters
4. 1030 divided by 10 three times: 1.030 kilograms
5. 12.6 times 10 once: 126 hectograms
6. 140 times 10 once: 1400 centimeters
7. 1,000,000 times 10 three times: 1,000,000,000 meters
8. The volume of the bathtub is $(2 \cdot 1) \cdot 0.5 = 1$ cubic meter. Now 1 meter = 100 centimeters. So, 1 cubic meter = $100 \cdot 100 \cdot 100$ = 1,000,000 cubic centimeters = 1,000,000 milliliters = 1000 liters.

Section 7.5

1. No. 1 gram = 0.035 ounces. So, 1.2 grams = $1.2 \cdot 0.035$ ounces = 0.042 ounces. Even a small hamburger weighs at least 2 ounces.
2. 4 bottles. 1 liter = 2.12 pints
   1 liter = 1000 milliliters
   1000 milliliters = 2.12 pints
   300 milliliters = 0.636 pints
   $4 \cdot 300$ milliliters = $4 \cdot 0.636$ = 2.544 pints
3. Yes. 3.5 meters = $3.5 \cdot 3.28$ feet = 11.48 feet > 9 feet

Section 7.6

1. 34,500
2. -0.002356
3. 12,345,000,000
4. 0.000689
5. $7.89 \cdot 10^{-7}$
6. $2.3459 \cdot 10^{3}$
7. $-6.7 \cdot 10^{25}$
8. $6.7 \cdot 10^{-25}$

Section 7.7

These are not the only Euler cycles possible for the graphs. Yours may be different, but yours should begin and end at the same vertex and include every edge exactly once.

1. (1, a, 2, b, 3, c, 4, d, 5, e, 3, f, 6, g, 1)
2. (1, a, 2, d, 3, e, 5, f, 6, g, 4, c, 2, b, 6, h, 7, i, 8, j, 6, k, 1)
3. Any cycle that includes all the edges would have to cross the bridge between vertices 2 and 3 twice.
4. Look at vertex 3. A potential Euler cycle that started at this vertex would leave the vertex along one of the edges b, c, or d. Let's suppose we leave vertex 3 along edge b. Then to include edges c and d in the cycle, we would have to come back to vertex 3 and then leave it again. But now there is no way to end up at vertex 3 (where we started) without going over b, c, or d a second time. The same problem would arise if we initially left vertex 3 along edge c or d or if we entered vertex 3 on some other potential Euler cycle.
5. Look at vertex 1 or 2 and argue as I did in Practice Exercise 4.
6. No Euler cycle. Vertex 6 has degree 1.
7. (1, a, 2, b, 3, c, 4, e, 5, d, 3, j, 2, k, 6, f, 5, g, 7, h, 6, i, 1)
8. No Euler cycle. The graph is not connected.
9. No Euler cycle. Vertex 2 has degree 3.

Chapter Eight

Mathematics of Numbers

What's the Point?

Numbers are probably the most obvious objects of mathematics in everyday life. You have a numerical home address, phone number, and social security number. If you cook, then your recipes list ingredients using fractions. Sporting events use numbers extensively. For example, numbers describe times, scores, and averages.

This chapter will help you to understand the variety of numbers, how they are used, and why mathematicians find them so fascinating.

Situation 25

At the Olympics

Fern and Zork are attending the Olympic Games. Just for fun, they decide to determine the many different kinds and uses of numbers during the track and field events. Fern notices that each runner in the 1600-meter race has a number on her back. Zork counters with the observation that there are three blocks on the awards platform and they are numbered. Fern responds by pointing out that the winning time for the 100-meter run was 8.93 seconds. Zork notices that the last contestant in the long jump leapt 8.75 meters. Fern, in desperation, remarks that one-fourth of the contestants in the shot put are from the Ukraine.

Describe the different uses of the numbers noticed by Fern and Zork.

____

____

____

____

____

____

8.1 COUNTING AND ORDERING

Counting is probably the most fundamental use of numbers. Whether we are counting the number of contestants in a race or sheep in a pasture, the purpose of counting is to determine how many things there are.

The numbers that are used for counting (1, 2, 3, 4, 5, 6, . . .) are called the **natural numbers.** The collection of natural numbers is denoted by *N*.

Before I go on, it is important to note the difference between a *number* and a *numeral.* A **numeral** is a symbol that names a **number.** For example, if you have three radios in your home, then the numeral 3 denotes the number of radios you have. In Chinese, the numeral 三 also denotes the number of radios you have. The number of radios is the same, but the numeral denoting that number may vary from culture to culture.

It is customary to blur the distinction between number and numeral when the context makes it clear which you are talking about.

The natural numbers are also used for ordering things. The winner of an Olympic event stands on the platform bearing the numeral 1. The second place finisher stands on a 2 and the third place finisher on a 3. These numbers are not used to count the medal winners. Instead, the numbers are used to order them.

When natural numbers are used to count things, they are called **cardinal numbers.** When natural numbers are used to order things, they are called **ordinal numbers.** Here are some examples.

Use of Natural Numbers	Cardinal or Ordinal?
The number of cows in a field	Cardinal
The steps in the instructions of a do-it-yourself book	Ordinal
The number of pages in this book	Cardinal
Using the page numbers to put the book back together if it falls apart	Ordinal

When 0 is included with the natural numbers, the new collection is called the **whole numbers.** The collection of whole numbers is denoted by *W*. Zero can be used for counting, but rarely for ordering. For example, the number of elephants in your room right now is probably 0, but it is odd to talk about the *zeroth* person in line.

Temperatures and bank balances sometimes go below 0. So, we have a whole series of negative natural numbers, -1, -2, -3, -4, and so on. When these negative natural numbers are included with the whole numbers, we have the **integers.** The collection of integers is denoted by *Z*.

Here is a summary of the collections of numbers we've defined so far.

Name of Collection	Numbers in Collection	Symbol of Collection
Natural numbers	1, 2, 3, 4, . . .	N
Whole numbers	0, 1, 2, 3, 4, . . .	W
Integers	. . . , -2, -1, 0, 1, 2, . . .	Z

Give three uses of natural numbers.

____

____

____

Give three uses of 0.

____

____

____

Give three uses of negative integers.

____

____

____

PRACTICE

Classify each of the following numbers as belonging to *N*, *W*, or *Z*. List all the collections that apply.

1. 3

2. -3

3. 0

4. 1890

5. $-8.67 \cdot 10^{23}$

8.2 MEASURING

When we move from counting to measuring, we immediately see the need for numbers other than integers because many measurements involve parts of whole units. You sometimes need only 1/4 teaspoon of salt in a recipe. Your room may measure 23 1/2 feet by 15 3/4 feet.

For measuring, then, we need to use a new collection of numbers called the **rational numbers.** A rational number is the ratio of two integers in which the bottom integer is not 0. The collection of rational numbers is denoted by Q. All these are rational numbers: 1/2, 3/4, −7/5, 3/1. Notice that 3/1 can be simplified to 3. Thus, the integers are also rational numbers.

Rational numbers can also be written as decimals by dividing the top number in the ratio by the bottom.

What is 7/5 when you divide it out? _____

What is 2/3 when you divide it out? _____

The two decimal numbers 1.4 and 0.6666666666 . . . are examples of the two kinds of decimals that arise from rational numbers. Rational numbers have decimal expansions that either terminate (like 7/5 = 1.4) or repeat (like 0.666666 . . .). Every rational number will either terminate or repeat when you divide it out.

PRACTICE

Find the decimal expansion of each of the following rational numbers. Tell if it is terminating or repeating. If it repeats, tell what string of digits repeats.

1. 3/4

2. 7/9

3. 11/7

4. 5/11

5. 7/20

Not all decimal numerals terminate or repeat. Here's an example, a number I'll call ♥.

$$♥ = 0.1010010001000010000010000001 \ldots$$

The number ♥ contains a pattern, but there is no repeating string of digits in it. What is the pattern of ♥?

Numbers whose decimal representations neither terminate nor repeat are called **irrational numbers.** The collection of irrational numbers is denoted by H.

Probably the most famous irrational number is π, the ratio of the circumference of a circle to its diameter. No matter how many decimal places you can calculate for π (the most calculated so far has been slightly over two billion!), it will never terminate or become repeating.

Here are some more irrational numbers.

$$\sqrt{2}, \quad \sqrt{3}, \quad 12.123112233111222333 \ldots$$

If we put the rational and irrational numbers into one big collection, we get the **real numbers.** The collection of real numbers is denoted by R. Every number that we normally encounter is a real number.

Here is the relationship between the various collections of numbers.

Every natural number is a whole number, but 0 is a whole number that isn't a natural number.

Every whole number is an integer, but the negative integers are not whole numbers.

Every integer is a rational number, but fractions that do not reduce to integers (e.g., 1/2, 2/3, 4/9) are rational numbers that are not integers.

Every rational number is a real number, but irrational numbers are real numbers that are not rational numbers.

The preceding relationships can be summarized quite nicely using subset notation:

$$N \subseteq W \subseteq Z \subseteq Q \subseteq R$$

and

$$H \subseteq R$$

PRACTICE

Name the collections that each of the following numbers belongs to:

6. 2.35

7. −56

8. 0.345345345 . . .

9. 0.9889888988889888889 . . .

10. 8

Situation 26

Unbreakable Codes

A considerable amount of information is transmitted electronically. Much of this information, like financial transfers, needs to be kept confidential. Because it is relatively easy for anyone with a computer to read the mass of information traveling back and forth, it is necessary that the confidential information be sent in code so that only the sender and intended receiver can understand it.

It may surprise you to know that one of the most widely used coding techniques relies on simple arithmetic and a specific class of natural numbers called the **prime numbers.** This coding technique is called the RSA Public-Key Cryptosystem.

The discussion in the next three sections is devoted to explaining how

13 14 20 16 07 27 06 14 16 27 21 28 02

can be deciphered as

SEND MORE DOUGH

8.3 PRIMES AND COMPOSITES

A certain kind of natural number that has long fascinated mathematicians and philosophers is the prime number. A **prime number** is a natural number greater than 1 that can be divided without remainder only by itself and 1. For example, 37 is prime because if you try to divide 37 by any natural number other than 37 and 1, you'll get a remainder. Try it!

Natural numbers greater than 1 that are not prime are called **composite numbers.** The number 18 is composite because it can be divided by 1, 2, 3, 6, 9, and 18 without remainder.

PRACTICE

Which of the following numbers are prime? Which are composite?

1. 35

2. 114

3. 17

4. 50

5. 8002

6. 29

What method did you use to answer Practice Exercises 1–6?

Devising a method for finding primes has occupied the energies of many mathematicians over many centuries. Many of the methods involve some fairly advanced mathematics, but one of the oldest is still effective and easy. This method of finding primes was invented by Eratosthenes of Cyrene (276–194 BC). Here is how the **Sieve of Eratosthenes** works.

To find all the primes up to a given number, say, 100, first write down all the natural numbers from 2 to that number. Like this:

	2	3	4	5	6	7	8	9	10	11	12	13	14	15
16	17	18	19	20	21	22	23	24	25	26	27	28	29	30
31	32	33	34	35	36	37	38	39	40	41	42	43	44	45
46	47	48	49	50	51	52	53	54	55	56	57	58	59	60
61	62	63	64	65	66	67	68	69	70	71	72	73	74	75
76	77	78	79	80	81	82	83	84	85	86	87	88	89	90
91	92	93	94	95	96	97	98	99	100					

We know that 2 is a prime (why?), but any multiple of 2 will not be a prime because that multiple can be divided by 2 without remainder. So, we put a slash through all the multiples of 2 to indicate that they are not primes.

	2	3	~~4~~	5	~~6~~	7	~~8~~	9	~~10~~	11	~~12~~	13	~~14~~	15
~~16~~	17	~~18~~	19	~~20~~	21	~~22~~	23	~~24~~	25	~~26~~	27	~~28~~	29	~~30~~
31	~~32~~	33	~~34~~	35	~~36~~	37	~~38~~	39	~~40~~	41	~~42~~	43	~~44~~	45
~~46~~	47	~~48~~	49	~~50~~	51	~~52~~	53	~~54~~	55	~~56~~	57	~~58~~	59	~~60~~
61	~~62~~	63	~~64~~	65	~~66~~	67	~~68~~	69	~~70~~	71	~~72~~	73	~~74~~	75
~~76~~	77	~~78~~	79	~~80~~	81	~~82~~	83	~~84~~	85	~~86~~	87	~~88~~	89	~~90~~
91	~~92~~	93	~~94~~	95	~~96~~	97	~~98~~	99	~~100~~					

The next *unslashed* number is 3. Now, 3 is a prime and all the multiples of 3 are not. Hence, we slash all the multiples of 3 that haven't been slashed already.

	2	3	~~4~~	5	~~6~~	7	~~8~~	~~9~~	~~10~~	11	~~12~~	13	~~14~~	~~15~~
~~16~~	17	~~18~~	19	~~20~~	~~21~~	~~22~~	23	~~24~~	25	~~26~~	~~27~~	~~28~~	29	~~30~~
31	~~32~~	~~33~~	~~34~~	35	~~36~~	37	~~38~~	~~39~~	~~40~~	41	~~42~~	43	~~44~~	~~45~~
~~46~~	47	~~48~~	49	~~50~~	~~51~~	~~52~~	53	~~54~~	55	~~56~~	~~57~~	~~58~~	59	~~60~~
61	~~62~~	~~63~~	~~64~~	65	~~66~~	67	~~68~~	~~69~~	~~70~~	71	~~72~~	73	~~74~~	~~75~~
~~76~~	77	~~78~~	79	~~80~~	~~81~~	~~82~~	83	~~84~~	85	~~86~~	~~87~~	~~88~~	89	~~90~~
91	~~92~~	~~93~~	~~94~~	95	~~96~~	97	~~98~~	~~99~~	~~100~~					

The next prime is 5. So, we slash all the multiples of 5.

	2	3	~~4~~	5	~~6~~	7	~~8~~	~~9~~	~~10~~	11	~~12~~	13	~~14~~	~~15~~
~~16~~	17	~~18~~	19	~~20~~	~~21~~	~~22~~	23	~~24~~	~~25~~	~~26~~	~~27~~	~~28~~	29	~~30~~
31	~~32~~	~~33~~	~~34~~	~~35~~	~~36~~	37	~~38~~	~~39~~	~~40~~	41	~~42~~	43	~~44~~	~~45~~
~~46~~	47	~~48~~	49	~~50~~	~~51~~	~~52~~	53	~~54~~	~~55~~	~~56~~	~~57~~	~~58~~	59	~~60~~
61	~~62~~	~~63~~	~~64~~	~~65~~	~~66~~	67	~~68~~	~~69~~	~~70~~	71	~~72~~	73	~~74~~	~~75~~
~~76~~	77	~~78~~	79	~~80~~	~~81~~	~~82~~	83	~~84~~	~~85~~	~~86~~	~~87~~	~~88~~	89	~~90~~
91	~~92~~	~~93~~	~~94~~	~~95~~	~~96~~	97	~~98~~	~~99~~	~~100~~					

Now the multiples of 7.

	2	3	~~4~~	5	~~6~~	7	~~8~~	~~9~~	~~10~~	11	~~12~~	13	~~14~~	~~15~~
~~16~~	17	~~18~~	19	~~20~~	~~21~~	~~22~~	23	~~24~~	~~25~~	~~26~~	~~27~~	~~28~~	29	~~30~~
31	~~32~~	~~33~~	~~34~~	~~35~~	~~36~~	37	~~38~~	~~39~~	~~40~~	41	~~42~~	43	~~44~~	~~45~~
~~46~~	47	~~48~~	~~49~~	~~50~~	~~51~~	~~52~~	53	~~54~~	~~55~~	~~56~~	~~57~~	~~58~~	59	~~60~~
61	~~62~~	~~63~~	~~64~~	~~65~~	~~66~~	67	~~68~~	~~69~~	~~70~~	71	~~72~~	73	~~74~~	~~75~~
~~76~~	~~77~~	~~78~~	79	~~80~~	~~81~~	~~82~~	83	~~84~~	~~85~~	~~86~~	~~87~~	~~88~~	89	~~90~~
~~91~~	~~92~~	~~93~~	~~94~~	~~95~~	~~96~~	97	~~98~~	~~99~~	~~100~~					

The next prime is 11. Slash all the multiples of 11 in the list above. What did you notice?

Were you surprised to see that these numbers had already been slashed? Try the multiples of the next prime, 13. You'll see that these multiples have already been slashed as well. What's going on here?

We know that 100, the largest number in our list, can be written as $10 \cdot 10$. This means that every composite number less than 100 must be the product of two numbers, one of which is less than 10. Try some examples now to see this:

Thus, the composite numbers less than 100 can all be divided by primes less than 10. When we slashed out the multiples of 7, we were done because the next prime (11) is larger than 10.

In general, we continue the slashing process until we reach a prime bigger than the square root of the last number in our list. In this case, $\sqrt{100} = 10$ and the next prime is 11. So, we can quit after 7. The unslashed numbers are the primes. Here they are with the composite numbers replaced by asterisks:

	2	3	*	5	*	7	*	*	*	11	*	13	*	*
*	17	*	19	*	*	*	23	*	*	*	*	*	29	*
31	*	*	*	*	*	37	*	*	*	41	*	43	*	*
*	47	*	*	*	*	*	53	*	*	*	*	*	59	*
61	*	*	*	*	*	67	*	*	*	71	*	73	*	*
*	*	*	79	*	*	*	83	*	*	*	*	*	89	*
*	*	*	*	*	*	97	*	*	*					

PRACTICE

7. Use the Sieve of Eratosthenes to find all the prime numbers less than 200.

How many primes are there? Well, we know that there is a never-ending supply of prime numbers. No matter how large a natural number you pick, there will be a prime number bigger than it! The proof that there is a never-ending supply of primes is beyond the scope of this book, but we can see how the argument goes with an example.

Suppose that Zork tells you that 13 is the largest prime. To convince him that he is wrong, you need to come up with a prime larger than 13. (We know that 17 is a prime larger than 13, but we'll pretend that we don't.) Instead, take all the primes less than or equal to 13. List them here:

If 13 is the largest prime, then these are all the primes that there are. Now multiply these primes together. This gives you

$$2 \cdot 3 \cdot 5 \cdot 7 \cdot 11 \cdot 13 = 30{,}030$$

Now add 1 to get 30,031. If 13 is the largest prime, then 30,031 must be composite. If 30,031 is composite, then it must be divisible by either 2, 3, 5, 7, 11, or 13. Is 30,031 divisible by 2, 3, 5, 7, 11, or 13?

(Answer: no)

So, since 30,031 is not divisible by 2, 3, 5, 7, 11, or 13, it must be prime itself and 13 is not the largest prime.

This argument can be successfully repeated for any so-called *largest* prime.

PRACTICE

A pair of **twin primes** is a pair of prime numbers that are separated by only one other number. For example, 11 and 13 are twin primes, 17 and 19 are twin primes, but 19 and 23 are not because there are three numbers between them. It is unknown whether or not the supply of twin primes is unlimited.

8. Find all the twin prime pairs between 1 and 200.

9. The natural numbers 2 and 3 are both prime. Are there any other consecutive natural numbers that are both prime?

But what about the code? Where do primes enter into coding? Well, the coding system that we are going to examine rests on a basic fact about primes. Namely, although it is relatively easy to find really large primes with a computer, it is very difficult and often impossible (even with computers) to find the two primes whose product is a given number.

Here's a small example. Find a prime number that divides 8051 without remainder.

____

____

It may have taken you some time to discover that $83 \cdot 97 = 8051$, but we were able to easily discover that 97 and 83 were primes using the Sieve of Eratosthenes. It becomes virtually impossible for even a large computer to systematically discover the primes that divide a number that is the product of two 150-digit primes.

The coding system we will look at makes use of this difficulty in finding the primes that divide large numbers. This coding system also uses something called the **mod operation.**

8.4 THE MOD OPERATION

What is the remainder when you divide 4 by 3? _____
What is the remainder when you divide 5 by 3? _____
What is the remainder when you divide 6 by 3? _____
What is the remainder when you divide 7 by 3? _____
What is the remainder when you divide 8 by 3? _____
What is the remainder when you divide 9 by 3? _____
What is the remainder when you divide 10 by 3? _____
What is the remainder when you divide 11 by 3? _____
What is the remainder when you divide 12 by 3? _____
What is the remainder when you divide 13 by 3? _____
What is the remainder when you divide 14 by 3? _____
What is the remainder when you divide 15 by 3? _____
What can you say about the remainder when any number is divided by 3?

____

____

____

You have discovered that when you divide a number by 3, the remainder is always either 0, 1, or 2.

Try dividing several numbers by 4 and examine the remainders. What do you see?

____

____

____

____

____

____

Right! The remainder when you divided by 4 is either 0, 1, 2, or 3.

Mathematicians have invented the **modulus operator** to denote the remainder of a division.

The symbol for this operator is **MOD.**

$$5 \text{ MOD } 3 = 2$$

means that when you divide 5 by 3, you get a remainder of 2.

To say that the remainder obtained when you divide 18 by 5 is 3, we write 18 MOD 5 = 3.

PRACTICE

Calculate each of the following:

1. 6 MOD 4

2. 8 MOD 11

3. 7 MOD 7

4. 234 MOD 7

5. 98,007 MOD 3

6. 87 MOD 11

8.5 RSA PUBLIC-KEY CRYPTOSYSTEM

You are probably familiar with **linear numeric codes.** In these codes, each letter in the alphabet is matched with a number. The simplest is this one (code 1).

Code 1:

A	B	C	D	E	F	G	H	I	J	K	L	M	N	O	P	Q
01	02	03	04	05	06	07	08	09	10	11	12	13	14	15	16	17
R	S	T	U	V	W	X	Y	Z								
18	19	20	21	22	23	24	25	26								

To encode the message LOOK OUT FOR DUCKS, simply replace each letter by its numeric counterpart. That is, the L is replaced by 12, each O by 15, the K by 11, and so on. The resulting coded message is

12 15 15 11 15 21 20 06 15 18 04 21 03 11 19

To decode the message, simply replace each number by its corresponding letter.

Now this code is probably too easy because we were not very imaginative in the way we matched letters to numbers. A better code might be this one (code 2).

Code 2:

A	B	C	D	E	F	G	H	I	J	K	L	M	N	O	P	Q
12	13	07	14	05	02	11	23	09	17	26	10	20	04	19	24	06
R	S	T	U	V	W	X	Y	Z								
01	18	08	22	15	25	03	21	16								

The message LOOK OUT FOR DUCKS becomes

10 19 19 26 19 22 08 02 19 01 14 22 07 26 18

in code 2.

PRACTICE

1. Encode this message using code 2.

OPEN THE LAST CAN

2. Decode this message using code 2.

11 19 19 14 20 19 01 04 09 04 11

There are two major problems with linear numeric codes like code 1 and code 2. First, the sender and receiver of the message must have the alphabet–number matching list. This makes changing the code difficult and expensive. Second, these codes are relatively easy to break.

The RSA Public-Key Cryptosystem (invented in 1977 by Ronald Rivest, Adi Shamir, and Leonard Adelman) allows the sender to encode a message without really knowing the code. Thus, the receiver is the only person with complete knowledge of the code. The receiver can change the code at will and announce part of it *publicly*. The code is secure from potential code-breakers because of the difficulty in breaking down large composite numbers into a product of primes. Here is how an RSA Public-Key Cryptosystem works.

First, the message sender puts her message into numeric form using the linear numeric code that I've called code 1 (the simple alphabet numbering code). She then looks up the receiver's public information in a public code directory. There will be three numbers in the directory.

1. b that denotes the size of the blocks into which the message should be put. For example, if the numeric version of the message is

12 15 15 11 15 21 20 06 15 18 04 21 03 11 19

and $b = 4$ (digits), then the code will be written

1215 1511 1521 2006 1518 0421 0311 1900

If $b = 6$, then the code will be written

121515 111521 200615 180421 031119

If $b = 2$, the code is left unchanged as

12 15 15 11 15 21 20 06 15 18 04 21 03 11 19

In this text, I will use $b = 2$, so that no reblocking will be necessary.

2. A code number n that is the product of two prime numbers.
3. A code number e that depends on n in a particular way that is discussed later.

The sender then takes each block (call it M) in the code and transforms it using the equation $E = M^e$ MOD n. She then sends the transformed message. Here is how the word SEND would be encoded if the public numbers were $b = 2$, $e = 7$, and $n = 33$.

1. Write the numeric version of SEND using my code 1.

   19 05 14 04

2. Form blocks of two digits each (because $b = 2$).

   19 05 14 04

3. Convert each block using $E = M^7$ MOD 33.

$$19^7 \text{ MOD } 33 = 893{,}871{,}739 \text{ MOD } 33 = 13$$
$$5^7 \text{ MOD } 33 = 78{,}125 \text{ MOD } 33 = 14$$
$$14^7 \text{ MOD } 33 = 105{,}413{,}504 \text{ MOD } 33 = 20$$
$$4^7 \text{ MOD } 33 = 16{,}384 \text{ MOD } 33 = 16$$

4. Transmit the converted numbers:

   13 14 20 16

PRACTICE

3. Encode the message MORE using the public numbers $b = 2$, $e = 7$, $n = 33$.

The receiver has three secret numbers, d, p, and q, that enable him to decode the message. He uses the equation

$$D = E^d \text{ MOD } n$$

Thus, for the message SEND that we encoded as 13 14 20 16, if the receiver's $d = 3$, he would decode the message like this:

$$13^3 \text{ MOD } 33 = 2197 \text{ MOD } 33 = 19$$

$$14^3 \text{ MOD } 33 = 2744 \text{ MOD } 33 = 5$$

$$20^3 \text{ MOD } 33 = 8000 \text{ MOD } 33 = 14$$

$$16^3 \text{ MOD } 33 = 4096 \text{ MOD } 33 = 4$$

The decoded message is 19 5 14 4, which becomes SEND when we use code 1.

Notice that the secret number d must be just right. If I tried to decode 13 using $d = 5$, I'd get 13^5 MOD 33 = 371293 MOD 33 = 10. But 10 is the letter J, not the S that is correct.

PRACTICE

4. Decode the message you encoded in Practice Exercise 3 using $d = 3$. Make sure that you get MORE.

We can now see how the code in Situation 26 was made. The encoder began with the message SEND MORE DOUGH. He then gave each letter a numerical value using the simple code 1.

19 05 14 04 13 15 18 05 04 15 21 07 08

Next, using the RSA Public-Key Cryptosystem and the values $b = 2$, $e = 7$, and $n = 33$, he transformed the message to

13 14 20 16 07 27 06 14 16 27 21 28 02

We've already done SEND and MORE. You should check that DOUGH becomes 16 27 21 28 02.

PRACTICE

Here is another RSA Public-Key Cryptosystem code using an abbreviated alphabet. This code works only for the letters A through I. Suppose that $b = 2$, $e = 3$, $d = 7$, and $n = 10$.

5. Encode BED.

6. Decode 06 09 03.

7. Decode 07 01 08 08 09 05.

8. Encode DIG.

Now that you've seen how the RSA Public-Key Cryptosystem works, it's time to look at how all the various numbers, public and secret, are related. The reasons behind the relationships involve more mathematics than we want to get into here, but the relationships themselves are interesting.

First, n must be the product of two prime numbers p and q, and it must be larger than the number of symbols in your alphabet. In Situation 26, $p = 3$ and $q = 11$, so $n = 3 \cdot 11 = 33$. The alphabet has 26 letters, so $n = 33$ is big enough. Now e and d depend on p and q in a special way. First, we find the number $\alpha = (p - 1) \cdot (q - 1) = 2 \cdot 10 = 20$. Next, pick an e that is less than n, doesn't divide α, and is prime. I picked $e = 7$. Finally, d must be chosen so that $e \cdot d$ MOD $\alpha = 1$. In this case, I found $d = 3$ ($3 \cdot 7$ MOD $20 = 1$).

Here is another set of numbers for an RSA code. Check to see that they meet all the conditions specified in the last paragraph.

$$p = 5, q = 11, e = 13, d = 37.$$

PRACTICE

9. Find values for p, q, e, and d that would serve as an RSA code for the abbreviated alphabet A–N (14 symbols).

Situation 27

Do You Guys Really Do This for a Living?

Fern's Uncle Belfer is a mathematician. On her 28th birthday, Fern got a card from Uncle Belfer that read

CONGRATULATIONS ON REACHING AN AGE
THAT IS BOTH
TRIANGULAR
AND
PERFECT

What did Uncle Belfer mean when he said that Fern's age was *triangular* and *perfect*?

Natural numbers have been a continual source of amusement for people for centuries. Mathematicians, philosophers, and hobbyists have delighted in discovering the properties of and relationships between natural numbers. Some properties of natural numbers, like the property of being prime, have useful applications. Other properties are simply enjoyable for their elegance or curiosity.

Fern's Uncle Belfer was referring to two of the myriad of properties of the natural numbers that have been discovered. We'll examine triangular numbers in the next section and then turn to perfect numbers.

8.6 FIGURATIVE NUMBERS

Around 550 BC, the Greek mathematician and philosopher Pythagoras and his followers noticed that some natural numbers were related to geometric shapes. For example, you can arrange six dots in the shape of a triangle, like this:

```
    •
   • •
  • • •
```

We say that 6 is a **triangular number.** Ten is also a triangular number because 10 dots can also be arranged in the shape of a triangle, like this:

```
     •
    • •
   • • •
  • • • •
```

Other triangular numbers are 3, 15, 21, and 28. Here's 28.

```
        •
       • •
      • • •
     • • • •
    • • • • •
   • • • • • •
  • • • • • • •
```

Also, we can arrange 9 dots in the shape of a square, like this:

Nine is called a **square number.** Other square numbers are 4, 16, 25, 36, and so on.
There are also **rectangular numbers** like 12:

and **octagonal numbers** like 24 (an octagon is an eight-sided figure with all sides the same length):

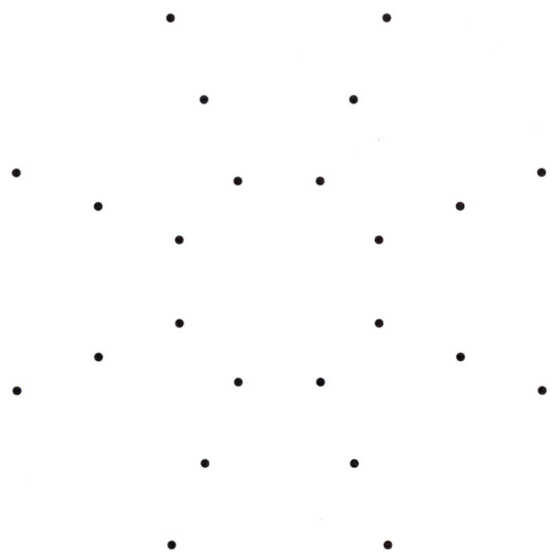

The general term for numbers that can be associated with geometric shapes is **figurative numbers.**

PRACTICE

1. Show that 3 is a triangular number.

2. Show that 16 is a square number.

3. Show that the octagonal number 24 is also a rectangular number.

4. Find a number that is triangular, rectangular, and hexagonal (a six-sided figure).

8.7 PERFECT NUMBERS

A **divisor** of a natural number is a natural number that divides it without remainder. For example, 5 is a divisor of 10 because 5 divides 10 without remainder, but 7 is not a divisor of 10 because 7 MOD 10 = 7. A **proper divisor** of a natural number is a divisor of the natural number that is not equal to the natural number. Thus, 2 is a proper divisor of 8, but 8 is not a proper divisor of 8 because 8 = 8.

PRACTICE

Find all the divisors of each of the following natural numbers. Which of the divisors are proper divisors?

1. 24

2. 28

3. 19

4. 30

5. 180

A **perfect number** is a natural number that is equal to the sum of its proper divisors. Thus, 6 is perfect because the proper divisors of 6 are 1, 2, and 3, and $6 = 1 + 2 + 3$. On the other hand, 4 is not perfect because the proper divisors of 4 are 1 and 2 and $4 \neq 1 + 2$.

PRACTICE

6. Show that 28 is a perfect number.

Perfect numbers are rare. The next perfect number after 28 is 496. The next one after that is 8128. The fifth perfect number is 33,550,336. The sixth perfect number is 8,589,869,056. After that, they really get big! In 1985, a perfect number with 130,100 digits was discovered.

Nobody knows if there is an unlimited supply of perfect numbers. We do know that every even perfect number must end with a 6 or an 8, but we don't know if there are any odd perfect numbers.

You have seen that 28 is triangular and perfect. To close this chapter, I'd like to mention one other interesting relationship that can be found among the natural numbers, **amicable pairs.**

8.8 AMICABLE PAIRS

Two natural numbers form an **amicable pair** if each is the sum of the proper divisors of the other. The numbers 220 and 284 form an amicable pair because the proper divisors of 220 are 1, 2, 4, 5, 10, 11, 20, 22, 44, 55, and 110, and $1 + 2 + 4 + 5 + 10 + 11 + 20 + 22 + 44 + 55 + 110 = 284$; and the proper divisors of 284 are 1, 2, 4, 71, and 142, and $1 + 2 + 4 + 71 + 142 = 220$.

Pythagoras (550 BC) knew about the amicable pair 220 and 284, but a second amicable pair was not found until 1636, a span of over 2100 years! The second amicable pair that was found (by Pierre de Fermat) is 17,296 and 18,416.

PRACTICE

1. Confirm that 17,296 and 18,416 are an amicable pair.

The third amicable pair that was found (by Descartes in 1638) is 9,363,584 and 9,437,056. As with the perfect numbers and the twin primes, it is unknown whether or not there is an unlimited supply of amicable numbers.

8.9 WHAT DO YOU KNOW?

If you have worked carefully through this chapter, then you know about the major collections of numbers, can encode and decode using the RSA Public-Key Cryptosystem, and are familiar with the ideas of prime, perfect, figurative, and amicable numbers. The exercises in this section are designed to test and refine your knowledge of the mathematics of numbers.

1. What is the difference between a number and a numeral?
2. Classify each of the following as a cardinal number or an ordinal number.
   (a) The number of hits a baseball player gets in a game
   (b) The position of a baseball player in the batting order
   (c) The number of cows grazing in a field
   (d) Your rank in your graduating class
   (e) The number of ants on a picnic blanket
3. Tell which collection of numbers each of the following numbers belongs to. List all collections that each number is in.
   (a) -3
   (b) 345
   (c) 19.06
   (d) $-7.8 \cdot 10^{13}$
   (e) $8.9 \cdot 10^{-56}$
   (f) 10^{100} (this number is called a *googol*)
   (g) 0.123123123123123 . . .
   (h) 0.223223322333223333223333 . . .
   (i) π
   (j) 1.00000000000000000000000000000000000000006
4. Find the decimal expansion of each of the following rational numbers. Tell if it is terminating or repeating. If it repeats, tell what string of digits repeats.
   (a) 1/4
   (b) 7/8
   (c) 14/9
   (d) 1/19
   (e) 4/5
   (f) 78/67
   (g) $-6/11$

5. Find the decimal expansion of 1/9, 2/9, 3/9, 4/9, 5/9, 6/9, 7/9, and 8/9. What pattern do you see? Does the pattern work for 11/9 and 67/9? Make up a rule for the decimal expansion of a rational number whose denominator is 9.
6. Find examples to illustrate each of the following propositions:
   (a) The sum of two odd natural numbers is even.
   (b) The product of two rational numbers is a rational number.
   (c) The product of two irrational numbers is not always irrational.
   (d) The sum of two irrational numbers is not always irrational.
7. The **Pythagorean theorem** says that the sum of the squares of the sides of a right triangle is equal to the square of the hypotenuse. Thus, for the right triangle shown below, $A^2 + B^2 = C^2$.

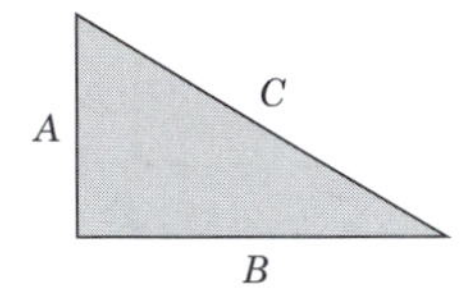

Use the Pythagorean theorem and a right triangle with both sides 1 centimeter long to show that it is possible to draw a line that has an irrational length.
8. A **Pythagorean triple** is a triple of three natural numbers in which the sum of the squares of the first two is equal to the square of the third. For example, (3, 4, 5) is a Pythagorean triple because $3^2 + 4^2 = 5^2$.
   (a) Show that (8, 15, 17) is a Pythagorean triple.
   (b) Show that (5, 12, 13) is a Pythagorean triple.
   (c) Find two other Pythagorean triples.
9. (a) Calculate $18 \cdot 11$, $12 \cdot 11$, $45 \cdot 11$, and $27 \cdot 11$.
   (b) Look at the answers to Exercise 9a. Do you see a pattern? What is it?
10. A **palindrome** is a number that reads the same backwards and forwards. For example, 11,311, 232, and 1,234,321 are palindromes. Find a palindrome that is also a square number.
11. Use the Sieve of Eratosthenes to find all the prime numbers less than 300.
12. How many twin prime pairs are there in the numbers less than 300?
13. **Goldbach's conjecture** states that any even number greater than 4 can be written as the sum of two odd prime numbers. Verify Goldbach's conjecture for each of the following:
   (a) 98
   (b) 38
   (c) 1120
   It is unknown whether or not Goldbach's conjecture is true.
14. Calculate each of the following:
   (a) 14 MOD 5
   (b) 24 MOD 8
   (c) 7 MOD 20
   (d) 35 MOD 2
   (e) 1,209,876 MOD 17
15. Suppose that n is an odd number. What is n MOD 2?
16. What is the smallest natural number that has all the numbers from 1 through 10 as divisors?
17. Using the RSA Public-Key Cryptosystem, encode the message

A DEAFENING FLAGMAN

with the code numbers $b = 2$, $e = 7$, $n = 15$.
18. Using the RSA Public-Key Cryptosystem, encode the message

CLIMB A BIG HILL

with the code numbers $b = 2$, $e = 7$, $n = 15$.
19. Using the RSA Public-Key Cryptosystem, encode the message

BIG DOG IN CAGE

with the code numbers $b = 2$, $e = 7$, $n = 22$ and the symbol list

A	B	C	D	E	F	G	I	N	O	P	R
01	02	03	04	05	06	07	08	09	10	11	12

20. Using the RSA Public-Key Cryptosystem, encode the message

POPCORN ON FRED

with the code numbers $b = 2$, $e = 7$, $n = 22$ and the symbol list

A	B	C	D	E	F	G	I	N	O	P	R
01	02	03	04	05	06	07	08	09	10	11	12

21. Using the RSA Public-Key Cryptosystem, decode the message

08 01 13 07 01 11 05

with the code numbers $b = 2$, $d = 7$, $n = 15$.
22. Using the RSA Public-Key Cryptosystem, decode the message

01 08 12 10 17 10 15 01 18 03 16

with the code numbers $b = 2$, $d = 3$, and $n = 22$. Use the symbol list from Exercise 19.

23. (a) Using the RSA Public-Key Cryptosystem, encode the message

CHECK THE BACK

with the code numbers $b = 2$, $e = 11$, $p = 3$, $q = 7$.

(b) What is a possible value for the code number d in this code?

24. Every natural number can be expressed as the sum of at most three triangular numbers. Confirm this statement for

(a) 20

(b) 30

(c) 178

25. Each of the following numbers is figurative. Tell whether each is triangular, rectangular, hexagonal, or pentagonal (five-sided figure). If a number belongs to more than one category, list them all. Draw figures to support your answers.

(a) 36

(b) 22

(c) 99

(d) 10

26. A number n is a **cubic number** if n dots can be arranged in the shape of a cube. The first cubic number is 8:

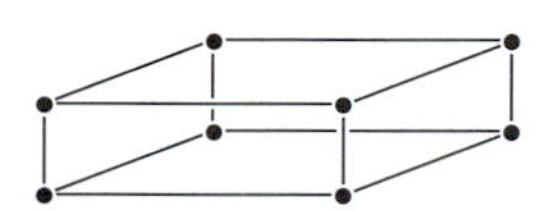

(a) Show that 27 is a cubic number.

(b) What is the next cubic number after 27?

27. A number k is a **crawler number** if k dots can be drawn to form a picture of a crawling insect. The first two crawler numbers are 10 and 13:

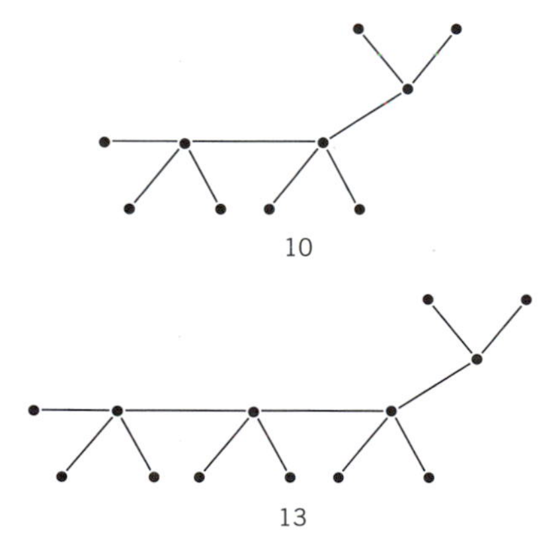

Find the next five crawler numbers.

28. Confirm that 496 and 8128 are perfect numbers.

29. A natural number is a **deficient number** if it is greater than the sum of its proper divisors. A natural number is an **abundant number** if it is less than the sum of its proper divisors. Complete the following table:

Natural Number	Sum of Proper Divisors	Type
2	1	Deficient
3	1	Deficient
4	3	Deficient
5	1	Deficient
6	6	Perfect
7	1	Deficient
8	7	Deficient
9	4	Deficient
10	8	Deficient
11	1	Deficient
12	16	Abundant
13		
14		
15		
16		
17		
18		
19		
20		

30. Are prime numbers deficient or abundant? Explain.

31. Explain why two times any perfect number is abundant.

32. In 1886, a 16-year-old Italian boy discovered that 1184 and 1210 formed an amicable pair. Verify this discovery.

33. A considerable amount of recreational mathematics centers on the year in the Western calendar considered as a natural number.

(a) 1991 was a palindromic year. What will the next palindromic year be?

(b) Find all the prime number years between 1990 and 2001.

(c) 1949 and 1951 are twin prime years. Are there any other twin prime years between 1951 and 2001?

(d) Is the current year abundant or deficient?

(e) What was happening in Europe during the last perfect year (496 AD)? Asia? Africa? The Americas?

34. In 1637, Pierre de Fermat conjectured that for all $n > 2$, there are no integer values for x, y, and z such that $x^n + y^n = z^n$. This conjecture is known as **Fermat's last theorem.** Write a report on recent efforts to prove this conjecture.

ANSWERS TO PRACTICE EXERCISES

Section 8.1

1. *N*, *W*, *Z*
2. *Z*
3. *W*, *Z*
4. *N*, *W*, *Z*
5. *Z*

Section 8.2

1. 0.75, terminating
2. 0.77777 . . . , repeating, 7
3. 1.571428571428 . . . , repeating, 571428
4. 0.454545 . . . , repeating, 45
5. 0.35, terminating
6. Rational, real
7. Integer, rational, real
8. Rational, real
9. Irrational, real
10. Natural, whole, integer, rational, real

Section 8.3

1. Composite; 35 can be divided by 7 and 5 without remainder.
2. Composite; 114 can be divided by 2 and 57 without remainder.
3. Prime.
4. Composite; 50 can be divided by 2, 5, 10, and 25 without remainder.
5. Composite; 8002 can be divided by 2 and 4001 without remainder.
6. Prime.
7. The prime numbers are circled. The last prime whose multiples are slashed is 13.

	(2)	(3)	~~4~~	(5)	~~6~~	(7)	~~8~~	~~9~~	~~10~~	(11)	~~12~~	(13)	~~14~~	~~15~~
~~16~~	(17)	~~18~~	(19)	~~20~~	~~21~~	~~22~~	(23)	~~24~~	~~25~~	~~26~~	~~27~~	~~28~~	(29)	~~30~~
(31)	~~32~~	~~33~~	~~34~~	~~35~~	~~36~~	(37)	~~38~~	~~39~~	~~40~~	(41)	~~42~~	(43)	~~44~~	~~45~~
~~46~~	(47)	~~48~~	~~49~~	~~50~~	~~51~~	~~52~~	(53)	~~54~~	~~55~~	~~56~~	~~57~~	~~58~~	(59)	~~60~~
(61)	~~62~~	~~63~~	~~64~~	~~65~~	~~66~~	(67)	~~68~~	~~69~~	~~70~~	(71)	~~72~~	(73)	~~74~~	~~74~~
~~76~~	~~77~~	~~78~~	(79)	~~80~~	~~81~~	~~82~~	(83)	~~84~~	~~85~~	~~86~~	~~87~~	~~88~~	(89)	~~90~~
~~91~~	~~92~~	~~93~~	~~94~~	~~95~~	~~96~~	(97)	~~98~~	~~99~~	~~100~~	(101)	~~102~~	(103)	~~104~~	

~~105~~	~~106~~	(107)	~~108~~	(109)	~~110~~	~~111~~	~~112~~	(113)	~~114~~	~~115~~	~~116~~
~~117~~	~~118~~	~~119~~	~~120~~	~~121~~	~~122~~	~~123~~	~~124~~	~~125~~	~~126~~	(127)	~~128~~
~~129~~	~~130~~	(131)	~~132~~	~~133~~	~~134~~	~~135~~	~~136~~	(137)	~~138~~	(139)	~~140~~
~~141~~	~~142~~	~~143~~	~~144~~	~~145~~	~~146~~	~~147~~	~~148~~	(149)	~~150~~	(151)	~~152~~
~~153~~	~~154~~	~~155~~	~~156~~	(157)	~~158~~	~~159~~	~~160~~	~~161~~	~~162~~	(163)	~~164~~
~~165~~	~~166~~	(167)	~~168~~	~~169~~	~~170~~	~~171~~	~~172~~	(173)	~~174~~	~~175~~	~~176~~
~~177~~	~~178~~	(179)	~~180~~	(181)	~~182~~	~~183~~	~~184~~	~~185~~	~~186~~	~~187~~	~~188~~
~~189~~	~~190~~	(191)	~~192~~	(193)	~~194~~	~~195~~	~~196~~	(197)	~~198~~	(199)	~~200~~

8. The twin prime pairs are underlined.

	2	3	4	5	6	7	8	9	10	11	12	13	14	15
16	17	18	19	20	21	22	23	24	25	26	27	28	29	30
31	32	33	34	35	36	37	38	39	40	41	42	43	44	45
46	47	48	49	50	51	52	53	54	55	56	57	58	59	60
61	62	63	64	65	66	67	68	69	70	71	72	73	74	75
76	77	78	79	80	81	82	83	84	85	86	87	88	89	90

91	92	93	94	95	96	97	98	99	100	101	102	103	104
105	106	107	108	109	110	111	112	113	114	115	116		
117	118	119	120	121	122	123	124	125	126	127	128		
129	130	131	132	133	134	135	136	137	138	139	140		
141	142	143	144	145	146	147	148	149	150	151	152		
153	154	155	156	157	158	159	160	161	162	163	164		
165	166	167	168	169	170	171	172	173	174	175	176		
177	178	179	180	181	182	183	184	185	186	187	188		
189	190	191	192	193	194	195	196	197	198	199	200		

9. No.

Section 8.4

1. 2
2. 8
3. 0
4. 3
5. 0
6. 10

Section 8.5

1. 19 24 05 04 08 23 05 10 12 18 08 07 12 04
2. GOOD MORNING
3. 07 27 06 14
4. 7^3 MOD 33 = 343 MOD 33 = 13 = M
27^3 MOD 33 = 19683 MOD 33 = 15 = O
6^3 MOD 33 = 216 MOD 33 = 18 = R
14^3 MOD 33 = 2744 MOD 33 = 5 = E
5. 08 05 04
6. FIG
7. CABBIE
8. 04 09 03
9. $p = 3, q = 5, e = d = 7$

Section 8.6

1.

  •
•   •

2.

• • • •
• • • •
• • • •
• • • •

3.

• • • • • •
• • • • • •
• • • • • •
• • • • • •

4. 6

Section 8.7

The proper divisors appear in bold print.

1. **1, 2, 3, 4, 6, 8, 12,** 24
2. **1, 2, 4, 7, 14,** 28
3. **1,** 19 (19 is a prime number. We could define a prime number as a natural number with only one proper divisor.)
4. **1, 2, 3, 5, 6, 10, 15,** 30
5. **1, 2, 3, 4, 5, 6, 9, 10, 12, 15, 18, 20, 30, 36, 45, 60, 90,** 180
6. 28 = 1 + 2 + 4 + 7 + 14

Section 8.8

1. Proper divisors of 17,296: 1, 2, 4, 8, 16, 23, 46, 47, 92, 94, 184, 188, 368, 376, 752, 1081, 2162, 4324, 8648.
Proper divisors of 18,416: 1, 2, 4, 8, 16, 1151, 2302, 4604, 9208.
1 + 2 + 4 + 8 + 16 + 23 + 46 + 47 + 92 + 94 + 184 + 188 + 368 + 376 + 752 + 1081 + 2162 + 4324 + 8648 = 18,416
and 1 + 2 + 4 + 8 + 16 + 1151 + 2302 + 4604 + 9208 = 17,296

Chapter Nine

Mathematics of Games

What's the Point?

In Chapter 4 you learned how to compute the probabilities of events and use those probabilities to make decisions. The decisions in that chapter all had one thing in common; nobody else was attempting to change the situation to their advantage and your disadvantage. In many situations, you will be faced with a human opponent whose goals may not be the same as yours. In some cases, your goals may be diametrically opposed.

Game theory is the mathematics of games. It offers suggestions on how best to make decisions when you are operating in competition with another person. Game theory is used extensively in negotiations, conflict resolution, and war. This chapter will help you to understand some of the principles of game theory, the mathematics of games.

Situation 28 If She Moves There, Then I Can Move Here and . . .

After several successful years as the CEO of a recycling company, you decide that you would like to run for mayor of your town. You successfully obtain your party's nomination and are now faced with a tough campaign against the incumbent mayor. Everyone agrees that there are only two basic issues in the campaign: the environment and the abilities of the candidates. You figure that you have strength in the environmental realm because of your years of experience in recycling and because the incumbent mayor had some trouble with the EPA a few years ago. You also know that you lack the experience in public office of your opponent, but are confident you can counteract that with your experience as a CEO and member of the school board.

Your sister, who is a mathematician and pollster, has prepared the following chart for you that shows the public's perceptions of you, the mayor, and the issues of environment and experience:

		The Mayor	
		Environment	Experience
You	Environment	4	−2
	Experience	−1	3

The numbers in the table represent *your* increase in popularity (positive numbers) or decrease in popularity (negative numbers) when each issue is discussed. For example, if you and the mayor both run commercials on the environment at the same time, then you will gain four popularity points with the voters. On the other hand, if the mayor talks about her experience while you are talking about the environment, the voters will be more inclined toward the mayor and you will lose two popularity points. If you both talk about experience, then you will gain three quality points because the voters are willing to remove an incumbent if they are convinced that the challenger is able to run the city. Finally, if you talk about your experience and the mayor highlights the environment, you will lose popularity points, as the voters are hesitant to support unexperienced administrators.

You know that the mayor knows what you know and she will be trying to make your popularity as low as possible. Thus, if you talk about the environment, the mayor will talk experience. Then you will shift to experience, and the mayor will shift to the environment, and on, and on. What percentage of your time should you spend talking about the environment to have the best chance of winning the election?

____

____

____

This is a fairly complicated situation. We will work our way up to a solution gradually. But first, it is appropriate to give some technical names to the situation at hand.

The competition between you and the mayor is called a **two-person 2 × 2 zero-sum game.** The *two-person* part is obvious. The 2 × 2 refers to the number of options each player has (in this case, to talk about the environment or experience). *Zero-sum* means that one player's gain is the other player's loss. Thus, a gain of 4 for you would be a loss of 4 for the mayor.

We would like to find the best strategy for playing this game. This strategy is called the **optimal strategy**. Our quest for it begins with a discussion of matrices.

9.1 MATRICES

A **matrix** (the plural is *matrices*) is a rectangular array of numbers. Here are some matrices.

Matrix A: $\begin{bmatrix} 2 & 3 & 5 \\ -3 & 1.5 & 9 \end{bmatrix}$

Matrix B: $[23 \quad 45 \quad 0.89]$

Matrix C: $\begin{bmatrix} -111 \\ 67 \\ 1.45 \end{bmatrix}$

Matrix D: $\begin{bmatrix} 3 & 4 & 5 \\ 8 & 1 & 7 \\ 2 & 0 & 8 \end{bmatrix}$

The **order of a matrix** is the two numbers that count the number of rows (horizontal) and columns (vertical) that a matrix has. The order is usually expressed as (number of rows) × (number of columns).

The order of matrix A is 2×3.
The order of matrix B is 1×3.
The order of matrix C is 3×1.
The order of matrix D is 3×3.

PRACTICE

Find the order of each of the following matrices:

1. $[5 \quad 6 \quad 7 \quad 8]$

2. $\begin{bmatrix} 4 & 5 \\ 9 & 10 \\ 1 & 0 \\ -3 & 8 \end{bmatrix}$

3. $\begin{bmatrix} 3 & 5 \\ 4 & 5 \\ 9.8 & 7 \\ 17 & -9.8 \\ 12 & 14 \end{bmatrix}$

A matrix that has order $1 \times n$ or $n \times 1$ (where n is any natural number) is called a **vector**. Matrices B and C are vectors. Matrix B is a **row vector**. Matrix C is a **column vector**.

Two matrices are equal if they have the same order and exactly the same numbers in the same positions. These two matrices are equal.

$$\begin{bmatrix} 2 & 3 & 4 \\ 0 & 7 & 9 \end{bmatrix}, \quad \begin{bmatrix} 2 & 3 & 4 \\ 0 & 7 & 9 \end{bmatrix}$$

These two matrices are not equal.

$$\begin{bmatrix} 2 & 3 & 4 \\ 0 & 7 & 9 \end{bmatrix}, \quad \begin{bmatrix} 2 & 3 & 4 \\ 0 & 8 & 9 \end{bmatrix}$$

These two matrices are not equal.

$$\begin{bmatrix} 2 & 3 & 4 \\ 0 & 7 & 9 \end{bmatrix}, \quad \begin{bmatrix} 2 & 3 \\ 4 & 0 \\ 7 & 9 \end{bmatrix}$$

Matrices are useful for organizing large amounts of information. Here are some examples.

Inventory Matrix

	Ducks	Cows	Horses
North pasture	140	200	30
South pasture	200	150	20

You can see at a glance how many horses are in the south pasture (20). How many ducks are in the north pasture? _____ How many cows are there altogether? _____

PRACTICE

4. Write the inventory matrix if three more cows are added to the south pasture.

Two-Person Game Matrix

		The Mayor	
		Environment	Experience
You	Environment	4	−2
	Experience	−1	3

This matrix shows your rating in the election campaign of Situation 28. What does the 3 represent?

____

____

____

A Probability Matrix

Change in Corn Futures Prices

		Up	Down	No change
	Hotter	.6	.1	.3
Weather	Colder	.4	.2	.4
	Normal	.1	.1	.8

The .8 means that the probability that corn futures prices will not change when the weather is normal is .8. What does the .6 mean?

____

____

____

9.2 MATRIX ADDITION

Two matrices may be added together only if they have the same order. To add together two matrices, simply add corresponding entries. Here's an example.

$$\begin{bmatrix} 3 \\ 2 \\ 4 \\ 14 \end{bmatrix} + \begin{bmatrix} 5 \\ 1 \\ 5 \\ 6 \end{bmatrix} = \begin{bmatrix} 8 \\ 3 \\ 9 \\ 20 \end{bmatrix}$$

Here's another.

$$\begin{bmatrix} 2 & 3 & 4 \\ 1 & 0 & 5 \\ 8 & 1 & 6 \end{bmatrix} + \begin{bmatrix} 9 & 0 & -5 \\ 0 & 1 & 5 \\ 2 & -9 & 4 \end{bmatrix} = \begin{bmatrix} 11 & 3 & -1 \\ 1 & 1 & 10 \\ 10 & -8 & 10 \end{bmatrix}$$

PRACTICE

1. Complete this matrix addition.

$$\begin{bmatrix} 2 & 4 & 6 & 8 \\ 4 & 6 & 3 & 1 \end{bmatrix} + \begin{bmatrix} 1 & 8 & 7 & 6 \\ 5 & 2 & 8 & 9 \end{bmatrix} = \begin{bmatrix} 3 & 12 & 13 & 14 \\ 9 & \_\_ & \_\_ & \_\_ \end{bmatrix}$$

Calculate:

2. $[2\quad 3\quad 4\quad 5] + [5\quad 9\quad 0\quad 8]$

3. $\begin{bmatrix} 1 & 0 \\ 2 & 3 \\ 1 & 5 \end{bmatrix} + \begin{bmatrix} 5 & 8 \\ 1 & -8 \\ 4 & 7 \end{bmatrix}$

One application of matrix addition is in the updating of inventory. Recall the inventory matrix given in the last section.

	Ducks	Cows	Horses
North pasture	140	200	30
South pasture	200	150	20

For convenience, I'll call this inventory matrix V. (It is customary to denote matrices by capital italic letters.)

Now suppose that we sell 5 of the horses from the north pasture, 50 of the cows from the south pasture, and buy 30 ducks for the north pasture. These transactions can be organized in a matrix like this.

	Ducks	Cows	Horses
North pasture	30	0	−5
South pasture	0	−50	0

Call this transaction matrix T. Then the new inventory N can be calculated by $V + T = N$, or

$$\begin{bmatrix} 140 & 200 & 30 \\ 200 & 150 & 20 \end{bmatrix} + \begin{bmatrix} 30 & 0 & -5 \\ 0 & -50 & 0 \end{bmatrix} = \begin{bmatrix} 170 & 200 & 25 \\ 200 & 100 & 20 \end{bmatrix}$$

PRACTICE

Here is an inventory matrix for a tire retailer.

	Passenger	Truck	Racing	Construction
Store No. 1	400	200	60	50
Store No. 2	500	300	40	0
Store No. 3	300	200	100	30

Yesterday, store no. 1 sold 60 passenger tires, 16 truck tires, 4 racing tires, and 0 construction tires. Store no. 2 sold 160 passenger tires, 70 truck tires, 0 racing tires, and 0 construction tires. Store no. 3 sold 55 passenger tires, 10 truck tires, 8 racing tires, and 4 construction tires.

4. Write a matrix S that shows yesterday's tire sales.

5. Use matrix addition to update the inventory.

Matrix subtraction is performed in a fashion similar to addition. Here's an example.

$$\begin{bmatrix} 2 & 5 \\ 3 & 7 \\ 6 & 2 \end{bmatrix} - \begin{bmatrix} 3 & 6 \\ 2 & 4 \\ 3 & 8 \end{bmatrix} = \begin{bmatrix} -1 & -1 \\ 1 & 3 \\ 3 & -6 \end{bmatrix}$$

9.3 MATRIX MULTIPLICATION

In the analysis of the two-person zero-sum game, we will need to be able to multiply together matrices. Matrix multiplication is not as obvious as matrix addition, but if you follow along carefully, you will soon be able to multiply matrices easily. I'll begin with vector multiplication.

To multiply a row vector ($1 \times n$ matrix) by a column vector ($n \times 1$ matrix), we compute the sum of the products of corresponding entries. Here are some examples.

$$[2 \quad 3 \quad 4] \cdot \begin{bmatrix} 6 \\ 5 \\ 7 \end{bmatrix} = 2 \cdot 6 + 3 \cdot 5 + 4 \cdot 7 = 55$$

$$[1 \quad 2 \quad 3 \quad 4] \cdot \begin{bmatrix} 5 \\ 7 \\ -4 \\ 8 \end{bmatrix} = 1 \cdot 5 + 2 \cdot 7 + 3 \cdot (-4) + 4 \cdot 8 = 39$$

Notice that the two vectors *must have the same number of entries* and the *row vector is always on the left.* The following two multiplications can't be done:

$$[2 \quad 3 \quad 4] \cdot \begin{bmatrix} 5 \\ 6 \\ 7 \\ 8 \end{bmatrix}$$ (Number of entries are not the same.)

$$\begin{bmatrix} 4 \\ 3 \\ 2 \end{bmatrix} \cdot [5 \quad 7 \quad 8]$$ (Row vector is not on the left.)

PRACTICE

Multiply or explain why you can't.

1. $[2 \quad 3 \quad 4] \cdot \begin{bmatrix} 3 \\ 1 \\ 6 \end{bmatrix}$

2. $[1 \quad 3 \quad -6 \quad 8] \cdot \begin{bmatrix} 7 \\ 3 \\ -8 \\ 0 \end{bmatrix}$

3. $[1 \quad 7 \quad 6 \quad 6] \cdot \begin{bmatrix} 4 \\ 5 \\ 2 \end{bmatrix}$

4. $\begin{bmatrix} 1 \\ 2 \\ 3 \end{bmatrix} \cdot [3 \quad 5 \quad 7]$

When we multiply two matrices that are not vectors, we perform several multiplications treating the rows and columns of the matrices as vectors. In short, we multiply the rows of the matrix on the left by the columns of the matrix on the right. Let's go slowly through an example.

We will multiply

$$\begin{bmatrix} 2 & 3 & 4 \\ 5 & 6 & 7 \end{bmatrix} \quad \text{by} \quad \begin{bmatrix} 8 & 9 \\ 1 & 0 \\ 10 & 11 \end{bmatrix}$$

Start by multiplying the first row of the left matrix by the first column of the right matrix, as if the row and column were vectors.

$$[2 \quad 3 \quad 4] \cdot \begin{bmatrix} 8 \\ 1 \\ 10 \end{bmatrix} = 2 \cdot 8 + 3 \cdot 1 + 4 \cdot 10 = 59$$

This result, 59, will be the entry in the first row and first column of the answer. Notice that the *first* row on the left times the *first* column on the right gives the entry in the *first* row and *first* column of the answer.

Next multiply the first row of the left matrix by the *second* column of the right matrix, as if the row and column were vectors.

$$[2 \quad 3 \quad 4] \cdot \begin{bmatrix} 9 \\ 0 \\ 11 \end{bmatrix} = 2 \cdot 9 + 3 \cdot 0 + 4 \cdot 11 = 62$$

This result, 62, will be the entry in the first row and second column of the answer. Notice that the *first* row on the left times the *second* column on the right gives the entry in the *first* row and *second* column of the answer.

The answer is now starting to take shape. We have the entries for its first row in the first and second columns:

$$\begin{bmatrix} 59 & 62 \\ & \end{bmatrix}$$

We have multiplied the first row of the left matrix by all the columns of the right matrix, so now we move to the second row.

Multiply the second row of the left matrix by the first column of the right matrix, as if the row and column were vectors.

$$[5 \quad 6 \quad 7] \cdot \begin{bmatrix} 8 \\ 1 \\ 10 \end{bmatrix} = 5 \cdot 8 + 6 \cdot 1 + 7 \cdot 10 = 116$$

This result, 116, will be the entry in the second row and first column of the answer. Notice that the *second* row on the left times the *first* column on the right gives the entry in the *second* row and *first* column of the answer.

The answer now looks like this.

$$\begin{bmatrix} 59 & 62 \\ 116 & \end{bmatrix}$$

Now multiply the second row of the left matrix by the second column of the right matrix, as if the row and column were vectors.

$$[5 \quad 6 \quad 7] \cdot \begin{bmatrix} 9 \\ 0 \\ 11 \end{bmatrix} = 5 \cdot 9 + 6 \cdot 0 + 7 \cdot 11 = 122$$

This result, 122, will be the entry in the second row and second column of the answer. Notice that the *second* row on the left times the *second* column on the right gives the entry in the *second* row and *second* column of the answer.

The answer now looks like this.

$$\begin{bmatrix} 59 & 62 \\ 116 & 122 \end{bmatrix}$$

The process is complete when we have multiplied every row in the left matrix by every column in the right matrix. We have done this in our example, so the final answer is

$$\begin{bmatrix} 2 & 3 & 4 \\ 5 & 6 & 7 \end{bmatrix} \cdot \begin{bmatrix} 8 & 9 \\ 1 & 0 \\ 10 & 11 \end{bmatrix} = \begin{bmatrix} 59 & 62 \\ 116 & 122 \end{bmatrix}$$

Here is another example without so much commentary. Make sure that you understand where *every* number in the calculation came from!

$$\begin{bmatrix} 2 & 3 & 4 & 5 \\ 1 & 2 & 3 & 6 \\ 2 & 0 & 4 & 6 \end{bmatrix} \cdot \begin{bmatrix} 3 & 4 \\ 5 & 2 \\ 1 & 0 \\ 4 & 1 \end{bmatrix} = \begin{bmatrix} 2 \cdot 3 + 3 \cdot 5 + 4 \cdot 1 + 5 \cdot 4 & 2 \cdot 4 + 3 \cdot 2 + 4 \cdot 0 + 5 \cdot 1 \\ 1 \cdot 3 + 2 \cdot 5 + 3 \cdot 1 + 6 \cdot 4 & 1 \cdot 4 + 2 \cdot 2 + 3 \cdot 0 + 6 \cdot 1 \\ 2 \cdot 3 + 0 \cdot 5 + 4 \cdot 1 + 6 \cdot 4 & 2 \cdot 4 + 0 \cdot 2 + 4 \cdot 0 + 6 \cdot 1 \end{bmatrix}$$

$$= \begin{bmatrix} 45 & 19 \\ 40 & 14 \\ 34 & 14 \end{bmatrix}$$

PRACTICE

5. Complete the multiplication.

$$\begin{bmatrix} 2 & 4 & 5 \\ 1 & 0 & 3 \end{bmatrix} \cdot \begin{bmatrix} 1 & 2 \\ -3 & 5 \\ 0 & 9 \end{bmatrix}$$

$$= \begin{bmatrix} 2 \cdot 1 + 4 \cdot (-3) + 5 \cdot 0 & 2 \cdot \underline{\qquad} + 4 \cdot \underline{\qquad} + \underline{\qquad} \cdot 9 \\ 1 \cdot 1 + 0 \cdot \underline{\qquad} + 3 \cdot \underline{\qquad} & \underline{\qquad\qquad\qquad\qquad} \end{bmatrix}$$

Multiply each of the following:

6. $\begin{bmatrix} 2 & 5 \\ 3 & 1 \\ 3 & 6 \end{bmatrix} \cdot \begin{bmatrix} 1 & 3 \\ 2 & 0 \end{bmatrix}$

7. $\begin{bmatrix} -1 & 34 \\ 3 & 5 \end{bmatrix} \cdot \begin{bmatrix} 8 & -9 & 4 \\ 2 & 5 & 0 \end{bmatrix}$

8. $[2 \quad 4] \cdot \begin{bmatrix} 3 & 5 & 8 & 7 \\ 2 & 5 & 4 & 8 \end{bmatrix}$

Because we use the rows and columns of matrices as vectors when we multiply together the matrices, it is necessary that the number of entries in each row in the left matrix be the same as the number of entries in each column in the right matrix. If this condition is not met, then the matrices cannot be multiplied together.

PRACTICE

Explain why each of the following multiplications is impossible:

9. $\begin{bmatrix} 1 & 0 & 5 \\ 2 & 3 & 4 \end{bmatrix} \cdot \begin{bmatrix} 3 & 5 \\ 2 & 6 \end{bmatrix}$

10. $\begin{bmatrix} 1 & 2 \\ 2 & 3 \\ 5 & 7 \end{bmatrix} \cdot \begin{bmatrix} 1 & 5 & 6 \\ 0 & 2 & 3 \\ 9 & 5 & 6 \end{bmatrix}$

11. What are the orders of the matrices being multiplied together in Practice Exercises 5–8? What is the order of the answer in each case? Can you see a pattern?

Let A be the matrix

$$\begin{bmatrix} 2 & 3 \\ 4 & 5 \end{bmatrix}$$

and B be the matrix

$$\begin{bmatrix} 1 & 4 \\ 3 & 0 \end{bmatrix}$$

Compute AB.

$$\begin{bmatrix} 2 & 3 \\ 4 & 5 \end{bmatrix} \cdot \begin{bmatrix} 1 & 4 \\ 3 & 0 \end{bmatrix} =$$

Now compute BA.

$$\begin{bmatrix} 1 & 4 \\ 4 & 5 \end{bmatrix} \cdot \begin{bmatrix} 2 & 3 \\ 3 & 0 \end{bmatrix} =$$

Are the two products equal? _____

You may have been surprised that, for matrices, AB is not necessarily equal to BA. This is an important result to remember. You cannot change the order of multiplication for matrices!

We now have all the tools necessary to analyze the election campaign game in Situation 28, but there is another kind of multiplication involving matrices that you may encounter. It is called **scalar multiplication.**

A **scalar** is a number. So, scalar multiplication is the multiplication of a matrix by a number. To do this, we simply multiply each entry in the matrix by the scalar. Here are three examples.

$$3 \cdot [7 \quad 8 \quad 9] = [21 \quad 24 \quad 27]$$

$$-5 \cdot \begin{bmatrix} 1 & 2 & 3 \\ 3 & -4 & 6 \\ 1 & 2 & 0 \end{bmatrix} = \begin{bmatrix} -5 & -10 & -15 \\ -15 & 20 & -30 \\ -5 & -10 & 0 \end{bmatrix}$$

$$0.7 \cdot \begin{bmatrix} 2 & 3 \\ 1 & 4 \end{bmatrix} = \begin{bmatrix} 0.14 & 0.21 \\ 0.7 & 0.28 \end{bmatrix}$$

PRACTICE

Perform the scalar multiplications.

12. $3 \cdot [3 \quad 5 \quad 7 \quad 9 \quad 10]$

13. $-5 \cdot \begin{bmatrix} 2 & 3 & 4 \\ 1 & 0 & -6 \end{bmatrix}$

Now onto the analysis of the campaign game of Situation 28.

9.4 TWO-PERSON ZERO-SUM GAMES

A **two-person game** is any conflict or competition between two people. A two-person game can be a board game like checkers or chess, a political campaign like that described in Situation 28, a contract negotiation, and so on. The two people involved in the game are called the **players**.

If a gain or advantage for one player results in an exactly equal loss or disadvantage to the other player, then the game is called **zero-sum**.

The playground game of **odds and evens** is an example of a two-person zero-sum game. In this game, each player simultaneously puts out one or two fingers. If the total of both players' fingers is an odd number, then one player wins. If the total is even, then the other player wins. Suppose that the winning player collects $1 from the loser. We can show the gain or loss for one of the players in a **payoff matrix**.

		Player *B* (odds)	
		1 Finger	2 Fingers
Player *A* (evens)	1 Finger	+1	−1
	2 Fingers	−1	+1

In this case, the entries in the matrix represent the payoff from player *A*'s point of view. Since player *A* wins on *evens*, she will gain a dollar when she puts out one finger and player *B* puts out one finger. The dollar gain is represented by the entry in the first row and first column of the payoff matrix.

The labels on the rows and columns represent the **moves** of players *A* and *B*, respectively. Thus, the first row gives the payoffs for player *A* when she takes the *1 finger* move.

This game is a zero-sum game because whenever player *A* gains a dollar, player *B* loses a dollar, and vice versa. The payoff matrix from player *B*'s point of view would be

		Player *B* (odds)	
		1 Finger	2 Fingers
Player *A* (evens)	1 Finger	−1	+1
	2 Fingers	+1	−1

To avoid confusion, all the payoff matrices in this book will be from the point of view of the player who has the row moves.

PRACTICE

Here is the payoff matrix for a two-person zero-sum game.

$$\begin{array}{ccc} & & \text{Player } B \\ & & \begin{array}{ccc} H & L & M \end{array} \\ \text{Player } A & \begin{array}{c} H \\ L \\ M \end{array} & \begin{bmatrix} 2 & -3 & 5 \\ 1 & 0 & -4 \\ 4 & 2 & -3 \end{bmatrix} \end{array}$$

1. If player A makes the move H and player B the move L, what is the gain or loss for player A? For player B?

2. If player A makes the move M and player B the move H, what is the gain or loss for player A? For player B?

3. Suppose that player A somehow knows that player B is going to make move M. What move should player A make?

Here is the payoff matrix of another two-person zero-sum game.

$$\begin{array}{ccc} & & \text{Player } B \\ & & \begin{array}{cc} \text{Pass} & \text{Play} \end{array} \\ \text{Player } A & \begin{array}{c} \text{Pass} \\ \text{Play} \end{array} & \begin{bmatrix} 4 & 5 \\ 3 & -3 \end{bmatrix} \end{array}$$

Look at the payoff matrix closely. Pretend that you are player A. If you have no idea what player B is going to do, should you *pass* or *play*?

____

____

____

Right! You should definitely pass because with that play you will gain no matter what player B does. We say that playing pass is player A's **best strategy**. Now pretend that you are Player B and you know that player A is an intelligent game player. What should you do?

____

Right again! Player B must also pass because he knows that player A will pass and he wants to minimize player A's inevitable gain. So, playing pass is player B's best strategy.

If both players play their best strategy, then player A will gain four. This gain is called the **value** of the game. If the value of a game is positive, then the game favors player A (row player). If the value of a game is negative, then the game favors player B (column player). If the value of a game is zero, then the game is **fair**.

In some games, the value of the game can easily be read from the payoff matrix. These games are called **strictly determined games.** Here is another strictly determined two-person zero-sum game's payoff matrix. See if you can find the value of the game and the best strategies for each player.

	Stop	Go
Stop	-2	-3
Go	6	-8

Right! The value of the game is -3. Player A's best strategy is to *stop*, whereas player B's is to *go*. Player A knows that player B will always go because player B cannot lose with that choice. As a result, player A must stop because that gives her the smallest loss (3).

Here are some more payoff matrices for strictly determined games. The value of the game is circled in each case.

Game I $\begin{bmatrix} 2 & -4 \\ -2 & \circled{-3} \end{bmatrix}$ Game II $\begin{bmatrix} 1 & -1 \\ \circled{2} & 8 \end{bmatrix}$ Game III $\begin{bmatrix} \circled{-3} & 6 \\ -7 & 3 \end{bmatrix}$

Can you make any general statement about the location of the value of the game in a strictly determined game's payoff matrix?

Did you notice that the value of the game in a strictly determined game is always the smallest entry in its row and the largest entry in its column? Player A will pick a row in which the smallest entry is larger than the smallest entry in any other row. So when player B picks a column, player A will get the best value in that column. At the same time, player B will pick the column whose largest entry is smaller than the largest entry in any other column. This will guarantee that player A will get a minimum advantage.

For example, in game I, player A sees that the worst she can do in picking row 1 is -4 and the worst she can do in picking row 2 is -3. She will pick row 2. Player B sees that the worst he can do in picking column 1 is 2 and the worst he can do in picking column 2 is −3. He will pick column 2. The value of the game is then −3, which is the smallest entry in its row and the largest in its column.

Explain why 2 is the value of game II.

If no entry is the smallest in its row and the largest in its column, then the game is not strictly determined. If a game is not strictly determined, then its value is called a **saddle point**.

PRACTICE

Find, when possible, the value of the game for each of the following payoff matrices:

4. $$\begin{bmatrix} 2 & 3 & 6 \\ 1 & -3 & 10 \\ -3 & 14 & 12 \end{bmatrix}$$

5. $$\begin{bmatrix} -1 & -3 & 14 \\ 2 & -7 & 10 \\ 17 & 6 & 7 \end{bmatrix}$$

6. $$\begin{bmatrix} 1 & 2 & 0 & 3 \\ -2 & -4 & -5 & 5 \\ 2 & 3 & -1 & 8 \\ -2 & 1 & -3 & 6 \end{bmatrix}$$

7. $$\begin{bmatrix} 2 & 3 & 9 & 11 \\ 2 & 4 & -4 & 7 \\ 1 & 2 & 3 & 4 \\ 9 & -78 & 2 & 6 \end{bmatrix}$$

8. Is the campaign game of Situation 28, whose payoff matrix is repeated below, a strictly determined game? If it is, what is the value of the game?

		The Mayor	
		Environment	Experience
You	Environment	4	−2
	Experience	−1	3

9.5 MIXED STRATEGIES

If a game is not strictly determined, then each player must vary his or her moves from game to game. For example, the odds and evens players will need to extend one finger sometimes and two fingers other times. When a player changes her move from game to game, she is using a **mixed strategy**. Mixed strategies are given in terms of the probability that a move will be chosen in any given game. Let's take a look at the election campaign game in Situation 28. The payoff matrix for this game is

		The Mayor	
		Environment	Experience
You	Environment	4	−2
	Experience	−1	3

Suppose that you decide to spend 100% of your time talking about the environment and the mayor decides to spend 100% of her time talking about experience. What will be the value of the game? _____.

Obviously, a decrease of two popularity points every time you run a commercial is not the best strategy for you. So, you decide to vary your approach and spend 70% of your time talking about the environment and the remaining 30% talking about experience. The mayor decides to spend 100% of her time talking about experience.

To compute the value of the game when you use this mixed strategy, we compute an average or **expected value** of the game. If you use the mixed strategy described in the last paragraph, 70% of the time the value of the game will be −2 because 70% of the time you will take the *environment* move while the mayor takes the *experience* move. Similarly, 30% of the time the value of the game will be 3. Thus, the expected value of the game will be

$$0.70 \cdot (-2) + 0.30 \cdot 3 = -0.5$$

If you use this mixed strategy, then you will lose an *average* of 0.5 popularity points every time you run a commercial.

Suppose that your strategy is to talk about the environment 55% of the time and about experience 45% of the time. Suppose the mayor continues to spend 100% of her time on experience. In this case, the expected value of the game is

$$0.55 \cdot (-2) + 0.45 \cdot 3 = 0.25$$

This new strategy will gain you an average of 0.25 popularity points per game (per commercial). The mayor, being a savvy politician, will not allow this. She will also use a mixed strategy to counteract yours. Suppose she decides to spend 80% of her time on experience and 20% on the environment. The calculation of the expected value of the game becomes a little trickier when both players use mixed strategies.

We can view the percentage of time each player plans to spend on each topic as probabilities that certain moves will be made. We will assume that neither candidate's decision depends on the other's. For example, if you spend 55% of your time on the environment and the mayor spends 20% of her time on it, then the probability that you will both be talking about the environment at the same time is $0.55 \cdot 0.20 = 0.11$.

The payoff is 4 when you both talk about the environment, so the expected payoff will be $0.11 \cdot 4 = 0.44$.

Here are the expected payoffs for the other possible combinations of moves under the two mixed strategies:

Your strategy: 55% environment (env), 45% experience (exp)

The Mayor's strategy: 20% environment, 80% experience

You	Mayor	Probability	Payoff	Expected payoff
env	env	$0.55 \cdot 0.20 = 0.11$	4	0.44
env	exp	$0.55 \cdot 0.80 = 0.44$	−2	−0.88
exp	env	$0.45 \cdot 0.20 = 0.09$	−1	−0.09
exp	exp	$0.45 \cdot 0.80 = 0.36$	3	1.08

The total of these expected payoffs is the expected value of the game. The expected value of the game under these mixed strategies is $0.44 + -0.88 + -0.09 + 1.08 = 0.55$, a gain of over a half a popularity point per game for you.

PRACTICE

1. Find the expected value of the campaign game when you use the strategy 60% on the environment and 40% on experience and the mayor uses the strategy 65% on experience and 35% on the environment.

The calculation of the expected value of a game where both players are using mixed strategies can be simplified somewhat by using matrices.

Write the strategy of the row player as a row vector. In the campaign, the row vector

$$A = [0.55 \quad 0.45]$$

indicates that you will spend 55% of your time on the environment and 45% of your time on experience.

Similarly, the *column* vector

$$B = \begin{bmatrix} 0.20 \\ 0.80 \end{bmatrix}$$

indicates that the mayor (column player) will spend 20% of her time on the environment and 80% on experience.

If P is the payoff matrix, then the expected value of the game E is

$$E = APB$$

In this example, we have

$$\begin{aligned} E &= [0.55 \quad 0.45] \cdot \begin{bmatrix} 4 & -2 \\ -1 & 3 \end{bmatrix} \cdot \begin{bmatrix} 0.20 \\ 0.80 \end{bmatrix} \\ &= [1.75 \quad 0.25] \cdot \begin{bmatrix} 0.20 \\ 0.80 \end{bmatrix} \\ &= 0.55 \end{aligned}$$

PRACTICE

2. Use matrices to calculate the expected value of the game described in Practice Exercise 1.

Calculate the expected value of each of the following games under the given mixed strategies:

3. $A = [0.3 \quad 0.7]$, $B = \begin{bmatrix} 0.5 \\ 0.5 \end{bmatrix}$, $P = \begin{bmatrix} 5 & -3 \\ -2 & 4 \end{bmatrix}$

4.

$$A = [0.8 \quad 0.2], \qquad B = \begin{bmatrix} 0.23 \\ 0.77 \end{bmatrix}, \qquad P = \begin{bmatrix} 1 & -2 \\ -2 & 8 \end{bmatrix}$$

5.

$$A = [0.4 \quad 0.6], \qquad B = \begin{bmatrix} 0.9 \\ 0.1 \end{bmatrix}, \qquad P = \begin{bmatrix} -8 & -2 \\ 3 & 1 \end{bmatrix}$$

6. The playground game of odds and evens is an example of a two-person zero-sum game. In this game, each player simultaneously puts out one or two fingers. If the total of both players' fingers is an odd number, then one player wins. If the total is even, then the other player wins. Suppose that the winning player collects \$1 from the loser. We can show the gain or loss for one of the players in a payoff matrix.

		Player B (odds)	
		1 Finger	2 Fingers
Player A (evens)	1 Finger	+1	−1
	2 Fingers	−1	+1

Find the expected value of this game if player A plays one finger 70% of the time and player B plays two fingers 80% of the time.

The expected value of a game changes with the changes in mixed strategies. However, both players would like to find the mixed strategy that gives them the best possible expected value. These strategies are called **optimal**. We have seen the optimal strategies for a strictly determined game; now we will find them for games not strictly determined.

9.6 OPTIMAL STRATEGIES

Let's look at how we might find the optimal strategies for you and the mayor in the election campaign game of Situation 28. Here is the payoff matrix again. You are player A (rows) and the mayor is player B (columns).

$$P = \begin{bmatrix} 4 & -2 \\ -1 & 3 \end{bmatrix}$$

Suppose that your optimal strategy is to play row 1 (the environment) with probability p. Then you will play row 2 (experience) with probability $1 - p$. Your optimal strategy can be given by the row vector

$$[p \quad 1-p]$$

If the mayor decides to play column 1 (environment), then your expected payoff is

$$\begin{aligned} E_A &= 4 \cdot p + (-1) \cdot (1-p) \\ &= 4p + -1 + p \\ &= 5p - 1 \end{aligned}$$

If the mayor decides to play column 2 (experience), then your expected payoff is

$$\begin{aligned} E_A &= -2 \cdot p + 3 \cdot (1-p) \\ &= -2p + 3 - 3p \\ &= -5p + 3 \end{aligned}$$

Notice that in both cases, the expected payoff E_A is linearly related to the probability p. We can draw the graph of these two linear relationships using p as the horizontal axis and E_A as the vertical axis.

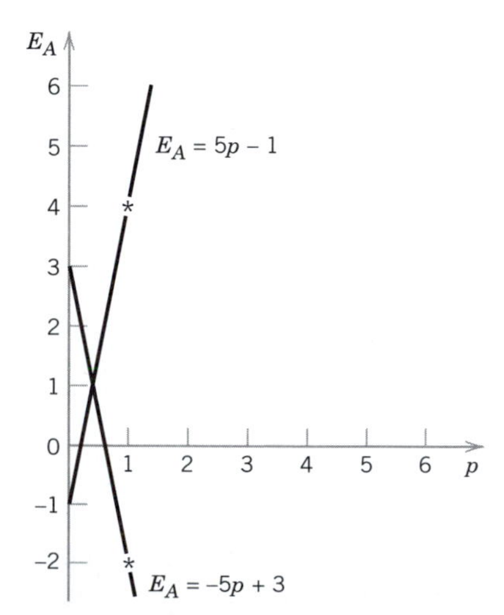

Focus your attention on the point where the two lines cross. If you change your p value in either direction from this point, one of the two lines will give you a lower E_A value than that at the crossing point. Thus, the p value at the crossing point is optimal. Any other p value will enable the mayor to decrease your expected payoff.

To find this p value, set the two E_A relationships equal to each other and solve for p.

$$-5p + 3 = 5p - 1$$
$$-5p = 5p - 4$$
$$-10p = -4$$
$$p = 0.4$$

Thus, your optimal strategy is to talk about the environment 40% of the time and experience 60% of the time. How did I get the 60% figure?

____

____

____

The row vector that represents this optimal strategy is

$$[0.4 \quad 0.6]$$

The mayor's optimal strategy is similarly calculated.

Suppose that her optimal strategy is to play column 1 (the environment) with probability q. Then she will play column 2 (experience) with probability $1 - q$. Her optimal strategy can be given by the column vector

$$\begin{bmatrix} q \\ 1 - q \end{bmatrix}$$

If you decide to play row 1 (environment), then the mayor's expected payoff is

$$E_B = 4 \cdot q + (-2) \cdot (1 - q)$$
$$= 4q + -2 + 2q$$
$$= 6q - 2$$

If you decide to play row 2 (experience), then the mayor's expected payoff is

$$E_B = -1 \cdot q + 3 \cdot (1 - q)$$
$$= -1q + 3 - 3q$$
$$= -4q + 3$$

Notice again that in both cases, the expected payoff E_B is linearly related to the probability q. We can draw the graph of these two linear relationships using q as the horizontal axis and E_B as the vertical axis.

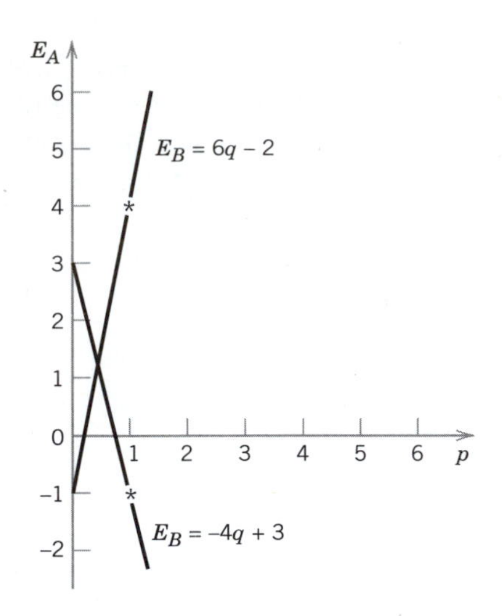

Again, the crossing point of the two lines represents the optimal strategy for the mayor because any change in the value of q from the crossing point will result in an increase in the value of E_B on one of the lines. The mayor wants to keep the E_B value as small as possible, so she will use the q value of the crossing point.

Therefore, we find the q value at the crossing point by setting the two E_B relationships equal to each other and solving for q.

$$\begin{aligned} 6q - 2 &= -4q + 3 \\ 6q &= -4q + 5 \\ 10q &= 5 \\ q &= 0.5 \end{aligned}$$

The mayor's optimal strategy is to talk about the environment (q) 50% of the time and experience ($1 - q$) 50% of the time. The column vector that represents the mayor's optimal strategy is

$$B = \begin{bmatrix} 0.50 \\ 0.50 \end{bmatrix}$$

If both you and the mayor employ your optimal strategies, then the expected value of the game is

$$E = APB = [0.4 \quad 0.6] \cdot \begin{bmatrix} 4 & -2 \\ -1 & 3 \end{bmatrix} \cdot \begin{bmatrix} 0.50 \\ 0.50 \end{bmatrix}$$

$$= [1 \quad 1] \cdot \begin{bmatrix} 0.50 \\ 0.50 \end{bmatrix} = 1$$

By employing your optimal strategy, you will gain one popularity point every time you run a commercial against the mayor. There is nothing she can do about it.

Here is another example of how to find the optimal strategies for a two-person 2×2 zero-sum game.

Fern and Zork are playing *hide and seek.* There are only two places to hide, under the bed or in the closet. Fern always hides and Zork always seeks. If the seeker (Zork) finds the hider (Fern) under the bed on the first try, then Fern must give Zork \$12. If Zork finds Fern in the closet on the first try, then she must give Zork \$14. If Zork looks under the bed and Fern isn't there, then Zork pays Fern \$15. If Zork looks in the closet and Fern isn't there, then Zork pays Fern \$10.

The payoff matrix for this game is

		Hider (Fern)	
		Under the bed	In the closet
Seeker (Zork)	Under the bed	12	−15
	In the closet	−10	14

This game is not strictly determined. Why not?

___

___

___

The expected payoff equations for Zork are

$$E_A = 12 \cdot p + -10 \cdot (1 - p) = 22p - 10$$

and

$$E_A = -15p + 14 \cdot (1 - p) = -29p + 14$$

The optimal p value is found by equating the two expected payoff equations and solving for p.

$$22p - 10 = -29p + 14$$
$$22p = -29p + 24$$
$$51p = 24$$
$$p = 24/51 \approx 0.471$$

The optimal strategy for Zork (row player) then is to look under the bed 47.1% of the time and in the closet 52.9% of the time.

Zork's optimal strategy is given in the row vector

$$A = [0.471 \quad 0.529]$$

The expected payoff equations for Fern (column player) are

$$E_B = 12 \cdot q + -15 \cdot (1 - q) = 27q - 15$$

and

$$E_B = -10q + 14 \cdot (1 - q) = -24q + 14$$

The optimal q value is found by equating the two expected payoff equations and solving for q.

$$27q - 15 = -24q + 14$$
$$27q = -24q + 29$$
$$51q = 29$$
$$q = 29/51 \approx 0.569$$

The optimal strategy for Fern then is to hide under the bed 56.9% of the time and the closet 43.1% of the time.

Fern's optimal strategy is given in the column vector

$$B = \begin{bmatrix} 0.569 \\ 0.431 \end{bmatrix}$$

The expected value of the game is

$$E = [0.471 \quad 0.529] \cdot \begin{bmatrix} 12 & -15 \\ -10 & 14 \end{bmatrix} \cdot \begin{bmatrix} 0.569 \\ 0.431 \end{bmatrix}$$

$$= [0.362 \quad 0.341] \cdot \begin{bmatrix} 0.569 \\ 0.431 \end{bmatrix} = 0.352949$$

Zork has about a 35¢ per game advantage over Fern.

We found the optimal strategy for Zork in the last example. He is supposed to look under the bed 47.1% of the time and in the closet 52.9% of the time. But, how can he be sure he is apportioning his time correctly?

One way that Zork can be sure his decisions are random and still within the guidelines of the optimal strategy is to obtain 51 marbles, 24 red and 27 blue. If he shakes the marbles up in a bag and then closes his eyes and selects one, the probability that he will get a red marble is 24/51 (.471). Thus, if he gets a red marble, he should look under the bed. If he gets a blue marble, then he should look in the closet.

Fern can use a similar technique with 29 red marbles and 22 blue marbles.

PRACTICE

1. Find the optimal strategy for the players in the odds and evens game (described in Section 9.5 Practice Exercise 6). What is the expected value of this game? Which player has the advantage?

2. Find the optimal strategy for each player in the following game. What is the expected value of the game? Which player has the advantage?

$$\begin{bmatrix} 8 & -7 \\ -2 & 1 \end{bmatrix}$$

The optimal strategies for a two-person 2×2 zero-sum game can always be found using the method described in this section, so long as the game is *not* strictly determined. The method may fail for strictly determined games!

9.7 WHAT DO YOU KNOW?

If you have worked carefully through this chapter, you are now able to add and multiply matrices, and you are able to find the optimal strategies for both players in a two-person 2×2 zero-sum game. The exercises in this section are designed to test and refine your knowledge of the mathematics of games.

In Exercises 1–4, determine the order of the matrix.

1. $[2 \quad 3 \quad 4]$

2. $\begin{bmatrix} 2 & 5 & 7 & 9 \\ 1 & 4 & 5 & 7 \\ 1 & 2 & 3 & 6 \end{bmatrix}$

3. $\begin{bmatrix} 1 & 0 & 0 & 0 & 0 \\ 0 & 1 & 0 & 0 & 0 \end{bmatrix}$

4. $\begin{bmatrix} 2 & 3 \\ 1 & 5 \\ 4 & 6 \\ 9 & 10 \\ 3 & 5 \end{bmatrix}$

5. Sam and Janet are salespeople for Zork's Used Cars. In June, Sam sold 6 cars and Janet sold 12. In July, Sam sold 12 cars and Janet 14. In August, Sam sold 7 cars and Janet 10. Write a 3×2 matrix that presents these sales figures.

In Exercises 6–16, let

$$A = [2 \quad 3 \quad 4]$$

$$X = [3 \quad 6 \quad 9]$$

$$B = \begin{bmatrix} 1 & 5 & 7 & 0 \\ 1 & 0 & 2 & 5 \end{bmatrix}, \quad C = \begin{bmatrix} 1 & -2 & 5 & 8 \\ 0 & 7 & 7 & 0 \end{bmatrix},$$

$$D = \begin{bmatrix} 1 & 2 \\ 2 & 5 \\ 3 & 0 \end{bmatrix}, \quad E = \begin{bmatrix} 5 & 6 \\ 1 & 2 \\ 0 & 0 \\ 1 & 4 \end{bmatrix}$$

Calculate each of the following or explain why the calculation is impossible:

6. $A + X$

7. $5 \cdot A$
8. $B - C$
9. $A \cdot D$
10. $D \cdot A$
11. $B \cdot D$
12. $D \cdot B$
13. $B \cdot E$
14. $2 \cdot B + C$
15. $C - E$
16. $X \cdot D + D$
17. A trucking company has three trucks that carry fruit from orchards to the market. Truck A carries 20 bushels of apples and 40 bushels of peaches. Truck B carries 25 bushels of apples and 35 bushels of peaches. Truck C carries 40 bushels of apples and 20 bushels of peaches.
(a) Write a 2×3 matrix that shows the trucks and their loads.
(b) Rewrite the matrix if truck B spilled 5 bushels of apples.
18. ZorkMart has four stores. Store I has 30 lawnmowers and 12 sets of playground equipment. Store II has 20 lawnmowers and 20 sets of playground equipment. Store III has 35 lawnmowers and 10 sets of playground equipment. Store IV has 15 lawnmowers and 30 sets of playground equipment.
(a) Write a 2×4 matrix that shows the stores and their stocks. Call this matrix S.
(b) Each lawnmower is worth $200 and each set of playground equipment $89. Write a 1×2 matrix that shows the value of each product. Call this matrix V.
(c) Multiply $V \cdot S$ and interpret the resulting matrix.

Each of the following matrices is a payoff matrix for a two-person 2×2 zero-sum game. Decide whether or not the game is strictly determined. If it is strictly determined, find the value of the game.

19. $\begin{bmatrix} 2 & 3 \\ 1 & 4 \end{bmatrix}$

20. $\begin{bmatrix} -1 & -2 \\ 8 & -3 \end{bmatrix}$

21. $\begin{bmatrix} 14 & -19 \\ 20 & 8 \end{bmatrix}$

22. $\begin{bmatrix} 2 & -4 \\ -4 & 6 \end{bmatrix}$

23. $\begin{bmatrix} -2 & 3 \\ 1 & -4 \end{bmatrix}$

24. $\begin{bmatrix} -1 & 2 \\ -8 & -3 \end{bmatrix}$

25. $\begin{bmatrix} 15 & -18 \\ 20 & -8 \end{bmatrix}$

26. $\begin{bmatrix} 12 & -40 \\ -4 & -60 \end{bmatrix}$

Find the expected value of each of the following games under the indicated mixed strategies:

27. $A = [0.45 \quad 0.55]$, $B = \begin{bmatrix} 0.90 \\ 0.10 \end{bmatrix}$, $P = \begin{bmatrix} 12 & -9 \\ -10 & 14 \end{bmatrix}$

28. $A = [0.25 \quad 0.75]$, $B = \begin{bmatrix} 0.80 \\ 0.20 \end{bmatrix}$, $P = \begin{bmatrix} 1 & -5 \\ -3 & -1 \end{bmatrix}$

29. $A = [1 \quad 0]$, $B = \begin{bmatrix} 0.50 \\ 0.50 \end{bmatrix}$, $P = \begin{bmatrix} 10 & -9 \\ -9 & 12 \end{bmatrix}$

30. $A = [0.33 \quad 0.67]$, $B = \begin{bmatrix} 0.45 \\ 0.55 \end{bmatrix}$, $P = \begin{bmatrix} 1 & 5 \\ -10 & 1 \end{bmatrix}$

Find the optimal strategies and the expected value for the games whose payoff matrices are given.

31. $\begin{bmatrix} 2 & -5 \\ -1 & 2 \end{bmatrix}$

32. $\begin{bmatrix} -4 & -2 \\ 3 & -7 \end{bmatrix}$

33. $\begin{bmatrix} -2 & 6 \\ 6 & -2 \end{bmatrix}$

34. $\begin{bmatrix} 1 & 0 \\ -2 & 3 \end{bmatrix}$

35. The optimal strategies for a two-person 2×2 zero-sum game can be obtained by formulas. These formulas are valid but, I think, rather hard to remember. Unless you have a great memory, you are better off using the techniques described in Section 9.6. But, for completeness' sake, here are the formulas.
Suppose the payoff matrix is

$$P = \begin{bmatrix} a & b \\ c & d \end{bmatrix}$$

Then the optimal p for the row player is

$$p = (d - c)/(a + d - b - c)$$

The optimal q for the column player is

$$q = (d - b)/(a + d - b - c)$$

Use these formulas to find the optimal strategies for the games in Exercises 31–34.

36. (Optional: requires excellent algebra skills!) Derive the formulas in Exercise 19.

37. A baseball player at bat with two strikes can be viewed as being involved in a two-person zero-sum game with the pitcher. The batter is trying to guess whether the pitcher will throw a fastball or curveball. If the pitcher throws a fastball and the batter has guessed a fastball, then the batter will hit a homerun (+10). If the pitcher throws a curveball and the batter has guessed a curveball, then the batter will get a base hit (+3). If the batter guesses wrong, then she will miss the ball (−5).

(a) Write the payoff matrix for this game with the batter as the row player.

(b) Find the optimal strategies for the batter and pitcher.

(c) Find the expected value of the game. Interpret this value.

38. Fern writes either a circle or triangle on a piece of paper. She then puts the paper face down on the table. Zork tries to guess what she has written. If Zork guesses *circle* and it is a circle, then Fern must pay him \$5. If Zork guesses *circle* and it isn't a circle, then he must pay Fern \$6. If Zork guesses *triangle* and it is a triangle, then Fern must pay him \$4. If Zork guesses *triangle* and it isn't a triangle, then he must pay Fern \$3.

(a) Find the optimal strategies for Fern and Zork.

(b) Find the expected value of this game.

(c) Suggest how Fern might carry out her strategy.

ANSWERS TO PRACTICE EXERCISES

Section 9.1

1. 1×4

2. 4×2

3. 5×2

4. Inventory Matrix:

	Ducks	Cows	Horses
North pasture	140	200	30
South pasture	200	153	20

Section 9.2

1. $\begin{bmatrix} 3 & 12 & 13 & 14 \\ 9 & 8 & 11 & 10 \end{bmatrix}$

2. $[7 \quad 12 \quad 4 \quad 13]$

3. $\begin{bmatrix} 6 & 8 \\ 3 & -5 \\ 5 & 12 \end{bmatrix}$

4. $S = \begin{bmatrix} -60 & -16 & -4 & 0 \\ -160 & -70 & 0 & 0 \\ -55 & -10 & -8 & -4 \end{bmatrix}$

5. $\begin{bmatrix} 400 & 200 & 60 & 50 \\ 500 & 300 & 40 & 0 \\ 300 & 200 & 100 & 30 \end{bmatrix} + \begin{bmatrix} -60 & -16 & -4 & 0 \\ -160 & -70 & 0 & 0 \\ -55 & -10 & -8 & -4 \end{bmatrix}$

$= \begin{bmatrix} 340 & 184 & 56 & 50 \\ 340 & 230 & 40 & 0 \\ 245 & 190 & 92 & 26 \end{bmatrix}$

Section 9.3

1. $2 \cdot 3 + 3 \cdot 1 + 4 \cdot 6 = 33$

2. $1 \cdot 7 + 3 \cdot 3 + (-6) \cdot (-8) + 8 \cdot 0 = 64$

3. Can't be done because the vectors do not have the same number of entries.

4. Can't be done because the row vector is not on the left.

5. $\begin{bmatrix} 2 \cdot 1 + 4 \cdot (-3) + 5 \cdot 0 & 2 \cdot 2 + 4 \cdot 5 + 5 \cdot 9 \\ 1 \cdot 1 + 0 \cdot (-3) + 3 \cdot 0 & 1 \cdot 2 + 0 \cdot 5 + 3 \cdot 9 \end{bmatrix}$

$= \begin{bmatrix} -10 & 69 \\ 1 & 29 \end{bmatrix}$

6. $\begin{bmatrix} 12 & 6 \\ 5 & 9 \\ 15 & 9 \end{bmatrix}$

7. $\begin{bmatrix} 60 & 179 & -4 \\ 34 & -2 & 12 \end{bmatrix}$

8. $[14 \quad 30 \quad 32 \quad 46]$

9. Each row in the left matrix has three entries, but each column in the right matrix has only two.

10. Each row in the left matrix has two entries, but each column in the right matrix has three.

11.

	Order		
Exercise	Left matrix	Right matrix	Answer
5	2×3	3×2	2×2
6	3×2	2×2	3×2
7	2×2	2×3	2×3
8	1×2	2×4	1×4

If the left matrix has order $p \times q$ and the right matrix order $q \times r$, then the answer has order $p \times r$. Also note that if the left matrix has order $p \times q$, then the right matrix *must* have order $q \times$ some integer.

12. $[9 \quad 15 \quad 21 \quad 27 \quad 30]$

13. $\begin{bmatrix} -10 & -15 & -20 \\ -5 & 0 & 30 \end{bmatrix}$

Section 9.4

1. Player A loses 3. Player B gains 3.
2. Player A gains 4. Player B loses 4.
3. Player A should make move H because that will give her a gain of 5. All other moves will give her a loss.
4. 2
5. 6
6. 0, this is a fair game
7. This game is not strictly determined.
8. The campaign game is not strictly determined.

Section 9.5

1. Your strategy: 60% environment (env), 40% experience (exp)

The mayor's strategy: 35% environment, 65% experience

You	Mayor	Probability	Payoff	Expected payoff
env	env	$0.60 \cdot 0.35 = 0.21$	4	0.84
env	exp	$0.60 \cdot 0.65 = 0.39$	−2	−0.78
exp	env	$0.40 \cdot 0.35 = 0.14$	−1	−0.14
exp	exp	$0.40 \cdot 0.65 = 0.26$	3	0.78

Expected value of the game is $0.84 + -0.78 + -0.14 + 0.78 = 0.70$.

2.
$$E = [0.60 \quad 0.40] \cdot \begin{bmatrix} 4 & -2 \\ -1 & 3 \end{bmatrix} \cdot \begin{bmatrix} 0.35 \\ 0.65 \end{bmatrix}$$
$$= [2 \quad 0] \cdot \begin{bmatrix} 0.35 \\ 0.65 \end{bmatrix}$$
$$= 0.70$$

3. 1.00
4. 0.092
5. −1.28
6.
$$E = [0.70 \quad 0.30] \cdot \begin{bmatrix} 1 & -1 \\ -1 & 1 \end{bmatrix} \cdot \begin{bmatrix} 0.20 \\ 0.80 \end{bmatrix}$$
$$= [0.4 \quad -0.4] \cdot \begin{bmatrix} 0.20 \\ 0.80 \end{bmatrix}$$
$$= -0.24$$

Section 9.6

1. Equations for player A:

$$E_A = 1 \cdot p + -1(1 - p) = 2p - 1$$
$$= -1 \cdot p + 1(1 - p) = -2p + 1$$

Optimal p for player A:

$$2p - 1 = -2p + 1$$
$$4p - 1 = 1$$
$$4p = 2$$
$$p = 0.50$$

Optimal strategy for player A

$$A = [0.50 \quad 0.50]$$

Equations for player B:

$$E_B = 1 \cdot q + -1(1 - q) = 2q - 1$$
$$= -1 \cdot q + 1(1 - q) = -2q + 1$$

Optimal q for player B:

$$2q - 1 = -2q + 1$$
$$4q - 1 = 1$$
$$4q = 2$$
$$q = 0.50$$

Optimal strategy for player B:

$$B = \begin{bmatrix} 0.50 \\ 0.50 \end{bmatrix}$$

Expected value of the game:

$$[0.50 \quad 0.50] \cdot \begin{bmatrix} 1 & -1 \\ -1 & 1 \end{bmatrix} \cdot \begin{bmatrix} 0.50 \\ 0.50 \end{bmatrix} = 0$$

Neither player has an advantage.

2. Equations for player A:

$$E_A = 8 \cdot p + -2(1 - p) = 10p - 2$$
$$= -7 \cdot p + 1(1 - p) = -8p + 1$$

Optimal p for player A:

$$10p - 2 = -8p + 1$$
$$18p - 2 = 1$$
$$18p = 3$$
$$p = 0.167$$

Optimal strategy for player A:

$$A = [0.167 \quad 0.833]$$

Equations for player B:

$$E_B = 8 \cdot q + -7(1 - q) = 15q - 7$$
$$= -2 \cdot q + 1(1 - q) = -3q + 1$$

Optimal q for player B:

$$15q - 7 = -3q + 1$$
$$18q - 7 = 1$$
$$18q = 8$$
$$q = 0.444$$

Optimal strategy for player B:

$$B = \begin{bmatrix} 0.444 \\ 0.556 \end{bmatrix}$$

Expected value of the game:

$$[0.167 \quad 0.833] \cdot \begin{bmatrix} 8 & -7 \\ -2 & 1 \end{bmatrix} \cdot \begin{bmatrix} 0.444 \\ 0.556 \end{bmatrix} = -0.333$$

Player B has an advantage.

Chapter Ten

Mathematics of Other Cultures and Other Times

What's the Point? This chapter is about the variety of counting systems that humans have used in the past or are now using. From a practical point of view, you may never need to read Chinese numerals, decipher Mayan hieroglyphic writing, or count in the Delaware language. Nevertheless, I feel that it is important to your understanding of mathematics and of the diversity of human culture to be aware of the alternatives to English numeration and the English counting system. This chapter will take you on a brief tour of the mathematics of other cultures and other times.

Situation 29 What Are All Those Funny Squiggles?

What do all the following have in common?

1. ||||| ∩∩ 𓍢𓍢𓍢𓍢𓍢𓍢 𓆼𓆼𓆼

2. ▾ ▴▴ ⟙⟙ << ▾▾▾▾▾

3. ͵γχκε

4. 三千六百二十五

5. ═ · ─

If you said that they all represent the number 3625, you were right! You have a great knowledge of numeration systems. If you didn't know that they all represented the number 3625, then the next few sections will show you why.

We'll begin with a review of our own numeration system.

10.1 BASE 10 NUMERATION

The numeration system that is used worldwide for business and science is called a **base 10 positional numeration system**. A **numeration system** is a systematic way of writing numerals. A numeration system is **positional** if the position of a symbol in a numeral is meaningful.

Our system is positional because 345 and 543 represent distinct numbers. The position of the 3 has different meanings in 345 and 543.

In a base 10 numeration system, each position represents a power of 10. For example, the 3 represents 300 in 345, 30 in 35, and 3 in 3. We can write 345 in an expanded form that illustrates the meaning of the positions.

$$\begin{aligned} 345 &= 3 \cdot 100 + 4 \cdot 10 + 5 \cdot 1 \\ &= 3 \cdot 10^2 + 4 \cdot 10^1 + 5 \cdot 10^0 \end{aligned}$$

Here are some more expanded base 10 numerals.

$$1234 = 1 \cdot 10^3 + 2 \cdot 10^2 + 3 \cdot 10^1 + 4 \cdot 10^0$$

$$3625 = 3 \cdot 10^3 + 6 \cdot 10^2 + 2 \cdot 10^1 + 5 \cdot 10^0$$

$$\begin{aligned} 10{,}190{,}456 = {} & 1 \cdot 10^7 + 0 \cdot 10^6 + 1 \cdot 10^5 + 9 \cdot 10^4 + 0 \cdot 10^3 \\ & + 4 \cdot 10^2 + 5 \cdot 10^1 + 6 \cdot 10^0 \end{aligned}$$

Notice that the 0 has an important role in a positional system. In the numeral 305, the 0 keeps the 3 in the place for 100. Without the 0, 35 could mean 35, 305, 3005, 30,000,005, 350, and so on.

PRACTICE

Write each of the following in expanded form:

1. 234

2. 14,567

3. 678,000,000

Not all numeration systems that humans have invented are positional, nor are they all base 10.

10.2 EGYPTIAN NUMERATION

The ancient Egyptian civilization had developed a nonpositional base 10 numeration system by about 3500 BC. The Egyptians used special pictographs for each power of 10. These pictographs are shown in the table below.

EGYPTIAN NUMERALS

Pictograph	Meaning	Number Represented
I	A staff	1
∩	A heel bone	10
𓍢	Curved rope	100
𓆼	Lotus flower	1000
𓂭	Bent finger	10,000
𓆐	Tadpole	100,000
𓁨	Astonished man	1,000,000
𓇳	Rising sun(?)	10,000,000

The Egyptian numeration system was **additive**. To express a number like 14, for example, the Egyptians simply wrote IIII∩ , repeating a symbol as many times as was necessary. Similarly, 24 would be written IIII∩∩ , and 230 would be written

∩∩∩ 𓍢𓍢

Because the system was additive, the order of writing the symbols really didn't matter. II∩ was the same as ∩II. However, the Egyptians were generally systematic and wrote the numerals from right to left with the largest units first. Thus, they would write 2,345,608 like this:

PRACTICE

Write each of the following using the ancient Egyptian numeration system:

1. 1492

2. 3625

3. 90,003

4. 23,452,721

10.3 GREEK NUMERATION

Another additive numeration system was that of the Ionic Greeks (c. 400 BC). The Ionic system used a greater number of symbols than the Egyptian system and consequently required more memorization. However, the use of more symbols did make writing some numerals a little easier. For example, the number 90 was written ∩∩∩∩∩∩∩∩∩ by the Egyptians but ϙ by the Ionic Greeks.

Here are the symbols of the Ionic Greek numeration system. The symbols used are the 24 letters of the Greek alphabet and three symbols borrowed from Phoenician (6, 90, and 900).

1	α	10	ι	100	ρ
2	β	20	κ	200	σ
3	γ	30	λ	300	τ
4	δ	40	μ	400	υ
5	ε	50	ν	500	ϕ
6	ϛ	60	ξ	600	χ
7	ζ	70	o	700	ψ
8	η	80	π	800	ω
9	θ	90	ϙ	900	ϡ

To name a number using the Ionic numeration system, select the appropriate symbols whose values add up to the number. The numerals were written left to right with the larger values first. For example,

$$84 = 80 + 4 = \pi\delta$$

and

$$245 = 200 + 40 + 5 = \sigma\mu\varepsilon$$

PRACTICE

Write each of the following using the Ionic Greek numeration system:

1. 888

2. 504

3. 279

For numbers larger than 999, the Greeks used special multiplier symbols. The accent mark (͵) placed to the left and below a symbol multiplied that symbol by 1000. Thus, ͵ε means 1000 · 5 or 5000. The letter M placed after the symbols for a number from 1 to 9999 multiplied the number by 10,000. Thus,

$$\beta\text{M} = 2 \cdot 10{,}000 = 20{,}000$$

and

$$\phi\delta\mathrm{M} = 504 \cdot 10{,}000 = 5{,}040{,}000$$

and

$$\phi\delta\mathrm{M}{,}\varepsilon = 5{,}040{,}000 + 5000 = 5{,}045{,}000$$

PRACTICE

Write each of the following using the Ionic Greek numeration system:

4. 8,880,000

5. 1,234,509

6. 3625

7. 146,298,789

10.4 ROMAN NUMERATION

You have probably seen Roman numerals on clock faces, building cornerstones, or at the end of movie credits. The Roman numeration system (50 BC) is an additive system with some subtractive and multiplicative aspects. The basic symbols of Roman numeration are:

I	V	X	L	C	D	M
1	5	10	50	100	500	1000

The numerals are written left to right with the highest values first. Thus, 1726 is written as 1000 + 500 + 100 + 100 + 10 + 10 + 5 + 1 = MDCCXXVI, and 3625 is written 1000 + 1000 + 1000 + 500 + 100 + 10 + 10 + 5 = MMMDCXXV.

PRACTICE

Write each of the following using the Roman numeration system:

1. 156

2. 8732

3. 3127

The Roman numeration system used a subtractive principle to express the digits 4 and 9 (except for 4000). Under this subtractive principle, the letter for the lower unit was placed *before* the letter for the higher unit. For example, 40 was conceived as $50 - 10$ and written XL, *10 less than 50.* Similarly, 900 was conceived as $1000 - 100$ and written CM, *100 less than 1000.*

In Roman numerals, 1994 is viewed as $1000 + (1000 - 100) + (100 - 10) + (5 - 1)$ and written MCMXCIV, and 1492 is written MCDXCII.

There was no problem with ambiguity in this system because the subtractive principle only applied at the highest possible level. That is, 99 could not be represented as IC, because it could be represented as XCIX. Only 10 could be subtracted from 50 or 100. Only 100 could be subtracted from 500 or 1000. So, 499 is CDXCIX, not ID.

PRACTICE

Write each of the following using the Roman numeration system:

4. 4498

5. 1999

6. 1944

For larger numbers, the Roman numeration system employed a multiplicative principle. The multiplicative principle said that a bar over a numeral multiplied the entire numeral by 1000, and a double bar multiplied it by 1000^2 (1,000,000). Thus, $\overline{\text{XIX}}$ was $19 \cdot 1000 = 19{,}000$, $\overline{\overline{\text{VII}}}$ was $7 \cdot 1000^2 = 7{,}000{,}000$, and $\overline{\text{XC}}\text{VI}$ was $90 \cdot 1000 + 6 = 90{,}006$.

PRACTICE

Write each of the following using the Roman numeration system:

7. 55,003

8. 23,003,674

9. 9,090,909

10.5 CHINESE NUMERATION

The traditional Chinese numeration system is a vertical multiplicative grouping system. The basic symbols of the system are symbols representing the digits 1 through 9 and symbols representing powers of 10 from 10^1 to 10^5. These symbols are:

一	二	三	四	五
1	2	3	4	5
六	七	八	九	
6	7	8	9	
十	百	千	萬	
10	100	1000	10,000	
億				
100,000				

Chinese numerals are written from the top down with the highest units first and the count symbol preceding its power of 10 symbol. For example, 2487 would be conceived as $2 \cdot 1000 + 4 \cdot 100 + 8 \cdot 10 + 7$ and written as

二 2
千 1000
四 4
百 100
八 8
十 10
七 7

If the count for a particular power of 10 is only 1, then the symbol for 1 is omitted. Thus, 1492 is conceived as $1 \cdot 1000 + 4 \cdot 100 + 9 \cdot 10 + 2$ and written

千 1000
四 4
百 100
九 9
十 10
二 2

PRACTICE

Write each of the following using the Chinese numeration system:

1. 728

2. 718

3. 708

4. 1468

5. 3625

6. 1999

In the last four sections, you have seen how additive numeration systems (Egyptian, Greek, Roman) and multiplicative grouping (Chinese) systems can be used to represent numbers systematically. The traditional Chinese system may have struck you as similar to our modern system in the way the numerals are conceived. Both our system and the Chinese system would see 237 as $2 \cdot 100 + 3 \cdot 10 + 7$. The difference lies in how the powers of 10 are expressed. The Chinese system explicitly writes the power of 10, whereas our system infers the power of 10 from the position of the numeral. This difference is why our system is called *positional* and the Chinese system is called multiplicative *grouping*.

The essential requirement of a positional system is a symbol for 0 to indicate that none of a particular power of 10 is present. The 0 distinguishes 203 from 23 or 230. Notice that in the Chinese system, the 0 is not needed because the power of 10 is written.

二	2	二	2	二	2
百	100	百	100	十	10
三	3	三	3	三	3
十	10				

We got our zero symbol and positional system from the Hindus (about 500 AD), but there were at least two other cultures who had positional numeration systems and the 0. These two cultures, the Mayan and Babylonian, are the subjects of the next two sections.

10.6 MAYAN NUMERATION

The Mayan civilization (300–1000 AD) of Mexico and Guatemala used a positional numeration system that was essentially base 20. There were only three basic symbols in this system:

· 1 (dot) — 5 (bar) 0

The numerals 1 through 19 were formed using the most economical combination of bars and dots. The numerals could be written vertically or horizontally. If the numeral was written vertically, then the bars were horizontal and the dots were above the bars. If the numeral was written horizontally, then the bars were vertical and the dots were to the left of the bars. Thus,

13 or 13

The Mayan system was essentially a base 20 positional system. It begins as expected, but in the third position where we would expect 400 (20^2), we get 360 ($18 \cdot 20^1$). This irregularity in the base 20 system is believed to exist because the Maya used their numeration system almost exclusively for astronomical, calendrical, and ritual computations. The number 360 figures prominently in their calendar. The Mayan place value system had the following form.

The first (bottom or right) position had the place value 1 (20^0), the next was 20 (20^1), the next was 360 ($18 \cdot 20^1$), followed by 7200 ($18 \cdot 20^2$), 144,000 ($18 \cdot 20^3$), 2,880,000 ($18 \cdot 20^4$), and so on.

For example, the number 1492 would be conceived as $4 \cdot 360 + 2 \cdot 20 + 12$ and written

••••	$4 \cdot 360$	=	1440
••	$2 \cdot 20$	=	40
•• over two bars	$12 \cdot 1$	=	12
			1492

Here's another example.

•• over one bar	$7 \cdot 144{,}000$	=	1,008,000
••••	$4 \cdot 7200$	=	28,800
shell (zero)	$0 \cdot 360$	=	0
•• over three bars	$17 \cdot 20$	=	340
one bar	$5 \cdot 1$	=	5
			1,037,145

PRACTICE

Write the value of each of the following Mayan numerals:

1. • over three bars / shell / •

2. •• over three bars / • / shell / shell

3. •• over two bars / • / •• / • / three bars

4. • over three bars / shell / one bar / one bar / ••

Converting one of our numerals into the Mayan numeration system can be a little tricky because we are not accustomed to thinking in base 20. Here are some suggestions on how to convert one of our numerals into the equivalent Mayan numeral. I'll use 246,873 as an example.

First divide the numeral by the highest possible Mayan place value (i.e., 1, 20, 360, 7200, 144,000, 2,880,000, etc.). In this case, 246,873 is more than 144,000 but less than 2,880,000, so I'll divide 246,873 by 144,000.

$$\begin{array}{r} 1 \\ 144{,}000 \overline{)\,246{,}783} \\ \underline{144{,}000} \\ 102{,}783 \end{array}$$

So, we have 246,873 = 1 · 144,000 + 102,783.

Now divide the remainder, 102,783, by the next lowest Mayan place value, 7200.

$$\begin{array}{r} 14 \\ 7200 \overline{)\,102{,}783} \\ \underline{100{,}800} \\ 1{,}983 \end{array}$$

Then we have 246,873 = 1 · 144,000 + 14 · 7200 + 1983.

Now divide 1983 by 360.

$$\begin{array}{r} 5 \\ 360 \overline{)\,1983} \\ \underline{1800} \\ 183 \end{array}$$

This gives us 246,873 = 1 · 144,000 + 14 · 7200 + 5 · 360 + 183.

Finally, divide 183 by 20.

$$\begin{array}{r} 9 \\ 20 \overline{)\,183} \\ \underline{180} \\ 3 \end{array}$$

The complete Mayan representation is then 246,873 = 1 · 144,000 + 14 · 7200 + 5 · 360 + 9 · 20 + 3, and the Mayan numeral for 246,873 is

PRACTICE

Write each of the following as Mayan numerals:

5. 365

6. 819

7. 260

8. 3625

9. 9999

10. 362,577

11. 23,567,001

10.7 Babylonian Numeration

The Babylonian (300 BC) numeration system was a base 60 (sexagesimal) positional system. The primary symbols of the Babylonian numeration system were

▾	<	▴▴
1	10	Place holder (0)

A Babylonian numeral is written left to right with the highest place value first. Within place values the symbols for 10 precede the symbols for 1. Thus,

< <▾▾ ▴▴ ▾▾▾▾

represented $10 \cdot 60^3 + 12 \cdot 60^2 + 0 \cdot 60^1 + 4 \cdot 60^0 = 2{,}203{,}204$.

And

▾▾▾▾ ▴▴ <<<<<▾▾ <<<▾▾▾▾ <<▾▾▾▾▾▾▾

represented $4 \cdot 60^4 + 0 \cdot 60^3 + 52 \cdot 60^2 + 34 \cdot 60^1 + 27 \cdot 60^0 = 52{,}029{,}267$.

The place holder symbol was only used between symbols. Thus, the numerals

▾▾▾▾ ($4 \cdot 60^1 = 240$) and ▾▾▾▾ ($4 \cdot 60^0 = 4$)

were written identically. The context in which the numeral appeared helped to distinguish ambiguous cases like these.

PRACTICE

Write the value of each of the following Babylonian numerals:

1. << ▾▾▾ <

2. <▴▴ ▾▾▾ <▾▾

3. <<▾▾▾▾▾ <<<<<▾ ▴▴ ▾▾

You can construct a Babylonian numeral using essentially the same method used to construct a Mayan numeral. Instead of dividing your number by Mayan place values, you will divide by powers of 60. For example, to write 36,254 in Babylonian numerals, first find the highest power of 60 that divides 36,254. That will be $60^2 = 3600$.

$$\begin{array}{r} 10 \\ 3600\overline{)36{,}254} \\ \underline{36{,}000} \\ 254 \end{array}$$

Now divide 254 by 60.

$$\begin{array}{r} 4 \\ 60\overline{)254} \\ \underline{240} \\ 14 \end{array}$$

Thus, 36,254 = 10 · 3600 + 4 · 60 + 14, and its Babylonian representation is

< ∷ <<∷

PRACTICE

Write each of the following using the Babylonian numeration system:

4. 1492

5. 1999

6. 3625

7. 144,000

8. 25,278,976

10.8 COMPUTER NUMERATION

The last numeration system that we will look at is the numeration system used by digital computers. This computer numeration system is a positional base 2 system. It is called **base 2** or **binary** numeration. The only symbols for the computer numeration system are 1 and 0.

Base 2 numerals are written left to right with the highest place values first. Thus, 1001010 represents $1 \cdot 2^6 + 0 \cdot 2^5 + 0 \cdot 2^4 + 1 \cdot 2^3 + 0 \cdot 2^2 + 1 \cdot 2^1 + 0 \cdot 2^0 = 64 + 0 + 0 + 8 + 0 + 2 + 0 = 74$.

PRACTICE

Find the value of each of the following binary numerals:

1. 10001

2. 101010101

3. 1011

4. 1010

Writing the binary numeral for a given number is a simple matter of dividing by 2 and recording the remainder. Let's write 234 as a binary numeral.

Begin by dividing 234 by 2. This gives 117 with a remainder of 0. Write down the remainder.

0

Now divide 117 by 2. This gives 58 with a remainder of 1. Write down the remainder to the left of the last remainder.

10

Now divide 58 by 2. This gives 29 with a remainder of 0. Write down the remainder to the left of the last remainder.

010

Now divide 29 by 2. This gives 14 with a remainder of 1. Write down the remainder to the left of the last remainder.

1010

Now divide 14 by 2. This gives 7 with a remainder of 0. Write down the remainder to the left of the last remainder.

01010

Now divide 7 by 2. This gives 3 with a remainder of 1. Write down the remainder to the left of the last remainder.

101010

Now divide 3 by 2. This gives 1 with a remainder of 1. Write down the remainder to the left of the last remainder.

1101010

Now divide 1 by 2. This gives 0 with a remainder of 1. Write down the remainder to the left of the last remainder.

11101010

When the quotient becomes 0, stop. The binary numeral representing 234 is 11101010. Check to see that this is the correct binary numeral!

The process of writing the binary representation of a number can be summarized this way:

1. Divide the number by 2 and write down the remainder.
2. Divide the quotient from the last division by 2 and write down the remainder to the left of the previous remainders.
3. Continue Step 2 until the quotient becomes 0.

Here's another example. We will write 167 as a binary numeral. The steps can be seen in the table below.

Number to be divided by 2	Quotient	Remainder	Binary numeral
167	83	1	1
83	41	1	11
41	20	1	111
20	10	0	0111
10	5	0	00111
5	2	1	100111
2	1	0	0100111
1	0	1	10100111

Thus, the binary representation of 167 is 10100111.

PRACTICE

Write each of the following as binary numerals:

5. 35

6. 1492

7. 88

8. 100

9. 3625

Situation 30

Counting Hats in Ixtapa

In an outdoor market in Ixtapa, you overhear a Nahuatl straw hat vendor counting his inventory. "*ce, ome, yey, naui,*" he begins. Then he continues, "*chica, chicace, chicome, chicuey.*" A mariachi band passes by and drowns his ninth count, but then you can hear him again. "*Matlactli, matlactlionce, matlactlionome, matlactlionyey,*" he counts and then stops.

What are the Nahuatl words for 9 and 14?

I really didn't expect you to know Nahuatl, a Native American language spoken in Mexico, when I posed the question in Situation 30. However, if you look closely at the patterns of the Nahuatl words, you may be able to guess how to say 9 and 14 in Nahuatl.

The next section takes a close look at the Nahuatl number words and counting system. As you examine this counting system, you will gain an appreciation for the range of possibilities when people use mathematics to organize their experiences. The exercises at the end of the chapter provide more examples of the variety of counting systems in the world.

10.9 COUNTING IN NAHUATL

Let's look a little closer at the counting words used by the Ixtapa straw hat vendor. Here are the numbers from 1 to 9 (my source for Nahuatl number words is Georges Ifrah's wonderful book, *From Zero to Infinity,* New York: Viking Penguin Inc., 1985):

1	*ce*	3	*yey*	5	*chica*	7	*chicome*	9	???????
2	*ome*	4	*nauī*	6	*chicace*	8	*chicuey*		

What do the Nahuatl words for the numbers 5 through 8 have in common?

Right! The Nahuatl words for the numbers 5 through 8 all begin with *chic*. Now look at the endings of the Nahuatl words for the numbers 6, 7, and 8. Notice that the word for 6, *chicace*, ends in *ce*, the word for 1. Thus, *chicace* is actually *chica-ce* or 5 + 1. Similar combinations form 7 and 8:

7 *chicome* = *chic*(a) + *ome* = 5 + 2

8 *chicuey* = *chic*(a) + (y)*ey* = 5 + 3

The linguistic rules of the Nahuatl language account for the loss of "a" in *chicome*, and the loss of "a" and "y" and introduction of "u" in *chicuey*. We will not concern ourselves with these rules here. It is sufficient to note that Nahuatl uses an additive principle to express the numbers 6, 7, and 8. If the same principle applies to 9, what would you guess the Nahuatl word for the number 9 is?

If you said *chicnaui*, you were exactly right. The Nahuatl word for the number 9 is formed from the words for 5 and 4. Thus, 9 = 5 + 4 = *chica* + *naui* = *chicnaui*.

The additive process used by the Nahuatl shouldn't seem too unusual to you because it is common in English. For example, the English word for 14 is *fourteen*, a combination of *four* and *ten*.

Now that we've gotten the Nahuatl word for the number 9, let's look carefully at the rest of the hat vendor's counting.

10	*matlactli*	13	*matlactlionyey*
11	*matlactlionce*	14	????
12	*matlactlionome*		

Look carefully at the beginnings and endings of the Nahuatl words for 11, 12, and 13. What do you see?

Right! The Nahuatl words for 11, 12, and 13 all begin with the Nahuatl word for 10 and end with those for 1, 2, and 3, respectively. The word *on* in each case is called a **particle**. Its purpose is to join *matlactli* with the endings. We can analyze the Nahuatl words like this:

11 = 10 + 1 = *matlactli-on-ce*

12 = 10 + 2 = *matlactli-on-ome*

13 = 10 + 3 = *matlactli-on-yey*

What is the Nahuatl word for 14? ______________

If you said *matlactlionnaui*, then you were right. The Nahuatl word for the number 14 is formed from 10 and 4.

Nahuatl introduces a new word at 15, *caxtulli*. The numbers 16 through 19 are then expressed as 15 + 1, 2, 3, and 4. A new term, *cempoualli*, appears at 20. Here is a longer list of Nahuatl number words (analyzed) beginning with 15.

15 *caxtulli*
16 *caxtulli-on-ce*
17 *caxtulli-on-ome*
18 *caxtulli-on-yey*
19 *caxtulli-on-naui*
20 *cem-poualli* (1 · 20)
21 *cem-poualli-on-ce* (1 · 20 + 1)
22 *cem-poualli-on-ome* (1 · 20 + 2) . . .
30 *cem-poualli-on-matlactli* (1 · 20 + 10) . . .
37 *cem-poualli-on-caxtulli-on-ome* (1 · 20 + 15 + 2) . . .
40 *ome-poualli* (2 · 20)
50 *ome-poualli-on-matlactli* (2 · 20 + 10)

60 *yey-poualli* (3 · 20)
100 *macuil-poualli* (5 · 20) (*macuilli* is another word for 5)
200 *matlactli-poualli* (10 · 20)
300 *caxtulli-poualli* (15 · 20)
400 *cen-tzuntli* (1 · 400)
8000 *cen-xiquipilli* (1 · 8000)

Nahuatl introduces new terms at the powers of 20 (20, 400, 8000), and numbers are expressed with reference to these powers of 20. We say that the Nahuatl counting system is essentially base 20. Note, however, that it is not purely base 20 because some lower numbers like 7 and 17 are formed without reference to 20. In a pure base 20 counting system, all the words for the numbers from 1–19 would be distinct and indivisible.

Thus, Nahuatl is a base 20 system with elements of base 5, base 10, and base 15. Essentially base 20 systems are found in several other languages, including Yucatec, Cakchiquel, Mam, Igbo, Irish, Welsh, and Breton.

PRACTICE

Write each of the following in Nahuatl:

1. 34

2. 29

3. 146

4. 854

5. 6392

10.10 WHAT DO YOU KNOW?

If you have worked carefully through this chapter, you have some understanding of the variety of written numeration systems and oral counting systems that exist or have existed in the world. The following exercises are designed to test and refine your knowledge of the mathematics of other cultures.

1. Write each of the following in expanded form in our base ten system:
(a) 45,678
(b) 1901
(c) 2,000,000,000,000

2. Write each of the following using the ancient Egyptian numeration system:
(a) 2089
(b) 17,888
(c) 27,045,431
(d) 1066

3. Find the number represented by each of the following Egyptian numerals:
(a) ||| ∩∩∩∩ 𓆼𓆼𓆼𓆼

(b)

(c)

(d)

4. Write each of the following using the Ionic Greek numeration system:
(a) 208
(b) 17,000
(c) 27,045,431
(d) 1066

5. Find the number represented by each of the following Ionic Greek numerals:
(a) ψμε
(b) δMφκα
(c) πηM͵βτλδ
(d) αMM͵αα

6. Write each of the following using the Roman numeration system:
(a) 208
(b) 17,000
(c) 27,045,431
(d) 1066

7. Find the number represented by each of the following Roman numerals:
(a) CXXIV
(b) MDCCXLVI
(c) DCCCLXIV
(d) $\overline{\text{X}}$CXX

8. Write each of the following using the Chinese numeration system:
(a) 208
(b) 17,000
(c) 245,431
(d) 1066

9. Find the number represented by each of the following Chinese numerals:
(a) 二千五百三十八
(b) 五千四十六
(c) 八億六千百二
(d) 千五百二十九

10. Write each of the following using the Mayan numeration system.
(a) 208
(b) 17,000
(c) 27,045,431
(d) 1066

11. Find the number represented by each of the following Mayan numerals:
(a)
(b)
(c)
(d)

12. Write each of the following using the Babylonian numeration system:
(a) 208
(b) 17,000
(c) 27,045,431
(d) 1066

13. Find the number represented by each of the following Babylonian numerals:
(a)
(b)
(c)
(d)

14. Write each of the following as binary numerals:
(a) 208
(b) 17,000
(c) 27,045,431
(d) 1066

15. Find the number represented by each of the following binary numerals:
(a) 100101
(b) 11111
(c) 1000100010001
(d) 1010111

16. Write each of the following in Nahuatl:
(a) 208
(b) 1066
(c) 7005
(d) 3400

17. Find the number represented by each of the following Nahuatl numerals:

(a) *cen-tzuntli-on-ome-poualli-on-caxtulli*

(b) *chica-ce-tzuntli-on-caxtulli-poualli-on-matlactli-on-naui*

(c) *cen-xiquipilli-on-yey-tzuntli-cem-poualli*

(d) *naui-xiquipilli-on-tzuntli-yey-poualli-on-chicace.*

18. The Mandarin Chinese number words form a system based on 10 that is very similar to that of English. Here is a list of Chinese number words.

1	*i*	11	*shih-i*	200	*erh-pai*
2	*er*	12	*shih-er*	300	*san-pai*
3	*san*	13	*shih-san*		
4	*si*	. . .			
5	*wu*	20	*erh-shih*		
6	*liu*	21	*erh-shih-i*		
7	*chi*	22	*erh-shih-erh*		
8	*pa*	. . .			
9	*chiu*	30	*san-shih*		
10	*shih*	100	*pai*		

How would each of the following numbers be said in Chinese?

(a) 35

(b) 124

(c) 674

(d) 987

19. This is how we count in Delaware, a Native American language of North America:

1	*kwati*	7	*niisaas*
2	*niisa*	8	*xaash*
3	*nakhaa*	9	*peskungk*
4	*neewa*	10	*telan*
5	*paleenkhk*	11	*telanokkwati*
6	*kwataash*	12	*telanookniisha*

Write each of the following in Delaware:

(a) 13

(b) 14

20. Here are some of the number words of Swahili, a language spoken in East Africa.

1	*moja*	11	*kumi na moja*
2	*mbili*	12	*kumi na mbili*
3	*tatu*	20	*ishirini*
4	*nne*	21	*ishirini na moja*
5	*tano*	30	*thelathini*
6	*sita*	40	*arubaini*
7	*saba*	100	*mia moja*
8	*nane*	200	*mia mbili*
9	*tisa*		
10	*kumi*		

Write each of the following in Swahili:

(a) 14

(b) 37

(c) 136

(d) 449

21. Here are some of the counting words in Welsh.

1	*un*	11	*un ar dec*
2	*dau*	12	*dau ar dec*
3	*tri*	15	*hymthec*
4	*petwar*	16	*un ar hymthec*
5	*pimp*	20	*ugeint*
6	*chwe*	30	*dec ar ugeint*
7	*seith*	40	*de-ugeint*
8	*wyth*	50	*dec ar de-ugeint*
9	*naw*	60	*tri-ugeint*
10	*dec* (or *deg*)	100	*cant*

Write each of the following in Welsh:

(a) 13

(b) 18

(c) 23

(d) 58

(e) 84

(f) 99

(g) 105

22. Here are some of the counting words in Yucatec, a Mayan language spoken in Mexico.

1	*hun*	11	*buluc*
2	*ca*	12	*lahca*
3	*ox*	13	*oxlahun*
4	*can*	14	*canlahun*
5	*ho*	20	*hun kal*
6	*uac*	30	*lahun ca kal*
7	*uuc*	40	*ca kal*
8	*uaxac*		
9	*bolon*		
10	*lahun*		

Write each of the following numbers in Yucatec:

(a) 15

(b) 16

(c) 80

(d) 280

23. Here are some of the number words in Yoruba, a language of West Africa.

1	*okan*	15	*meedogun*
2	*meji*	16	*meridogan*
3	*meta*	17	*metadilogun*
4	*merin*	18	*mejidilogun*
5	*matun*	19	*nokandilogun*
6	*mefa*	20	*ogun*
7	*meje*	21	*mokandilogun*

8	*mejo*	25	*meedogbon*
9	*mesan*	26	*meridogbon*
10	*mewa*	30	*ogbon*
11	*mokanla*	40	*ogoji*
12	*mejila*	41	*mokanlelogoji*
13	*metala*	50	*aadota*
14	*merinla*	56	*merindilogota*
		60	*ogota*

(a) Explain how the Yoruba word for 12 is formed.

(b) Explain how the Yoruba word for 18 is formed.

(c) Explain how the Yoruba word for 56 is formed.

(d) What is the Yoruba word for 57?

(e) What is the Yoruba word for 43?

24. Write an essay in which you compare and contrast the number words for 1–10 in each of the following languages: English, German, French, Spanish, Russian, Hindi.

25. Write an essay about the significance of the numbers 260, 819, and 584 to the Maya.

26. Write a critical summary of one of the articles in the book *Native American Mathematics* edited by Michael Closs (Austin, TX: University of Texas Press, 1986).

27. Write a review of the book *Ethnomathematics* by Marcia Ascher (Pacific Grove, CA: Brooks/Cole, 1991).

28. Write an essay on the Sumerian numeration system.

29. Write an essay about the mathematical aspects of North American medicine wheels.

ANSWERS TO PRACTICE EXERCISES

Section 10.1

1. $2 \cdot 10^2 + 3 \cdot 10^1 + 4 \cdot 10^0$

2. $1 \cdot 10^4 + 4 \cdot 10^3 + 5 \cdot 10^2 + 6 \cdot 10^1 + 7 \cdot 10^0$

3. $6 \cdot 10^8 + 7 \cdot 10^7 + 8 \cdot 10^6 + 0 \cdot 10^5 + 0 \cdot 10^4 + 0 \cdot 10^3 + 0 \cdot 10^2 + 0 \cdot 10^1 + 0 \cdot 10^0$

Section 10.2

1. 𓏺𓏺 𓎆𓎆𓎆𓎆𓎆𓎆𓎆𓎆𓎆 𓍢𓍢𓍢𓍢𓍢𓍢 𓆼

2. 𓏺𓏺𓏺𓏺𓏺 𓎆𓎆 𓍢𓍢𓍢𓍢𓍢𓍢 𓆼𓆼𓆼

3. 𓏺𓏺𓏺 𓂭𓂭𓂭𓂭𓂭𓂭𓂭𓂭𓂭

4. 𓏺 𓎆𓎆 𓍢𓍢𓍢𓍢𓍢𓍢𓍢 𓆼𓆼 𓂭𓂭𓂭𓂭𓂭 𓆐𓆐𓆐𓆐 𓁨𓁨𓁨 𓁨𓁨

Section 10.3

1. ωπη

2. ϕδ

3. σοθ

4. ωπηΜ

5. ρκτΜ͵δϕθ

6. ͵γχκε

7. αΜΜ͵δχκθΜ͵ηψπθ

Section 10.4

1. CLVI

2. MMMMMMMMDCCXXXII

3. MMMCXXVII

4. MMMMCDXCVIII (1000 + 1000 + 1000 + 1000 + (500 − 100) + (100 − 10) + 5 + 1 + 1 + 1)

5. MCMXCIX

6. MCMXLIV

7. $\overline{\text{LVIII}}$

8. $\overline{\overline{\overline{\text{XXIII}}}}$MMMDCLXXIV

9. $\overline{\overline{\text{IX}}}$XCCMIX

Section 10.5

1. 七百二十八

2. 七百十八

3. 七百八

4. 千四百六十八

5. 三千六百二十五

6. 千九百九十九

Section 10.6

1. 46,801

2. 86,760

3. 1,735,955

4. 18,721,902

5. 1 · 360
0 · 20
5 · 1

6. 2 · 360
4 · 20
19 · 1

7. 13 · 20
0 · 1

8. 10 · 360
1 · 20
5 · 1

9. 1 · 7200
7 · 360
13 · 20
19 · 1

10. 2 · 144,000
10 · 7200
7 · 360
2 · 20
17 · 1

11. 8 · 2,880,000
3 · 144,000
13 · 7200
3 · 360
16 · 20
1 · 1

Section 10.7

1. 790
2. 2,160,192
3. 5,367,602
4.
5.
6.
7.
8.

Section 10.8

1. 17
2. 341
3. 11
4. 10
5. 100011
6.

Number to be divided by 2	Quotient	Remainder	Binary Numeral
1492	746	0	0
746	373	0	00
373	186	1	100
186	93	0	0100
93	46	1	10100
46	23	0	010100
23	11	1	1010100
11	5	1	11010100
5	2	1	111010100
2	1	0	0111010100
1	0	1	10111010100

Thus, the binary representation of 1492 is 10111010100.

7. 1011000
8. 1100100
9. 111000101001

Section 10.9

1. *cem-poualli-on-matlactli-on-naui*
2. *cem-poualli-on-chic-naui*
3. *macuil-poualli-on-ome-poualli-on-chica-ce*
4. *ome-tzuntli-on-ome-poualli-on-matlactli-on-naui*
5. *caxtulli-tzuntli-on-caxtulli-poualli-on-naui-poualli-on-matlactli-on-ome*

Chapter Eleven

Mathematics of Reasoning

What's the Point? If you want to be able to understand tax forms, read contracts, and follow arguments in the workplace and public forums, then you need to have a basic understanding of **logic**, the mathematics of reasoning. The activities in this chapter will help you to learn how to understand basic logical statements and arguments.

Situation 31

Following IRS Instructions

Read the following quotation from the 1991 Instructions for the 1040 Tax Form. The quote refers to the procedure for checking on your refund.

> *If you call to find out about the status of your refund and do not receive a refund date, please wait 7 days before calling back.*

Do you understand it? If not, read it again until you do.

Now look at how three people responded to the IRS instructions.

1. Kyoko called the IRS to find out about her refund. She did not receive a refund date. She called back 10 days later. Did she follow the instructions correctly? _____
Explain.

____

____

2. Sam called the IRS to find out about his refund. He did not receive a refund date. Sam called back the next day. Did he follow the instructions correctly? _____
Explain.

____

____

3. Kenyatta called the IRS to find out about his refund. They told him that he would get a refund by May 3. He called back 6 days later. Did he follow the instructions correctly? _____
Explain.

____

____

Did you say that Kyoko followed instructions, but that Sam didn't because he didn't wait 7 days before calling back? Good! But what about Kenyatta? Actually, Kenyatta did follow the instructions! Let's see why.

We'll have to back up a little and work our way up to Kenyatta's response because this IRS rule is, in fact, fairly complicated.

11.1 PROPOSITIONS AND TRUTH VALUE

The IRS instructions are intended to be unambiguous. However you respond to an instruction, it should be clear whether or not you responded correctly. A well-written contract has the same features. You should be able to decide whether or not a promise or commitment made in the contract has been kept. Our first task in logic is to distinguish sentences that can be true or false from those that are neither.

Look at these two sentences.

1. Lansing is the capital of Michigan.

2. Look out for that mosquito!

Is sentence 1 true? _____

Is sentence 2 true? _____

Sentence 1 is true. You either knew that Lansing was the capital of Michigan, or you could look it up. However, sentence 2 is not true, nor is it false. The whole idea of calling sentence 2 true or false is a strange one.

Sentences like 1, which are true or false, are called **propositions**. Sentences that are neither true nor false (like 2) are not propositions. Propositions always assert something. Sentences that are not propositions do not assert anything.

PRACTICE

Try to decide whether or not the sentences are propositions. If they are propositions, are they true or false?

1. John Adams was the president of the United States in 1992.

2. Pass the salt, please.

3. Are you ready for Freddy?

4. The earth is the third planet from the sun.

We say that the **truth value** of practice sentence 1 is *false* and the truth value of sentence 4 is *true*. Practice sentences 2 and 3 do not have a truth value.

Why isn't practice sentence 3 a proposition? Practice sentence 3 is a question and questions don't assert anything. Compare practice sentence 3 to the next two examples.

3(a). I am ready for Freddy.

3(b). I am not ready for Freddy.

Sentences 3(a) and 3(b) are propositions because they assert something that is either true or false. For example, if the speaker in 3(a) *is* ready for Freddy, then 3(a) is true.

Practice sentence 3 is not a proposition because a question doesn't assert anything.

Sometimes, you know that a sentence is a proposition even if you don't know its truth value. For example,

5. The current President of the United States can count to 40 in Norwegian.

Sentence 5 may be true or false, but it is certainly one or the other. Therefore, sentence 5 is a proposition.

Here are three more propositions.

(a) You called to find out about the status of your refund.

(b) You did not receive a refund date.

(c) You waited 7 days before calling back.

Do these look familiar? They are, in fact, the three propositions that make up the IRS instructions in Situation 31. Each one of these propositions can be either true or false. Collectively, there are eight distinct possibilities for this group of three propositions.

	Truth Value for		
Possibility	(a)	(b)	(c)
1	True	True	True
2	True	True	False
3	True	False	True
4	True	False	False
5	False	True	True
6	False	True	False
7	False	False	True
8	False	False	False

Go back and read how Kyoko, Sam, and Kenyatta responded to the IRS instructions. You should see that Kyoko's situation fits into possibility 1, Sam's into possibility 2, and Kenyatta's into possibility 4.

Now we already had decided that Kyoko followed instructions and Sam didn't. In more mathematical terms, for Kyoko, the IRS statement was true, for Sam it wasn't. The next task of logic is to show how the truth value of a complicated proposition (like the IRS's) can be calculated from the truth values of the smaller propositions (a, b, c) from which it is made.

11.2 SIMPLE AND COMPOUND PROPOSITIONS

A proposition that asserts a single fact is called a **simple proposition**. These propositions are simple.

1. Alexander Hamilton could read Chinese.

2. The earth has one moon.

3. The electronic digital computer was invented in 1219 AD.

A proposition that is not simple is called **compound**. These are compound propositions.

4. The earth has one moon *and* the earth's surface is more than 50% water.

5. Alexander Hamilton could read Chinese *or* he could read English.

6. *If* Alexander Hamilton could read Chinese, *then* the earth has one moon.

7. *If* you call to find out about the status of your refund *and* do *not* receive a refund date, *then* please wait 7 days before calling back.

You can make a compound proposition by connecting two or more propositions with any of the connecting words *and*, *or*, and *if . . . then*. These connecting words are called **logical connectives**.

Another logical connective is *not*. It can be used to assert the negation of any proposition. For example,

8(a) Thomas Fuller was a calculating prodigy.

8(b) Thomas Fuller was *not* a calculating prodigy.

PRACTICE

Underline the connectives in each of the following propositions:

1. Big Bird went to the fair and Oscar stayed home.

2. If you want your company's future protected, then contact our office.

3. A natural disaster does not have to be a business disaster.

4. If you are over 18 years of age and you can read, then you are eligible for our life and casualty plan.

5. Fred is a duck or my name is not Rumpelstiltskin.

Here is that IRS sentence again.

> *If you call to find out about the status of your refund and do not receive a refund date, please wait 7 days before calling back.*

Can you see that it is a compound proposition made up of three simple propositions? Sometimes changing the wording helps. Here is the IRS sentence with the logical connectives in capital letters and with some minor word changes.

> *IF you call to find out about the status of your refund AND you do NOT receive a refund date, THEN you wait 7 days before calling back.*

Logical connectives show up everywhere. Watch for them! To help you out, I'll underline all the logical connectives in the next paragraph.

11.3 SYMBOLS

Now we will see how the truth value of a compound proposition can be calculated from those of the simple propositions from which it is constructed. The calculation will be a little easier to do if we start using some symbols. Symbols are at the heart of mathematics and are used to make things easier, not harder. If you start to have problems with the symbols, then stop and ask yourself if you really understand what they stand for. If you don't, go back and make sure you do!

It is customary to use lowercase letters to stand for propositions. For example, if I say, "let p be the proposition 'Thomas Fuller was a calculating prodigy,' " then, until I tell you otherwise, whenever I write p, I mean, "Thomas Fuller was a calculating prodigy."

If we write the symbol $\neg$ in front of a letter representing a proposition, then we are asserting the negation of that proposition. Thus, $\neg p$ means, "Thomas Fuller was *not* a calculating prodigy."

The logical connective *and* will be represented by the symbol $\wedge$. For example, if p is "Thomas Fuller was a calculating prodigy" and q is "Thomas Fuller was a slave," then $p \wedge q$ means, "Thomas Fuller was a calculating prodigy *and* Thomas Fuller was a slave."

The logical connective *or* will be represented by the symbol $\vee$. Thus, $p \vee q$ means, "Thomas Fuller was a calculating prodigy or Thomas Fuller was a slave."

Here are some more examples! Let p be "The Governor of Illinois is a woman," and q be "Canadian geese fly south for the winter." Complete the table.

Symbolic Proposition	English Proposition
$\neg p$	The Governor of Illinois is not a woman.
$p \wedge q$	The Governor of Illinois is a woman and Canadian geese fly south for the winter.
$p \vee \neg p$	The Governor of Illinois is a woman or the Governor of Illinois is not a woman.
$\neg q \wedge p$	Canadian geese do not fly south for the winter and the Governor of Illinois is a woman.
______	The Governor of Illinois is not a woman and Canadian geese fly south for the winter.
$p \vee \neg q$	______________________

The last basic logic symbol is $\rightarrow$. We will use the $\rightarrow$ to represent the logical connective *if . . . then*. For example, if p is "Thomas Fuller was a calculating prodigy" and q is "Thomas Fuller could add large numbers in his head," then $p \rightarrow q$ is "If Thomas Fuller was a calculating prodigy, then Thomas Fuller could add large numbers in his head."

Here are some more examples! Let p be "The Governor of Illinois is a woman," and q be "Canadian geese fly south for the winter." Complete the table.

Symbolic Proposition	English Proposition
$p \rightarrow q$	If the Governor of Illinois is a woman, then Canadian geese fly south for the winter.
$q \rightarrow p$	If Canadian geese fly south for the winter, then the Governor of Illinois is a woman.
$p \rightarrow \neg p$	If the Governor of Illinois is a woman, then the Governor of Illinois is not a woman.
______	If Canadian geese do not fly south for the winter, then the Governor of Illinois is a woman.
$\neg p \rightarrow q$	______________________

In English, the conditional is not always written or stated directly as *if* ______ *then* ______. Sometimes, the word *then* is omitted. And sometimes, the word order is switched around.

Take the conditional proposition, "If I study, then I am happy," as an example. This proposition could be expressed in English as

"If I study, I am happy."

or as

"I am happy, if I study."

Equivalent variations of *if p then q* are

If p, q.

q, if p.

p only if q.

q is necessary for p.

p is sufficient for q.

PRACTICE

Write each of the following in symbols using the suggested **boldface** letters for the simple propositions:

1. If I am **h**appy, then I am **r**ich.

2. Sam is **h**appy, if he is **s**kiing.

3. Sam is **h**appy only if he is **s**kiing.

4. Having a **c**ar is necessary for **g**oing to the park.

5. **O**wning a VCR is sufficient for **w**atching movies late at night at home.

We can use parentheses to group our symbols to make more complex compound propositions. Like $(\neg p \vee q) \rightarrow q$, that is, ''If the Governor of Illinois is not a woman or Canadian geese fly south for the winter, then Canadian geese fly south for the winter.''

Did that last proposition make any sense to you? Probably not! That's OK. We are building up mathematical symbolism for propositions here and are not concerned about the meaning of the propositions. Play with the symbolism and don't worry that some of these propositions are fairly silly.

Here are some more examples. Complete the table.

Let p be ''Lansing is the capital of Michigan,'' q be ''Detroit is the largest city in Michigan,'' and r be ''Plymouth Voyagers are made in Detroit.''

Symbolic Proposition	English Proposition
$(p \vee q) \rightarrow r$	If Lansing is the capital of Michigan or Detroit is the largest city in Michigan, then Plymouth Voyagers are made in Detroit.
$q \rightarrow (p \wedge r)$	If Detroit is the largest city in Michigan, then Lansing is the capital of Michigan and Plymouth Voyagers are made in Detroit.

Symbolic Proposition	English Proposition
$(\neg p \lor p) \to q$	If Lansing is not the capital of Michigan or Lansing is the capital of Michigan, then Detroit is the largest city in Michigan.
$(q \lor \neg r) \to p$	______
______	If Plymouth Voyagers are not made in Detroit, then Lansing is the capital of Michigan or Detroit is the largest city in Michigan.

Here is one more example. Let p be "You called to find out about the status of your refund," q be "You received a refund date," and r be "You waited 7 days before calling back."

The IRS sentence

If you call to find out about the status of your refund and do not receive a refund date, please wait 7 days before calling back.

is symbolized by $(p \land \neg q) \to r$.

Before we analyze the truth value of this IRS proposition, take time to familiarize yourself with the logical notation we have covered so far.

PRACTICE

Write each of the following compound propositions in symbols. Use the letters in **bold** to represent the simple propositions. For example in Practice Exercise 6, let

p = "The price of corn is higher this month than last month"
q = "The quality of Sam's investments is better"

6. If the **p**rice of corn is higher this month than last month, then the **q**uality of Sam's investments is better.

7. **P**otatoes are vegetables or **r**aisins are not fruit.

8. If Warsaw is the capital of **P**oland, then **q**oph is the 19th letter of the Hebrew alphabet and summer is the **r**ainy season in Belize.

9. If **p**ickling carrots is illegal in Kansas and **r**insing onions is not illegal in Nebraska, then **t**rimming okra is not illegal in Colorado.

10. If **m**athematics is fun for you, then you will love playing **P**archeesi or you will not love playing **P**archeesi.

11.4 BASIC TRUTH TABLES

NEGATION

Let p be "The current president of the United States can count to 40 in Spanish." Is $\neg p$ true or false?

If you said, "It depends on whether p is true or false," then you were exactly right. If the current president of the United States can count to 40 in Spanish, then it is false to say that he can't. (If p is true, then $\neg p$ will be false.) On the other hand, if he can't count to 40 in Spanish, it is true to say that he can't. (If p is false, then $\neg p$ will be true.)

We can summarize this analysis in a **truth table**.

	p	$\neg p$
1.	T	F
2.	F	T

Line 1 of the truth table says that if p is true (T), then $\neg p$ is false (F). Line 2 says that if p is false, then $\neg p$ is true. Since p must be either true or false, the truth table shows all the possibilities for truth values of p and $\neg p$.

What is even more wonderful about the truth table is that it will be the same no matter what p represents! Thus, if p is the true proposition, "Lansing is the capital of Michigan," we can see from the table (line 1) that $\neg p$ is false.

Because the truth table is the same for any p, we say that this table is **the truth table for negation**.

CONJUNCTION

Now what about $\wedge$? It may be helpful to view its truth value from the point of view of promising. Suppose I promise that "I will pick you up at 8 o'clock and I will buy you dinner at Stoney's."

What do I have to do to keep my promise?

So, my promise is kept only when I do *both* of the things I promised to do. The truth table for $\wedge$ reflects this idea. We will say that the proposition $p \wedge q$ is true only when both p and q are true. The proposition $p \wedge q$ is called a **conjunction**, and the propositions p and q the **conjuncts** of the conjunction. A conjunction is true only when both conjuncts are true. Otherwise, it is false.

The truth table for conjunction is:

	p	q	$p \wedge q$
1.	T	T	T
2.	T	F	F
3.	F	T	F
4.	F	F	F

Notice that all the possible combinations of truth values for p and q are listed in the table.

"July 4 is a U.S. holiday and Aristotle was an Australian" is a false proposition because although July 4 is a U.S. holiday, Aristotle was not an Australian; this is an instance of line 2 of the truth table for conjunction.

"July 4 is a U.S. holiday and December 25 is a U.S. holiday" is a true proposition because both conjuncts are true; this is an instance of line 1 of the truth table for conjunction.

PRACTICE

True or false?

1. The capital of Michigan is Lansing and Napoleon was born in Kansas.

2. The moon is made of green cheese and my brother's name is Sam.

3. The capital of Michigan is Lansing and Detroit is the largest city in Michigan.

DISJUNCTION

There are two distinct ways of viewing the compound proposition $p \vee q$. For example, let p be "I had a hamburger for lunch" and q be "I had a falafel sandwich for lunch." Now consider $p \vee q$ ("I had a hamburger for lunch or I had a falafel sandwich for lunch"). Everyone will agree that this last proposition is true if I had a hamburger for lunch. Everyone would also agree that it is true if I had a falafel sandwich for lunch. And everyone would also agree that it is false if I had neither a hamburger nor a falafel sandwich for lunch.

Thus, a partial truth table for $p \vee q$ is:

	p	q	$p \vee q$
1.	T	T	?
2.	T	F	T
3.	F	T	T
4.	F	F	F

But what is the truth value of $p \vee q$ if I had *both* a hamburger and falafel sandwich for lunch? Many people would say, "False, because you said that you had one or the other for lunch." Indeed, we tend to use *or* in an exclusive fashion in everyday speech. For example, "Do your homework or no more TV," "I'll buy the Dodge or the Honda," "We're going on vacation to Colorado or Delaware."

However, in mathematics we use *or* in an inclusive fashion. Thus, $p \vee q$ will be true even when both p and q are true. This approach actually makes sense if we view the proposition as a promise. Suppose I promise that "I will buy you a hamburger or I will buy you a falafel." I will only break my promise if I fail to buy you either item. If I am generous enough to buy you both, I haven't gone back on my promise!

The proposition $p \vee q$ is called a **disjunction**. The propositions p and q are called the **disjuncts** of the disjunction. A disjunction is false only when both disjuncts are false. Otherwise, it is true.

Thus, the complete truth table for **disjunction** is

	p	q	$p \vee q$
1.	T	T	T
2.	T	F	T
3.	F	T	T
4.	F	F	F

PRACTICE

True or false?

4. The capital of Michigan is Lansing or Napoleon was born in Kansas.

5. The moon is made of green cheese or my brother's name is Sam.

6. The capital of Michigan is Lansing or Detroit is the largest city in Michigan.

CONDITIONAL

The discussion of the truth value of $p \vee q$ pointed out something very important about propositions. Propositions are either true or they are false. If a proposition is not false, then it must be true. We will make use of this important point in the construction of the truth table for $\rightarrow$, the **conditional**.

Again, it is helpful to view the proposition as a promise. Suppose that I make the following promise to you:

"If you order a hamburger for lunch, then I will pay for your lunch."

Now let's look at some possible scenarios. Let p be "You order a hamburger for lunch" and q be "I'll pay for your lunch." The promise is $p \rightarrow q$.

1. If you do order a hamburger (p is true) and I pay for your lunch (q is true), then I have kept my promise ($p \rightarrow q$ is true).

2. If you do order a hamburger (p is true) and I don't pay for your lunch (q is false), then I have broken my promise ($p \rightarrow q$ is false).

3. If you don't order a hamburger (p is false), then I am relieved of any obligation to pay for your lunch. I could pay for your lunch (q could be true). On the other hand, I could decide to not pay for your lunch (q could be false). Whatever I do, I *have not broken my promise* ($p \rightarrow q$ is true).

Notice that in scenario 3, $p \to q$ is true because it is not false.

We conclude that the only time the proposition $p \to q$ is false is when p is true and q is false.

The proposition $p \to q$ is called a **conditional**. Proposition p is called the **antecedent** of the conditional. Proposition q is called the **consequent** of the conditional. The conditional is false only when it has a true antecedent and false consequent. Otherwise, it is true.

This gives us the truth table for the **conditional**.

	p	q	$p \to q$
1.	T	T	T
2.	T	F	F
3.	F	T	T
4.	F	F	T

PRACTICE

True or false?

7. If the capital of Michigan is Lansing, then Napoleon was born in Kansas.

8. If the moon is made of green cheese, then my brother's name is Sam.

9. If the capital of Michigan is Lansing, then Detroit is the largest city in Michigan.

Complete the statements.

10. A disjunction is false only when

11. A conjunction is true only when

12. A conditional proposition is false only when

13. Complete the dialogue between you and your good friend Fern.

Fern: This math book is crazy!
You: Why?
Fern: Look at this! It says that the sentence ''If the moon is made of green cheese, then Fern is the King of England'' is true.
You: That's right. It is true.
Fern: But how can it be true when nothing in it is true?
You: Because in the truth table . . .
Fern: Don't give me that truth table mumbo jumbo. Explain it to me in words I can understand.

You:

A **tautology** is a proposition that is always true. How can that be? Is anything *always* true? Yes, there are propositions that are always true. Here is one of them:

$$p \vee \neg p$$

Draw its truth table, and you will see that it is always true.

Here are some more tautologies. You should construct their truth tables to confirm that they are always true.

$$p \rightarrow (p \vee q)$$

$$\neg(p \vee q) \rightarrow (\neg p \wedge \neg q)$$

11.5 BIGGER TRUTH TABLES

Remember the quotation from the IRS's 1991 Instructions for the 1040 Tax Form? The quote, with which we began this chapter, refers to the procedure for checking on your refund.

> *If you call to find out about the status of your refund and do not receive a refund date, please wait 7 days before calling back.*

Earlier, we symbolized this instruction as follows: Let p be "You called to find out about the status of your refund," q be "You received a refund date," and r be "You waited at least 7 days before calling back."

$$(p \wedge \neg q) \rightarrow r$$

We are now ready to analyze the truth value of this instruction from the point of view of the three taxpayers Kyoko, Sam, and Kenyatta. Here are their cases again.

1. Kyoko called the IRS to find out about her refund. She did not receive a refund date. She called back 10 days later.
2. Sam called the IRS to find out about his refund. He did not receive a refund date. Sam called back the next day.
3. Kenyatta called the IRS to find out about his refund. They told him that he would get a refund by May 3. He called back 6 days later.

Let's make a truth table for $(p \wedge \neg q) \rightarrow r$. The first thing we notice is that there are eight different possible arrangements of truth values for the three simple propositions p, q, and r. This is because for three propositions, either they are all true, all false, two of them are true and the third is false, or two of them are false and the third is true.

	p	q	r	$(p$	$\wedge$	$\neg q)$	$\rightarrow$	r
1.	T	T	T					
2.	T	T	F					
3.	T	F	T					
4.	T	F	F					
5.	F	T	T					
6.	F	T	F					
7.	F	F	T					
8.	F	F	F					

In order to complete the table, we gradually build up the truth value of $(p \wedge \neg q) \rightarrow r$.

First, copy in the values of p and r.

	p	q	r	$(p$	$\wedge$	$\neg q)$	$\rightarrow$	r
1.	T	T	T	T				T
2.	T	T	F	T				F
3.	T	F	T	T				T
4.	T	F	F	T				F
5.	F	T	T	F				T
6.	F	T	F	F				F
7.	F	F	T	F				T
8.	F	F	F	F				F

Next, since $\neg q$ is part of the compound proposition, enter in its values by reversing the values of q.

	p	q	r	$(p$	$\wedge$	$\neg q)$	$\rightarrow$	r
1.	T	T	T	T		F		T
2.	T	T	F	T		F		F
3.	T	F	T	T		T		T
4.	T	F	F	T		T		F
5.	F	T	T	F		F		T
6.	F	T	F	F		F		F
7.	F	F	T	F		T		T
8.	F	F	F	F		T		F

Now, use the truth table for $\wedge$ to fill in the truth value of $(p \wedge \neg q)$. Recall that $(p \wedge \neg q)$ is true *only when both p and* $\neg q$ *are true.*

	p	q	r	$(p$	$\wedge$	$\neg q)$	$\rightarrow$	r
1.	T	T	T	T	F	F		T
2.	T	T	F	T	F	F		F
3.	T	F	T	T	T	T		T
4.	T	F	F	T	T	T		F
5.	F	T	T	F	F	F		T
6.	F	T	F	F	F	F		F
7.	F	F	T	F	F	T		T
8.	F	F	F	F	F	T		F

Finally, using the values of $(p \wedge \neg q)$ as the antecedent and the values of r as the consequent, complete the table. Recall that $(p \wedge \neg q) \rightarrow r$ is false only when $(p \wedge \neg q)$ is true and r is false.

	p	q	r	$(p$	$\wedge$	$\neg q)$	$\rightarrow$	r
1.	T	T	T	T	F	F	**T**	T
2.	T	T	F	T	F	F	**T**	F
3.	T	F	T	T	T	T	**T**	T
4.	T	F	F	T	T	T	**F**	F
5.	F	T	T	F	F	F	**T**	T
6.	F	T	F	F	F	F	**T**	F
7.	F	F	T	F	F	T	**T**	T
8.	F	F	F	F	F	T	**T**	F

Now if we locate Kyoko's, Sam's, and Kenyatta's situations on the table, we will have the answers to our questions.

		p	q	r	$(p$	$\wedge$	$\neg q)$	$\rightarrow$	r
	1.	T	T	T	T	F	F	**T**	T
Kenyatta	2.	T	T	F	T	F	F	**T**	F
Kyoko	3.	T	F	T	T	T	T	**T**	T
Sam	4.	T	F	F	T	T	T	**F**	F
	5.	F	T	T	F	F	F	**T**	T
	6.	F	T	F	F	F	F	**T**	F
	7.	F	F	T	F	F	T	**T**	T
	8.	F	F	F	F	F	T	**T**	F

From the truth table, we can see that Kenyatta and Kyoko followed the IRS's instructions, but Sam did not.

PRACTICE

1. Go back and reread the IRS quote in Situation 31. Write the IRS quote on a blank sheet of paper. Close your book and try to re-create the truth table for the IRS quote.

Big truth tables can be a little overwhelming, so let's look at another one as we try to answer this question:

For what values of p, q, and r is the proposition $(p \rightarrow r) \vee \neg q$ true?

First, we set up the truth table with all the possible truth values for the simple propositions p, q, and r.

	p	q	r	$(p$	$\rightarrow$	$r)$	$\vee$	$\neg q$
1.	T	T	T					
2.	T	T	F					
3.	T	F	T					
4.	T	F	F					
5.	F	T	T					
6.	F	T	F					
7.	F	F	T					
8.	F	F	F					

Next, copy the values of p and r into the table.

	p	q	r	$(p$	$\rightarrow$	$r)$	$\vee$	$\neg q$
1.	T	T	T	T		T		
2.	T	T	F	T		F		
3.	T	F	T	T		T		
4.	T	F	F	T		F		
5.	F	T	T	F		T		
6.	F	T	F	F		F		
7.	F	F	T	F		T		
8.	F	F	F	F		F		

Next, put in the values of $\neg q$.

	p	q	r	$(p$	$\rightarrow$	$r)$	$\vee$	$\neg q$
1.	T	T	T	T		T		F
2.	T	T	F	T		F		F
3.	T	F	T	T		T		T
4.	T	F	F	T		F		T
5.	F	T	T	F		T		F
6.	F	T	F	F		F		F
7.	F	F	T	F		T		T
8.	F	F	F	F		F		T

Now put in the values of $(p \rightarrow r)$. Recall that $(p \rightarrow r)$ is false only when p is true and r is false.

	p	q	r	$(p$	$\rightarrow$	$r)$	$\vee$	$\neg q$
1.	T	T	T	T	**T**	T		F
2.	T	T	F	T	**F**	F		F
3.	T	F	T	T	**T**	T		T
4.	T	F	F	T	**F**	F		T
5.	F	T	T	F	**T**	T		F
6.	F	T	F	F	**T**	F		F
7.	F	F	T	F	**T**	T		T
8.	F	F	F	F	**T**	F		T

Finally, put in the values of $(p \rightarrow r) \vee \neg q$. Recall that $(p \rightarrow r) \vee \neg q$ is false only when both $(p \rightarrow r)$ and $\neg q$ are false.

	p	q	r	$(p$	$\rightarrow$	$r)$	$\vee$	$\neg q$
1.	T	T	T	T	T	T	**T**	F
2.	T	T	F	T	F	F	**F**	F
3.	T	F	T	T	T	T	**T**	T
4.	T	F	F	T	F	F	**T**	T
5.	F	T	T	F	T	T	**T**	F
6.	F	T	F	F	T	F	**T**	F
7.	F	F	T	F	T	T	**T**	T
8.	F	F	F	F	T	F	**T**	T

We can see that $(p \rightarrow r) \vee \neg q$ is true except when p and q are true and r is false (line 2).

PRACTICE

Construct truth tables for each of the following propositions:

2. $p \rightarrow (r \wedge q)$

3. $\neg p \vee (p \wedge \neg q)$

4. $(p \to r) \to q$

5. $(p \vee \neg p) \wedge q$

6. Here is a quote from the IRS's 1991 *Tax Guide for Small Businesses* (p. 54).

If the IRS finds that your travel allowance practices are not based on reasonably accurate estimates of travel costs, your employees will not be considered to have accounted to you and they may be required to prove their expenses to the IRS.

(a) Represent this IRS statement symbolically. (*Hint:* In everyday speech, we often omit the word *then* in a conditional proposition.)

(b) Construct a truth table for your symbolic proposition.

(c) Suppose that the IRS statement is true. Also, suppose that the IRS has told you that "Your employees will not be considered to have accounted to you." Can you conclude the IRS has found that your travel allowance practices are not based on reasonably accurate estimates? _____

Why or why not?

Situation 32 Making Sense of Arguments

Consider the following excerpt from a letter to the editor of a newspaper. The writer of the letter is offering his or her view on a proposed property tax increase.

> *If this tax increase is passed, then many citizens will have to reduce their spending for other things. Local businesses will see a decline in income. Help your local businesses. Don't pass this tax increase!*

Does the writer favor or oppose the tax increase? _____

Do the writer's reasons for opposing the tax increase make sense? _____

Why or why not?

____

____

____

____

Those were fairly difficult questions. One of the useful aspects of logic is that it gives us a way of deciding what it means to ''make sense.''

11.6 PREMISES AND CONCLUSION

Let's look more closely at the letter to the editor. What is the writer arguing against?

____

Why does the writer say that we should not support the tax increase?

____

____

Notice that the letter has two main parts. First, the writer tells us what the results of the tax increase would be. Then the writer tells us that *because of those results,* we should be opposed to the tax increase.

The proposition that asserts the point of the argument, or that tells you what the author wants you to do, is called the **conclusion** of the argument. The propositions that are offered as reasons for the conclusion are called the **premises** of the argument.

11.7 VALIDITY

An argument *makes sense* if the conclusion follows logically from the premises. The mathematical term for makes sense is **valid**. We are concerned then with how to decide whether or not an argument is valid. Fortunately, you already have all the tools necessary to test an argument for validity.

We'll begin by trying to put the letter to the editor in symbols. The first sentence is relatively easy. Let t be "This tax increase is passed," and r be "Many citizens will have to reduce their spending for other things." So, the first sentence is $t \rightarrow r$.

The second sentence is a simple proposition. So, let b be "Local businesses will see a decline in income." Now, the letter obviously suggests that the decline in business is a result of the reduced spending. Therefore, we should really write this second sentence as $r \rightarrow b$, because that is what the author is asserting.

We then have the two premises:

1. $t \rightarrow r$

2. $r \rightarrow b$

The last two sentences actually go together as one compound proposition. Suppose that the author is saying that if the tax increase is not passed, then local businesses will not be hurt. For purposes of discussion, we can assume that helping is the same as not hurting. Thus, the author's conclusion is $\neg t \rightarrow \neg b$.

The complete argument of the letter to the editor is:

Premise 1	$t \rightarrow r$
Premise 2	$r \rightarrow b$
Conclusion	$\neg t \rightarrow \neg b$

Now we must decide what it means for the conclusion to *follow* from the premises. Or, in mathematical terms, what it means for an argument to be valid. Suppose that every premise in the argument is true. In that case, if the conclusion actually follows from the premises, then the conclusion must be true.

We say that an argument is valid *if its conclusion is true whenever all its premises are true.*

We can use truth tables to look at all the possible ways for an argument to have true premises. We will build a table that contains each premise and the conclusion.

				Premise 1			**Premise 2**			**Conclusion**		
	t	r	b	t	$\rightarrow$	r	r	$\rightarrow$	b	$\neg t$	$\rightarrow$	$\neg b$
1.	T	T	T	T	**T**	T	T	**T**	T	F	**T**	F
2.	T	T	F	T	**T**	T	T	**F**	T			
3.	T	F	T	T	**F**	F						
4.	T	F	F	T	**F**	F						
5.	F	T	T	F	**T**	T	T	**T**	T	T	**F**	F
6.	F	T	F	F	**T**	T	T	**F**	F			
7.	F	F	T	F	**T**	F	F	**T**	T	T	**F**	F
8.	F	F	F	F	**T**	F	F	**T**	F	T	**T**	T

Why didn't I complete the table? Because I didn't need to! Remember that all we care about when checking for validity are those cases where *all the premises are true.* So, we only need to complete the lines in which both $t \rightarrow r$ and $r \rightarrow b$ are true. These are lines 1, 5, 7, and 8.

Now look at the values of the conclusion $\neg t \rightarrow \neg b$ in lines 1, 5, 7, and 8. We see that the conclusion is true in lines 1 and 8, but false in lines 5 and 7. We must then say that this conclusion *does not follow from the premises.* Why?

____

____

In lines 5 and 7, we have the possibility that both premises could be true, but the conclusion would be false. Therefore, the conclusion does not follow from the premises. This argument is not valid. Wait a minute! This argument certainly seems reasonable. Why do the mathematics say that it isn't?

Let's look at the conclusion from a different perspective. Suppose the author is trying to conclude that ''If businesses are not to be hurt, then the tax increase should not be passed.'' In this case, the conclusion is $\neg b \rightarrow \neg t$. The argument has the symbol representation:

Premise 1	$t \rightarrow r$
Premise 2	$r \rightarrow b$
Conclusion	$\neg b \rightarrow \neg t$

The truth table for the argument is shown below. (I've left off some of the columns because we can now work columns like $t \rightarrow r$ without rewriting the t and r columns. Fill in the missing columns if you like.)

				Premise 1			Premise 2			Conclusion		
	t	r	b	t	$\rightarrow$	r	r	$\rightarrow$	b	$\neg b$	$\rightarrow$	$\neg t$
1.	T	T	T		T			T			T	
2.	T	T	F		T			F				
3.	T	F	T		F							
4.	T	F	F		F							
5.	F	T	T		T			T			T	
6.	F	T	F		T			F				
7.	F	F	T		T			T			T	
8.	F	F	F		T			T			T	

Here we can see that the conclusion is true whenever all the premises are true. This argument is valid. Its conclusion follows from its premises.

We can see from this that $\neg b \rightarrow \neg t$ is not the same as $\neg t \rightarrow \neg b$. If the premises are true, then $\neg b \rightarrow \neg t$ is true. We know that if the businesses are doing well, then the tax increase must not have been passed. On the other hand, $\neg t \rightarrow \neg b$ may not necessarily be true because even if the tax increase is not passed, the businesses could still be hurt for reasons totally unrelated to the tax increase (like a tornado).

11.8 ANOTHER EXAMPLE

Is the following argument valid?

> *If the price of coffee goes down, then Colombia's economy suffers a setback. If Colombia's economy suffers a setback, then the value of my vacation home in Colombia declines. The value of my vacation home in Colombia did not decline. Thus, the price of coffee did not go down.*

We can decide whether or not the argument is valid by constructing its truth table. But first we need to symbolize the argument.

Let p be ''the price of coffee goes down,'' e be ''Colombia's economy suffers a setback,'' and h be ''the value of my vacation home in Colombia declines.'' Then we have

Premise 1	$p \rightarrow e$
Premise 2	$e \rightarrow h$
Premise 3	$\neg h$
Conclusion	$\neg p$

You can usually tell what the conclusion of an argument is because it is often marked by a **conclusion marker word** like *thus*, *therefore*, *so*, or *hence*.

Now we make the truth table for the argument. Remember that we only are concerned with those lines of the table for which all the premises are true. So, whenever a premise is false in a line, we can stop investigating that line.

				Premise 1	**Premise 2**	**Premise 3**	Conclusion
	p	e	h	$p \to e$	$e \to h$	$\neg h$	$\neg p$
1.	T	T	T	T	T	F	
2.	T	T	F	T	F		
3.	T	F	T	F			
4.	T	F	F	F			
5.	F	T	T	T	T	F	
6.	F	T	F	T	F		
7.	F	F	T	T	T	F	
8.	F	F	F	T	T	T	**T**

The only line in which all the premises are true is line 8. The conclusion is true in line 8. Thus, this argument is valid. Its conclusion follows from its premises.

PRACTICE

1. Use truth tables to decide whether or not the following argument is valid:

If Sam doesn't walk to work, then he rides his bicycle. If it is raining, then Sam doesn't ride his bicycle. It isn't raining. Therefore, Sam didn't walk to work.

2. (a) Write the following argument in symbols. Use the suggested **boldface** letters to represent the simple propositions.

If there are **t***wo apples on the table, then one of them* **b***elongs to Ralph. If one of the apples belongs to Ralph, then Ralph is* **h***appy. Ralph is not happy. Therefore, there are not two apples on the table.*

(b) Compare the symbolic argument you wrote in part (a) to the symbolic argument at the beginning of this section.

(c) Is the argument in this exercise valid? Explain.

11.9 BASIC FORMS AND FALLACIES

Using truth tables, we have the tools to analyze a wide variety of arguments. However, much of the reasoning that we have to do in daily life involves a handful of related argument types. These types are so common that you should learn to recognize them.

MODUS PONENS (VALID)

This argument is called ***modus ponens***.

Premise 1	$p \rightarrow q$
Premise 2	p
Conclusion	q

The truth table shows that *modus ponens* is valid.

			Premise 1	**Premise 2**	**Conclusion**
	p	q	$p \rightarrow q$	p	q
1.	T	T	T	T	**T**
2.	T	F	F		
3.	F	T	T	F	
4.	F	F	T	F	

MODUS TOLLENS (VALID)

This argument is called ***modus tollens***.

Premise 1	$p \to q$
Premise 2	$\neg q$
Conclusion	$\neg p$

The truth table shows that *modus tollens* is valid.

			Premise 1	**Premise 2**	**Conclusion**
	p	q	$p \to q$	$\neg q$	$\neg p$
1.	T	T	T	F	
2.	T	F	F		
3.	F	T	T	F	
4.	F	F	T	T	**T**

AFFIRMING THE CONSEQUENT (NOT VALID)

This argument is called **affirming the consequent**.

Premise 1	$p \to q$
Premise 2	q
Conclusion	p

The truth table shows that affirming the consequent is not valid. Arguments that are not valid are called **fallacies**.

			Premise 1	**Premise 2**	**Conclusion**
	p	q	$p \to q$	q	p
1.	T	T	T	T	**T**
2.	T	F	F		
3.	F	T	T	T	**F**
4.	F	F	T	F	

DENYING THE ANTECEDENT (NOT VALID)

This argument is called **denying the antecedent**.

Premise 1	$p \to q$
Premise 2	$\neg p$
Conclusion	$\neg q$

The truth table shows that denying the antecedent is not valid.

			Premise 1	Premise 2	Conclusion
	p	q	$p \rightarrow q$	$\neg p$	$\neg q$
1.	T	T	T	F	
2.	T	F	F		
3.	F	T	T	T	**F**
4.	F	F	T	T	**T**

PRACTICE

Name the type of argument and tell whether or not it is valid.

1. If I study all night, then I will pass the test. I didn't pass the test. Thus, I didn't study all night.

Type? Valid?

2. If the elevator is working, then Mary will be on time. Mary was on time. Thus, the elevator is working.

Type? Valid?

3. If the bait is a newt, then the raccoon will reject it. The bait was a newt. Thus, the raccoon rejected it.

Type? Valid?

4. If the lichens grow in Illinois, then they have rubixanthin in their thalli. The lichens don't grow in Illinois. Therefore, they do not have rubixanthin in their thalli.

Type? Valid?

Situation 33

Some and All

Consider the following classic argument:

> *All politicians are liars. Some liars are eloquent speakers. Therefore, some politicians are eloquent speakers.*

Is this argument valid?

Hey! These propositions are different! Where are the logical connectives? We can't use truth tables because we don't know how to symbolize the propositions. We need to increase our knowledge of propositions to handle arguments like these.

Propositions that assert something about all or some kinds of things are called **quantified propositions**. The mathematical treatment of quantified propositions is a little different from the truth value approach we've used so far. We'll begin with some terminology and then get right into the analysis of arguments with quantified propositions.

11.10 UNIVERSAL AND PARTICULAR

There are four basic types of quantified propositions.

1. The **universal affirmative** proposition asserts that every member of a class of things has a certain property. Here are some universal affirmative propositions.

All politicians are liars.

All ducks are yellow.

All bachelors are unmarried.

All Plymouths are made by Chrysler Corporation.

Universal affirmative propositions are called **type A** propositions.

2. The **universal negative** proposition asserts that no member of a class of things has a certain property. Here are some universal negative propositions.

No politicians are liars.

No ducks are yellow.

No bachelors are unmarried.

No Plymouths are made by Chrysler Corporation.

Universal negative propositions are called **type E** propositions.

3. The **particular affirmative** proposition asserts that *at least one* member of a class of things has a certain property. Here are some particular affirmative propositions.

Some politicians are liars.

Some ducks are yellow.

Some bachelors are unmarried.

Some Plymouths are made by Chrysler Corporation.

Particular affirmative propositions are called **type I** propositions.

4. The **particular negative** proposition asserts that *at least one* member of a class of things does not have a certain property. Here are some particular negative propositions.

Some politicians are not liars.

Some ducks are not yellow.

Some bachelors are not unmarried.

Some Plymouths are not made by Chrysler Corporation.

Particular negative propositions are called **type O** propositions.

PRACTICE

Identify the type (A, E, I, or O) of each of the following quantified propositions:

1. All libraries have books.

2. Some cows are not brown.

3. No perfect number is odd.

4. Some liars are eloquent speakers.

5. All coins are valuable.

6. Some coins are old.

7. Some old things are not valuable.

There is a nice relationship between the universal and particular propositions. If we wish to deny the truth of an A proposition, we can state an O proposition. That is, to deny ''All ducks are white'' we need only find one duck that isn't white. In other words, ''Some ducks are not white.'' Furthermore, we can deny an E proposition with an I proposition. For example, to deny ''No cows are brown'' we need only find one cow that is brown. In other words, ''Some cows are brown.'' Here are some more propositions and their **negations**, the propositions that deny them.

Proposition (type)	**Negation (type)**
All lawyers are honest. (A)	Some lawyers are not honest (O)
No carpet is made of grass. (E)	Some carpets are made of grass. (I)
Some bugs are harmful. (I)	No bugs are harmful. (E)
Some cats are not clever. (O)	All cats are clever. (A)

PRACTICE

Write the negation of each of the following propositions:

8. All doctors are rich.

9. Some dogs have fleas.

10. No man is an island.

11. Some prisoners are not guilty.

11.11 ENGLEBRETSEN LINES AND DIAGRAMS

A, E, I, and O propositions make assertions about the properties of collections of things. A very nice way to depict the assertions of these propositions is to use Englebretsen lines.

Englebretsen lines represent a collection of things by a line segment with a dot on its right end.

————•

The dot is labeled with a letter or letters naming the collection and the line segment represents the members of that collection. Thus, we could represent the collection of libraries like this

Library
————•

or in a shorthand version like this

l
————•

Englebretsen lines are combined to make diagrams that represent quantified propositions. It will take two Englebretsen lines for each proposition because the A, E, I, and O propositions each mention two collections of things.

Consider the A proposition, ''All libraries have books.'' The Englebretsen lines for the collections in this proposition are

Library and Have books
————• ————•

We can combine these two Englebretsen lines to express the meaning of the proposition ''All libraries have books'' by hooking together the two Englebretsen lines.

l hb
————•————•

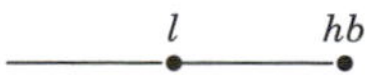

The dot *hb* marks the collection of things that have books. All points on the line segment to the left of *hb* represent things that have books. Similarly, the dot marked *l* and all points on the line segment to the left of *l* represent things that are libraries. From this **Englebretsen diagram**, we can see that anything that is a library is also a thing that has books, because any point to the left of dot *l* is also to the left of dot *hb*.

Here are some more examples:

Proposition	Englebretsen Diagram
All men are mortal.	Men Mortals
All cows are ungulates.	*c* *u*
All ungulates are cows.	*u* *c*
All taxi drivers are rich.	*td* *r*

PRACTICE

Draw Englebretsen diagrams for each of the following A propositions:

1. All politicians are liars.

2. All pizza deliverers are reckless drivers.

3. All ducks are yellow.

4. All math teachers are weird.

Let *p* represent "politicians," *l* "liars," *d* "ducks," and *y* "yellow things." Write the A proposition that corresponds to each of the following Englebretsen diagrams:

5. 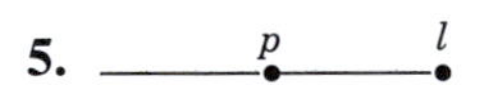

6. *l* *p*

7. *d* *y*

8. *y* *p*

Here is the Englebretsen diagram of the E proposition, "No man is an island."

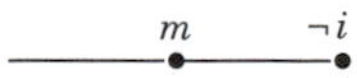

$m \quad \neg i$

This diagram makes use of the fact that when we assert no member of a collection has a property, then we are also saying that every member of the collection fails to have that property. That is, everything that is a man is a thing that is not an island.

Here are some more examples of A and E propositions depicted by Englebretsen diagrams. Complete the table.

Proposition	Type	Englebretsen diagram
All emus have feathers.	A	$e \quad f$
No emus have feathers.	E	$e \quad \neg f$
All coffee is Colombian.	A	$c \quad Co$
No coffee is Colombian.	E	$c \quad \neg Co$
All peaches are juicy.	A	
No peaches are juicy.	E	

The I and O propositions do not assert anything about entire collections of things, but only some members of the collections. For example, "Some ducks are yellow" makes no claim about all ducks. Instead, "Some ducks are yellow" merely asserts there is at least one duck that is yellow.

The Englebretson diagram for the I proposition, "Some ducks are yellow," is

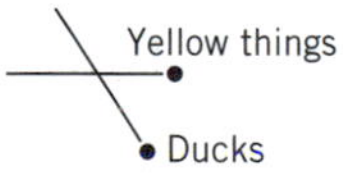

Yellow things
Ducks

Notice that the Englebretsen line for ducks crosses the Englebretsen line for yellow things. That point of intersection means there is at least one yellow duck. Also notice that it doesn't matter how the two lines are oriented so long as they intersect. Thus,

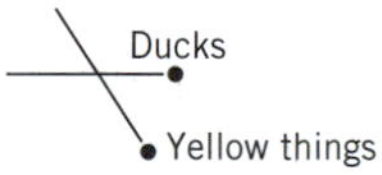

Ducks
Yellow things

is also an Englebretsen diagram for "Some ducks are yellow."

Similarly, the Englebretsen diagram for the O proposition, "Some ducks are not yellow," looks like this

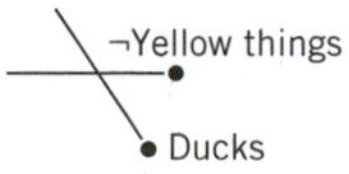

¬Yellow things
Ducks

PRACTICE

Draw Englebretsen diagrams for each of the following propositions:

9. Some politicians are liars.

10. Some pizza deliverers are not reckless drivers.

11. Some yellow things are ducks.

12. Some math teachers are not weird.

Let p represent "politicians," l "liars," d "ducks," and y "yellow things." Write the proposition that corresponds to each of the following Englebretsen diagrams:

13. 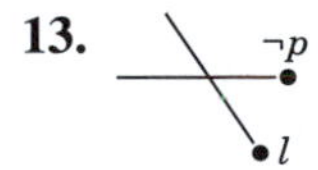

14. 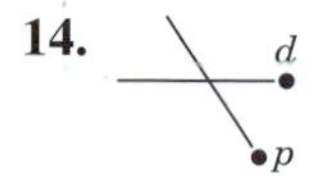

15. $\neg l$

p

11.12 ENGLEBRETSEN DIAGRAMS OF ARGUMENTS

Englebretsen diagrams can be used to decide whether or not an argument involving A, E, I, and O propositions is valid. Consider this argument.

All ducks are yellow. Some ducks are mallards. Therefore, some mallards are yellow.

The first sentence, an A proposition, has the Englebretsen diagram

Ducks Yellow

The second sentence, an I proposition, has the Englebretsen diagram

Mallards

Ducks

If we combine these two diagrams while maintaining the relationships they assert, we get

Ducks Yellow

Mallards

This last Englebretsen diagram represents the premises of the argument. The argument is valid if the conclusion is already shown in the Englebretsen diagram of the premises. Thus, we look for the conclusion

Yellow

Mallards

in the diagram of the premises. And we can see it! The conclusion requires that the Englebretsen line for mallard cross the Englebretsen line for yellow. And in the Englebretsen diagram for the premises

Ducks Yellow

Mallards

the mallard line does cross the yellow things line! Thus, this argument is valid.

Here's another example.

1. No petunias are edible flowers.
2. All Bilzter reds are petunias.
Therefore, no Bilzter reds are edible flowers.

Englebretsen diagram for premise 1:

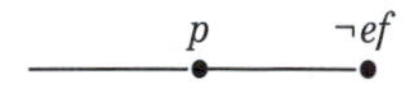

Englebretsen diagram for premise 2:

BR *p*

Combined Englebretsen diagram for premises:

BR *p* *¬ef*

Englebretsen diagram for conclusion:

BR *¬ef*

Is the Englebretsen diagram for the conclusion already present in the combined Englebretsen diagram for the premises? Yes. Therefore, this argument is valid.

Now here's an example of an invalid argument.

1. Some mice are pizza lovers.
2. Some pizza lovers are lawyers.
Therefore, some mice are lawyers.

Englebretsen diagram for premise 1:

pl
Mice

Englebretsen diagram for premise 2:

Lawyers
pl

Combined Englebretsen diagram for premises:

Mice
Lawyers
pl

Englebretsen line for conclusion:

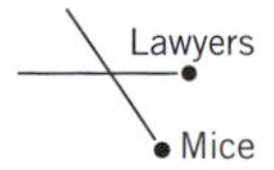

We can see that the conclusion is not present in the combined Englebretsen diagram of the premises. (Note that the mouse line and lawyer line do not intersect in the combined premise diagram, but they do in the conclusion diagram.) Thus, the argument is not valid.

Now let's look at the argument of Situation 33.

> *All politicians are liars. Some liars are eloquent speakers. Therefore, some politicians are eloquent speakers.*

The Englebretsen diagram for the first premise is

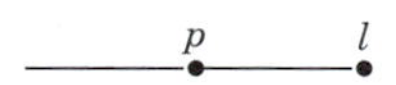

The Englebretsen diagram for the second premise is

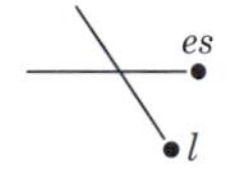

The combined Englebretsen diagram for the premise is a little tricky in this case. We know that the lawyer (1) line and eloquent speaker (*es*) line intersect (premise 2). But since the politician (*p*) line is part of the lawyer line, we could draw the intersection of the lawyer line and eloquent speaker line in two ways.

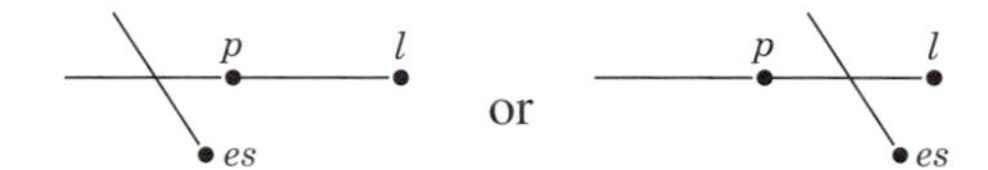

The premises do not tell us which of these possibilities is correct, so we must use both of them. If the argument is valid, then the Englebretsen diagram of the conclusion will be included in *both* of the combined Englebretsen diagrams of the premises.

The Englebretsen diagram of the conclusion is

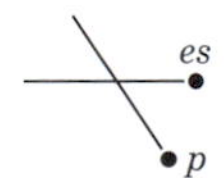

Now although this diagram appears in one of the combined premises diagram, it doesn't appear in the other. Thus, the argument is not valid.

PRACTICE

Use Englebretsen diagrams to determine whether or not the following arguments are valid:

1. All dirty dogs have fleas.
Some terriers are dirty dogs.
Therefore, some terriers have fleas.

2. All dolphins are intelligent.
Some cats are intelligent.
Therefore, some cats are dolphins.

3. Some birds can't fly.
All birds are winged animals.
Therefore, some winged animals can't fly.

11.13 WHAT DO YOU KNOW?

If you have worked carefully through this chapter, you can now symbolize propositions and propositional arguments, and you can test an argument's validity using either truth tables or Englebretsen diagrams. The exercises in this section are designed to test and refine your logical skills.

Decide whether or not each of the sentences in Exercises 1–10 is a proposition.

1. The capital of Honduras is Tegucigalpa.
2. All marigolds are yellow.
3. Stop that!
4. Please pass the salt.
5. My dog is a fox terrier.
6. Some cats are not Persian.
7. Turn left at the stoplight.
8. I don't like green eggs and ham.
9. If you shake that can, then it will explode when you open it.
10. Do you know the muffin man?

Write each of the following propositions in symbols using the suggested **bold** letters:

11. The **s**ky is blue and her **e**yes are green.
12. If the **l**ast letter is a vowel, then the **f**inal syllable is stressed.
13. Sam is **s**eventy years old and his **b**rother is eighty.
14. The money is in the **s**afe or it is hidden in the **m**attress.
15. I can eat **br**ead if I don't like tuna **p**izza.
16. If you have a **t**ouch-tone phone, then **f**ollow the recorded instructions, or if you have a **r**otary phone, then **w**ait for an operator to take your call.
17. You can **d**educt the actual cost of running your truck or you can take the **s**tandard mileage rate.
18. If you **t**ook an investment credit on property or the property use **c**hanged, you may have to **r**efigure the credit.
19. If you do not want to be able to **u**nderstand tax forms, then you do not need to have a **b**asic understanding of logic.

For Exercises 20–32, let p be "The bear ate our pizza," s be "The raccoons stole our shoes," and t be "There are bear tracks around our campsite." Write each proposition in English.

20. $\neg p \vee s$
21. $t \rightarrow p$
22. $s \wedge \neg p$
23. $\neg t \rightarrow p$
24. $s \rightarrow \neg p$
25. $s \vee t$
26. $t \rightarrow \neg s$

27. $\neg t \wedge s$

28. $(t \vee s) \rightarrow \neg p$

29. $\neg s \rightarrow (p \vee t)$

30. $t \rightarrow (t \wedge p)$

31. $(p \rightarrow s) \rightarrow t$

32. $p \rightarrow (s \rightarrow t)$

For Exercises 33–41, let c be ''Lansing is the capital of Michigan,'' m be ''The moon weighs 32 pounds,'' and n be ''The Nile river flows north.'' True or false?

33. $c \rightarrow m$

34. $m \rightarrow c$

35. $\neg n \vee c$

36. $\neg(\neg c)$

37. $c \wedge (m \rightarrow n)$

38. $\neg c \rightarrow (m \rightarrow \neg n)$

39. $(c \vee n) \wedge (n \vee m)$

40. $c \rightarrow (n \wedge n)$

41. $\neg(c \rightarrow n) \vee (m \rightarrow \neg n)$

Construct truth tables for the following propositions:

42. $p \vee \neg p$

43. $p \rightarrow (p \vee q)$

44. $p \vee (q \rightarrow p)$

45. $\neg q \vee (p \wedge q)$

46. $(\neg p \vee q) \rightarrow r$

47. $(p \wedge q) \rightarrow (r \vee p)$

48. $p \rightarrow (p \rightarrow q)$

49. $(\neg p \rightarrow r) \rightarrow \neg q$

50. $\neg(p \vee q) \rightarrow (\neg p \wedge \neg q)$

51. $(p \rightarrow (p \rightarrow q)) \rightarrow q$

A **tautology** is a proposition that is always true.

52. Show that $(p \rightarrow q) \rightarrow (\neg p \vee q)$ is a tautology.

53. Which of the propositions in Exercises 42–51 are tautologies?

Use truth tables to decide whether or not each argument is valid. The conclusions are marked with a ❍.

54. 1. $p \vee q$
2. $\neg p$
❍. q

55. 1. $p \rightarrow \neg q$
2. q
❍. $\neg p$

56. 1. $p \vee q$
2. $\neg p \rightarrow q$
❍. $\neg q$

57. 1. $\neg(p \rightarrow q)$
2. q
❍. r

58. 1. $(\neg p \rightarrow r) \wedge (p \rightarrow q)$
2. $\neg r$
❍. q

59. 1. $(r \rightarrow p) \rightarrow q$
2. $\neg p$
❍. $r \vee q$

60. 1. $p \rightarrow q$
2. $q \rightarrow r$
❍. $p \rightarrow r$

61. 1. $p \rightarrow (r \vee q)$
2. $\neg r$
3. $\neg q$
❍. p

62. If I eat a hearty breakfast, then I feel good all day. If I eat a big lunch, then I did not eat a hearty breakfast. I did not feel good all day. Therefore, I did not eat a big lunch.

63. I went to the beach. If I didn't see my friend Al, then Al was playing golf. Either I didn't go to the beach or Al wasn't playing golf. Therefore, I saw my friend Al.

64. If I vote for Martha and she wins, then my taxes will increase. If I vote for Claude and Martha wins, then my taxes will increase. I'll either vote for Claude or Martha. Therefore, Martha will win.

65. Sam went to Chicago or Yohei went to Austin. If Yohei went to Austin, then Myrna went to Lahore. Myrna didn't go to Lahore. Therefore, Sam went to Chicago.

66. If we painted the living room blue, then we needed to buy new carpeting. If we needed to buy new carpeting, then we found a good deal on carpeting. We found a good deal on carpeting. Therefore, we painted the living room blue.

Draw Englebretsen diagrams for each of the following quantified propositions:

67. Some seals are horn players.

68. Some even numbers are prime.

69. All rock guitarists have beards.

70. No accordion players have beards.

71. All polar bears are cold.

72. Some Inuits are computer programmers.

73. Some gorillas cannot read.

74. No orangutans can read.

75. Some Aztecs could write.

76. All Vulcans have pointy ears.

77–86. Write the negation of each of the propositions in Exercises 67–76.

Use Englebretsen lines to determine whether or not each of the arguments in Exercises 87–95 is valid.

87. Some pizzas have anchovies. All things that have anchovies are disgusting. Therefore, some pizzas are disgusting.

88. All cowboys sing the blues. No blues singer can ride a bronco. Therefore, no cowboy can ride a bronco.

89. Some cats are not vegetarians. No vegetarians eat tuna fish. Therefore, some cats eat tuna fish.

90. All popcorn is natural. All grapes are natural. Therefore, all grapes are popcorn.

91. Some math books are fat. All fat books are hard to read. Therefore, some math books are not hard to read.

92. No man is an island. All islands have coconut trees. Therefore, no man has a coconut tree.

93. Some ducks like rain. Some penguins like rain. Therefore, some penguins are ducks.

94. All dogs will have their day. Some things that have their day are happy. Therefore, some dogs are happy.

95. No cows can sing arias. Some gorillas can sing arias. Therefore, some gorillas are not cows.

96. Here is another quote from the *1991 Instructions for Tax Form 1040* (p. 77).

> *If you do not fully understand the answer you receive, or you feel that our representative may not fully understand your question, our representative needs to know this.*

(a) Write this instruction symbolically.

(b) Suppose Fern did not fully understand the answer she received, but she did feel that the representative did fully understand her question. Does the IRS representative need to know this? Explain.

(c) Under what conditions does the IRS representative *not* need to know this?

97. Here is another quote from the *1991 Instructions for Tax Form 1040* (p. 17).

> *If your IRA distribution is fully taxable, enter it on line 16b and do not make an entry on line 16a. If only part is taxable, enter the total distribution on line 16a and the taxable part on line 16b.*

(a) Write the two sentences of this instruction symbolically.

(b) Suppose that Kim followed the instructions correctly. Kim made an entry on line 16a. Was Kim's IRA distribution fully taxable? Explain.

ANSWERS TO PRACTICE EXERCISES

Section 11.1

1. False proposition.

2. Not a proposition.

3. Not a proposition.

4. True proposition.

Section 11.2

1. Big Bird went to the fair *and* Oscar stayed home.

2. *If* you want your company's future protected, *then* contact our office.

3. A natural disaster does *not* have to be a business disaster.

4. *If* you are over 18 years of age *and* you can read, *then* you are eligible for our life and casualty plan.

5. Fred is a duck *or* my name is *not* Rumpelstiltskin.

Section 11.3

1. $h \rightarrow r$

2. $s \rightarrow h$

3. $h \rightarrow s$

4. $g \rightarrow c$

5. $o \rightarrow w$

6. $p \rightarrow q$

7. $p \vee \neg r$

8. $p \rightarrow (q \wedge r)$

9. $(p \wedge \neg r) \rightarrow \neg t$

10. $m \rightarrow (p \vee \neg p)$

Section 11.4

1. False (line 2).

2. False (line 3 or line 4). You don't need to know the truth value of "my brother's name is Sam."

3. True (line 1).

4. True (line 2).

5. True (line 3) if my brother's name is Sam. False (line 4) if my brother's name is not Sam.

6. True (line 1).

7. False (line 2).

8. True (line 3 or line 4).

9. True (line 1).

10. Both disjuncts are false.

11. Both conjuncts are true.

12. The antecedent is true and the consequent is false.

13. ⟨For class discussion⟩

Section 11.5

1. See page 383.

2.

	p	q	r	p	$\rightarrow$	$(r$	$\wedge$	$q)$
1.	T	T	T	T	**T**	T	T	T
2.	T	T	F	T	**F**	F	F	T
3.	T	F	T	T	**F**	T	F	F
4.	T	F	F	T	**F**	F	F	F
5.	F	T	T	F	**T**	T	T	T
6.	F	T	F	F	**T**	F	F	T
7.	F	F	T	F	**T**	T	F	F
8.	F	F	F	F	**T**	F	F	F

3.

	p	q	$\neg p$	$\vee$	$(p$	$\wedge$	$\neg q)$
1.	T	T	F	**F**	T	F	F
2.	T	F	F	**T**	T	T	T
3.	F	T	T	**T**	F	F	F
4.	F	F	T	**T**	F	F	T

4.

	p	q	r	$(p$	$\rightarrow$	$r)$	$\rightarrow$	q
1.	T	T	T	T	T	T	**T**	T
2.	T	T	F	T	F	F	**T**	T
3.	T	F	T	T	T	T	**F**	F
4.	T	F	F	T	F	F	**T**	F
5.	F	T	T	F	T	T	**T**	T
6.	F	T	F	F	T	F	**T**	T
7.	F	F	T	F	T	T	**F**	F
8.	F	F	F	F	T	F	**F**	F

5.

	p	q	$(\neg p$	$\vee$	$p)$	$\wedge$	q
1.	T	T	F	T	T	**T**	T
2.	T	F	F	T	T	**F**	F
3.	F	T	T	T	F	**T**	T
4.	F	F	T	T	F	**F**	F

6. (a) Let f be "The IRS finds that your travel allowance practices are based on reasonably accurate estimates of travels costs," e be "Your employees will be considered to have accounted to you," and m be "your employees may be required to prove their expenses to the IRS." Then the IRS statement is $\neg f \rightarrow (\neg e \wedge m)$.

(b)

	f	e	m	$\neg f$	$\rightarrow$	$(\neg e$	$\wedge$	$m)$
1.	T	T	T	F	**T**	F	F	T
2.	T	T	F	F	**T**	F	F	F
3.	T	F	T	F	**T**	T	T	T
4.	T	F	F	F	**T**	T	F	F
5.	F	T	T	T	**F**	F	F	T
6.	F	T	F	T	**F**	F	F	F
7.	F	F	T	T	**T**	T	T	T
8.	F	F	F	T	**F**	T	F	F

(c) This situation describes those lines in the truth table for which $\neg f \rightarrow (\neg e \wedge m)$ is true and $\neg e$ is true. Those lines are lines 3, 4, and 7. However, only in line 7 is $\neg f$ true. Thus, you *cannot* conclude the IRS has found that your travel allowance practices are not based on reasonably accurate estimates of travel costs.

Section 11.8

1. Let w be "Sam walks to work," b be "Sam rides his bicycle," and r be "It is raining."
The argument in symbols:

Premise 1	$\neg w \rightarrow b$
Premise 2	$r \rightarrow \neg b$
Premise 3	$\neg r$
Conclusion	$\neg w$

Truth table:

	Premise 1	Premise 2	Premise 3	Conclusion
w b r	$\neg w \rightarrow b$	$r \rightarrow \neg b$	$\neg r$	$\neg w$
1. T T T	T	F		
2. T T F	T	T	T	F
3. T F T	T	T	F	
4. T F F	T	T	T	F
5. F T T	T	F		
6. F T F	T	T	T	T
7. F F T	F			
8. F F F	F			

The argument is *not valid* because it is possible for it to have all of its premises true with a false conclusion (see line 4).

2. (a)

Premise 1	$t \rightarrow b$
Premise 2	$b \rightarrow h$
Premise 3	$\neg h$
Conclusion	$\neg t$

(b) The two arguments have exactly the same form. Look!

Premise 1	$p \rightarrow e,$	$t \rightarrow b$
Premise 2	$e \rightarrow h,$	$b \rightarrow h$
Premise 3	$\neg h,$	$\neg h$
Conclusion	$\neg p,$	$\neg t$

You could turn either argument into the other by just interchanging the letters (p and t, e and b).

(c) This argument is valid because it has exactly the same form as the argument in Section 11.8 and that argument is valid.

Section 11.9

1. *Modus tollens*, yes.
2. Affirming the consequent, no.
3. *Modus ponens*, yes.
4. Denying the antecedent, no.

Section 11.10

1. A
2. O
3. E
4. I
5. A
6. I
7. O
8. Some doctors are not rich.
9. No dog has fleas.
10. Some men are islands.
11. All prisoners are guilty.

Section 11.11

1. p — l

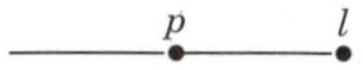

2. pd — rd

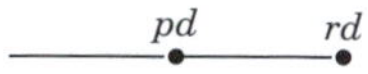

3. d — y

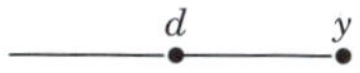

4. mt — w

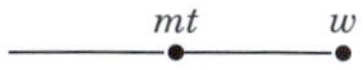

5. All politicians are liars.
6. All liars are politicians.
7. All ducks are yellow.
8. All yellow things are politicians.
9. l, p
10. $\neg rd$, pd
11. d, y
12. $\neg w$, mt
13. Some liars are not politicians.
14. Some politicians are ducks.
15. Some politicians are not liars.

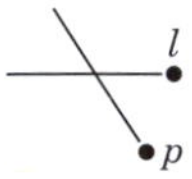

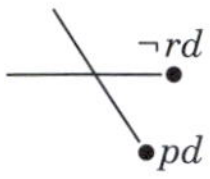

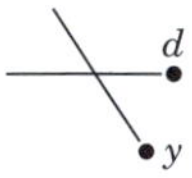

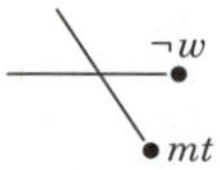

Section 11.12

1. dd — Have fleas; Terriers   Valid

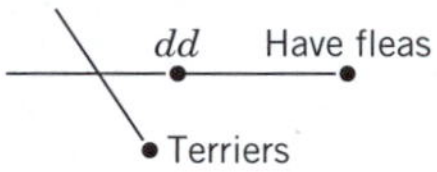

2. d — i

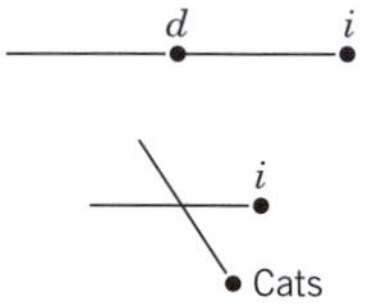

   i; Cats   Not Valid

3. Bird — wa; $\neg$Fly   Valid

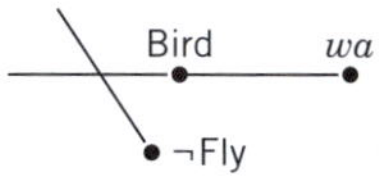

Chapter Twelve

Mathematics of Computers

What's the Point? Computers are an omnipresent part of modern life. There are computers in automobiles, microwave ovens, and toys. Computers manage banking accounts, process credit card purchases, and mail catalogs to prospective customers. This chapter is designed to help you understand the main features of computers, the operation of some popular computer applications, and the main ideas behind some of the more recent developments in computer science.

Situation 34

How Does It Work?

You have just come home with your brand new Cyberfrog 3000 personal computer. After plugging it in and following the manufacturer's *easy* installation instructions, you play a few games, send an e-mail to an old friend, and buy a new shirt from Ben's of Boston. Then your niece comes in and asks, "How does it work?" What do you tell her?

____

____

____

____

____

____

If you were able to give your niece an accurate account of the operation of your computer including things like RAM, CPU, CD-ROM, and compilers, then you probably don't need to read the next three sections. On the other hand, if you don't know how your computer works, then the next three sections will give you a basic overview of its operation and some of the mathematics behind it all. We'll begin with computer hardware.

12.1 COMPUTER ORGANIZATION AND ARCHITECTURE

A **computer** is a machine that can precisely follow a series of instructions called a **program**. The computer itself is called **hardware** and the instructions that the computer follows comprise the **software**. In this section, we'll look at the hardware components of a computer.

A rather typical-looking personal computer is sketched below. What you see in this illustration are the external parts of the major hardware components of a computer. The two most obvious components are the **keyboard**, which resembles a typewriter keyboard, and the **monitor**, which looks like a television screen.

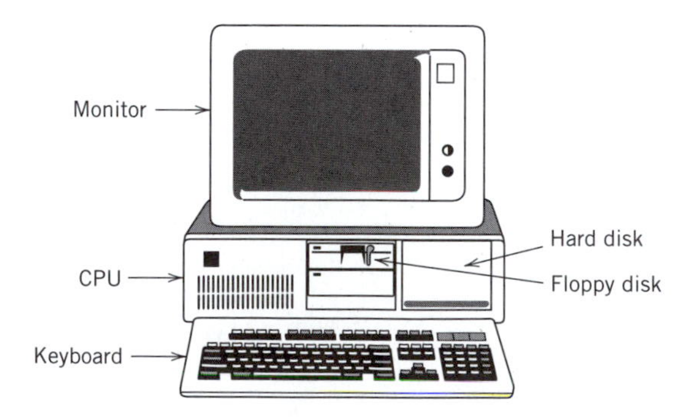

The keyboard is one kind of **input device**. Input devices provide a way of giving the computer instructions and information. These instructions or information, which are called **input**, can come directly from a human using the computer, directly from another computer, or indirectly from either a human or another computer. The keyboard is a direct way of giving your input to the computer. Although the keyboard is the most widespread type of direct input device, direct human input can also be accomplished with a computer **mouse**, **track ball**, **light pen**, or by voice command.

What kinds of direct input devices are available for the computers that you have access to or know about?

A computer can receive input indirectly via computer disks or magnetic tape. (We'll see how this works later on!) In the case of disk input, the input is first placed on the disk using a direct input device called a **disk drive** and then the computer obtains, or **reads**, the input from the disk. The process is similar for magnetic tape input. Indirect input is like a letter to the computer that can be read over and over again. Direct input is more like a conversation that happens once and is over.

The monitor is one kind of **output device**. Output devices provide a way for the computer to give information to a human user or another computer. This information, which is called **output**, can go directly from the computer to an output device, directly to another computer, or indirectly to either an output device or another computer. The monitor is a direct output device. Although the monitor is the most widespread type of direct output device, direct output can also be accomplished with a printer or by sound.

What kinds of direct output devices are available for the computers that you have access to or know about?

A computer can send output indirectly via computer disks or magnetic tape. In these cases, the output is first placed on the disk or tape by the computer by means of the disk drive. We say that the computer **writes** information to the disk or tape. The output can then be read by a human user or computer at their convenience.

The relationship between the computer, the human user, and the input and output devices can be illustrated with a diagram of boxes and lines called a **flowchart**. In the flowchart shown below, the boxes represent the hardware components and the arrow heads on the lines represent the flow of information. Thus, for example, information flows from the computer to the output device.

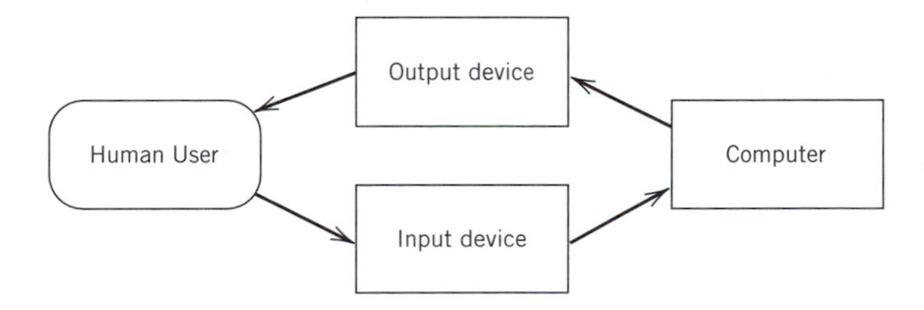

Now that we see how the human user is able to communicate with the computer, we need to examine the internal components of the computer itself. Every computer has three essential components: a **processing unit**, an **arithmetic and logic unit**, and an **internal memory**.

Collectively, these three components are called the **central processing unit** or **CPU**. The relationships between these three units are illustrated in the flowchart shown below.

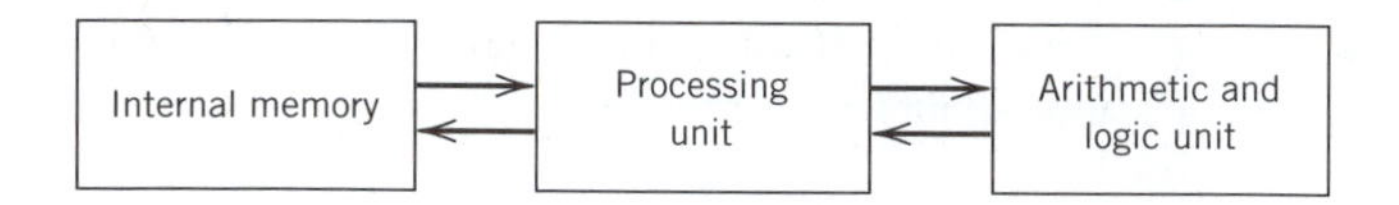

The processing unit (also called the **processor**) controls all the activity within the computer. The processor moves information in and out of the internal memory and carries out instructions supplied by the user. The arithmetic and logic unit performs arithmetic calculations and logical operations as instructed by the processor. The internal memory is a storage facility that shows intermediate calculations, data, instructions, and other information required by the processor.

The final major hardware component of a computer is the **external memory**. External memory is additional storage space on disks for information. Disks that cannot be removed from the computer are called **hard disks**. A **hard drive** is the input–output device that puts information on and takes information off a hard disk. Disks that can be removed and carried about are called **floppy disks**. Computers work with information expressed as binary numerals (see Section 10.8). Memory in a computer is measured in **kilobytes** (see Section 7.4 for a review of the metric system). A **byte** consists of eight binary digits. A binary digit is called a **bit**. Internal memory is called **random access memory** or **RAM**. If we can store 1024 kilobytes of information in internal memory, then we say that our CPU has 1024K RAM. External memory is usually much greater in capacity than internal memory and is frequently measured in **megabytes** (1 megabyte is 1 million bytes). External memory sometimes is constructed so that the computer may obtain information from it but may not put information into it. This kind of memory is called **read-only memory** or **ROM**. Compact disks that contain read-only memory are called **CD-ROMS**.

Finally, since input and output are so essential to using a computer, it is customary to include these units in the term *computer*. We can then present a flowchart of the typical computer:

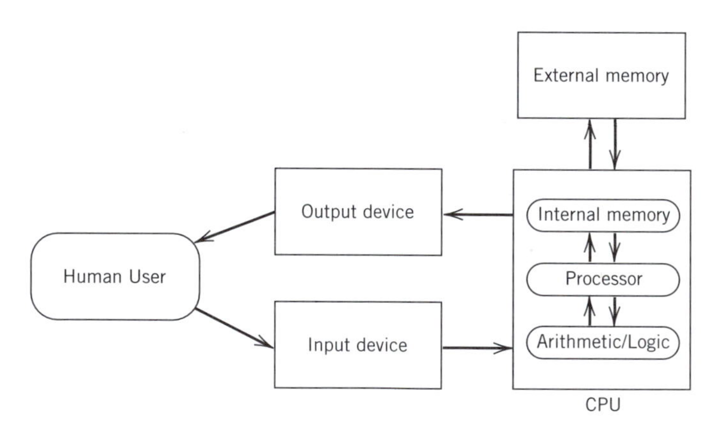

PRACTICE

1. Name the three components of the CPU.

2. What is the difference between internal and external memory?

3. A computer has 640K RAM and a 3-megabyte hard drive. What does that mean?

12.2 COMPUTER LANGUAGES

The term *software* refers to the instructions that the user provides to the CPU. The human user has an idea about what specific tasks she would like the computer to perform, and these tasks and the instructions for performing them must be given to the computer. The process of getting the computer to carry out your instructions is called **programming**.

Programming involves communication between the human user and the computer. This communication is accomplished by means of languages that can be jointly understood by the computer and the user.

Humans understand **natural language** like English and Spanish, but at its most basic level, the computer understands and works with only binary numerals (strings of zeros and ones). Each computer has its own **machine language**, which consists of strings of ones and zeros. If we wanted to give the computer instructions directly, we would need to give it instructions solely with strings of zeros and ones. A set of instructions in machine language might look like this:

01101010001010101011001100100001011

10101010000000101101010101010100001000

0000001110101001010100010001000010

It takes a long time to write a program in machine language and it is really easy to make a mistake. So, some intermediate languages have been developed to make it easier for the human user to communicate his instructions to the computer.

Assembly languages use some key words like ADD. A set of instructions called the **assembler** can take programs written in assembly language and translate them into machine language. Unfortunately, assembly languages, like machine languages, are different from computer brand to brand. The assembly language for a Macintosh® is not the same as that for the IBM PC®.

Most people use a **high-level language**, which has English-like words and phrases, to program their computer. High-level languages can be used on any computer that has a **compiler** for that language. Compilers are computer programs that translate a high-level language into machine language. Some popular high-level languages are BASIC, Pascal, COBOL, FORTRAN, LISP, Ada, C, and Prolog.

The flowchart below shows the relationship of the kinds of languages used by humans to communicate with computers. The languages appear in ovals and the translating programs in rectangles.

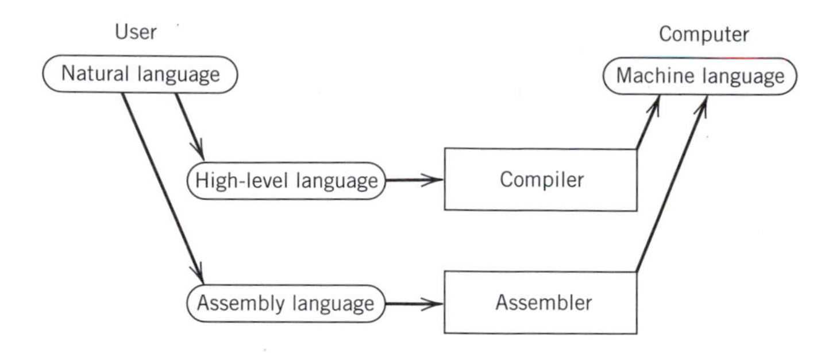

12.3 COMPUTER PROGRAMS

Computers can do only a handful of tasks, but they can do them very fast. Computers can do arithmetic, make logical comparisons, follow logical instructions, and move information within memory, from input devices to memory, and from memory to output devices. As a result, all high-level computer languages have the ability to instruct the computer to do each of the following tasks:

1. Follow a sequence of instructions in order.
2. Repeat an instruction until told to stop.
3. Choose from alternative instructions based on specified information.
4. Send results to and receive information from specified output and input devices.

Any ordered set of these types of instructions is a **computer program**. When the computer has performed the instructions of a program, it has **executed** the program.

Although it is not my intention to teach you how to program a computer, I think that you will better understand how programming works if we look at a very small example. In fact, we'll only look at a piece of a computer program.

Suppose that we are writing a pricing program for a supermarket. The supermarket is having a sale on laundry detergent. The detergent is being sold for $3.00 a box, but if you buy five or more boxes, it is only $2.80 a box. Let's look at the part of the computer program that will figure out how much to charge for the laundry detergent.

The first thing we notice is that we can't tell the computer ahead of time how many boxes of detergent a customer will want to buy. So, in our computer program, we will use a **variable** to represent the number of boxes of detergent a customer buys. Most high-level languages will let us call this variable BOXES. So, when the customer gets to the checkout line, the cashier will give the computer a value of BOXES by means of some input device.

Now the computer cannot know the amount to charge the customer until it knows the value of BOXES, so we will use another variable called COST to represent the cost of detergent based on the number of boxes. The computer needs to reason as follows:

1. If the number of boxes (i.e., the value of BOXES) is less than five, then the cost is $3.00 times the number of boxes.
2. If the number of boxes is five or more, then the cost is $2.80 times the number of boxes.

Stop! Let's make sure that you understand the preceding paragraphs. Suppose that the customer has two boxes of detergent. What should he be charged? _____
What should he be charged if he has six boxes of detergent? _____

Correct! The customer is charged $6.00 for two boxes of detergent and $16.80 for six boxes of detergent.

The instructions for this decision in the high-level language Pascal are written this way:

```
IF BOXES < 5 THEN
  COST := BOXES*3.00
ELSE
  COST := BOXES*2.80;
```

After these instructions are compiled, the computer can execute them. The computer will compare the value given to it for BOXES with 5. If BOXES is less than 5, then the computer will compute the value of COST to be BOXES times 3.00. Otherwise, the value of COST will be computed to be BOXES times 2.80.

Here are the same instructions in some other high-level languages. Look at the similarities and differences.

FORTRAN
```
IF (BOXES.LT.5) THEN
  COST = BOXES*3.00
ELSE
  COST = BOXES*2.80
```

C
```
IF (BOXES < 5) THEN
  COST = BOXES*3.00;
ELSE
  COST = BOXES*2.80;
```

COBOL
```
IF BOXES IS LESS THAN 5
  COMPUTE COST = BOXES*3.00
ELSE
  COMPUTE COST = BOXES*2.80.
```

LISP
```
(IF (< BOXES 5) (SETQ COST (* BOXES 3.00))
                (SETQ COST (* BOXES 2.80)))
```

PRACTICE

1. Suppose the supermarket decides that a customer must buy at least seven boxes of detergent before she can get the $2.80 per box price. Rewrite the programs in this section to make this change.

Each of the preceding sets of instructions directs the computer to do the same tasks. The instructions all mean the same thing. However, they express that meaning in different ways. The way in which the words and symbols of a language must be put together to express meaning is called the **syntax** of the language. Computer programmers must learn the syntax of the high-level language they use. The syntax of computer languages is an application of a branch of mathematics called **formal language theory** that also has uses in codes, biology, and computer animation. The next section examines some of the basics of formal language theory.

12.4 FORMAL LANGUAGES

Take a look at the following two sentences:

1. The dog ate the ice cream.
2. Ice dog the the cream ate.

Which of the two sentences is written in correct English? Why is the other one wrong?

____

____

____

If you said that the words in sentence 2 were in the wrong order, then you were correct. The English language requires a certain order between the various parts of a sentence. Word order is a part of English grammar or syntax. It is so important that we can recognize sentences even if we don't know the words in them. Look at this sentence.

3. The felmers flooped the grelxt.

What is the verb in sentence 3? _____ Right! You can tell that the verb is *flooped* even though you probably don't know what *flooped* means. You probably also identified *the felmers* as the subject of the sentence and *the grelxt* as the object of the sentence.

In a simple declarative English sentence, the subject comes first, then the verb, and then the object. We can express this fact using a simple mathematical notation like

$$\textit{Sentence} \rightarrow \textit{subject verb object}$$

This says that a *sentence* consists of the sequence *subject verb object*, in that order. It is an example of a **production rule**. Production rules are an important feature of formal languages. **Formal language theory** is the mathematical study of formal languages.

A **formal grammar** consists of two sets of symbols, T and NT, and a set of production rules. Here's an example of a formal grammar that we will call G_1:

G_1:

$T = \{a, b\}$
$NT = \{S, C, D\}$
Production rules:

1. $S \rightarrow aC$
2. $C \rightarrow bD$
3. $D \rightarrow a$

The set T is called the set of **terminal symbols**. The set NT is called the set of **nonterminal symbols**. The symbol S, which stands for *start*, is always a member of the set NT. The production rules provide instructions for constructing **words** in the **formal language** for which G_1 is the grammar. Here's how it works.

How Production Rules Construct Words

1. Begin with the symbol S. Find any production rule with S on its left-hand side. Replace S by the right-hand side of that rule. The result is called the **current string**.
2. Select any nonterminal in the current string. Find any production rule with this nonterminal on its left-hand side. Replace the nonterminal by the right-hand side of the production rule. The result is called the current string.
3. Repeat step 2 until there are no nonterminals in the current string or you cannot find a production rule that applies. If there are no nonterminals in the current string, then the current string is a word in the language.

In this example, only rule 1 has S on the left-hand side, so we can replace S by aC. The current string is now aC. We will show this replacement like this:

$$S \Rightarrow_1 aC$$

The subscript indicates the number of the rule used.

Now, C is a nonterminal symbol, so we can replace it by bD using rule 2. The current string is now abD.

$$S \Rightarrow_1 aC \Rightarrow_2 abD$$

Notice that the a is not altered.

Finally, D is a nonterminal symbol, so we can replace it by a using rule 3.

$$S \Rightarrow_1 aC \Rightarrow_2 abD \Rightarrow_3 aba$$

Now, aba has no nonterminal symbols in it, so aba is in the set of strings that can be obtained using the production rules of G_1. This set is called the **language generated by $\mathbf{G_1}$**. The language generated by G_1 is denoted $L(G_1)$. The strings in a language are called the **words** of the language. aba is a word in $L(G_1)$.

In this example, aba is the only word of $L(G_1)$, so $L(G_1) = \{aba\}$. Why is aba the only word of $L(G_1)$?

____

____

____

Some languages have many words. Consider the language generated by the grammar G_2 shown below.

G_2:

$T = \{a, b\}$
$NT = \{S, C\}$
Production rules:

1. $S \rightarrow aC$
2. $S \rightarrow aS$
3. $C \rightarrow b$

One word in $L(G_2)$ is ab, which is generated by

$$S \Rightarrow_1 aC \Rightarrow_3 ab$$

But, we can also generate aab by using rule 2 first:

$$S \Rightarrow_2 aS \Rightarrow_1 aaC \Rightarrow_3 aab$$

Also, we can generate the word $aaaab$ by reusing rule 2:

$$S \Rightarrow_2 aS \Rightarrow_2 aaS \Rightarrow_2 aaaS \Rightarrow_1 aaaaC \Rightarrow_3 aaaab$$

Notice that the production rules can be used in any order so long as a nonterminal symbol is replaced each time a rule is used.

The words of $L(G_2)$ are strings that begin with one or more a and end in b.

PRACTICE

1. Show that $aaaaab$ is generated by G_2.

2. Let G_3 be the grammar defined by

$T = \{a, b, c\}$
$NT = \{S, D\}$
Production rules:

1. $S \to aDb$
2. $D \to cD$
3. $D \to c$

(a) Show that $acccb$ is a word of $L(G_3)$.

(b) Show that acb is a word of $L(G_3)$.

(c) Describe $L(G_3)$.

Here, as another example, is the grammar G_4.

G_4:

$T = \{the, dog, ate, pizza, \#\}$
$NT = \{S, NP, VP, VA, N\}$

Production rules:

1. $S \rightarrow NP\#VP$
2. $VP \rightarrow V\#NP$
3. $VP \rightarrow V$
4. $NP \rightarrow A\#N$
5. $A \rightarrow$ the
6. $N \rightarrow$ dog
7. $N \rightarrow$ pizza
8. $V \rightarrow$ ate

Let's generate a word in $L(G_4)$:

$$S \Rightarrow_1 NP\#VP \Rightarrow_4 A\#N\#VP \Rightarrow_5 \text{the}\#N\#VP \Rightarrow_6 \text{the\#dog}\#VP$$

$$\Rightarrow_2 \text{the\#dog}\#V\#NP \Rightarrow_8 \text{the\#dog\#ate}\#NP \Rightarrow_4 \text{the\#dog\#ate}\#A\#N$$

$$\Rightarrow_5 \text{the\#dog\#ate\#the}\#N \Rightarrow_7 \text{the\#dog\#ate\#the\#pizza}$$

So, the#dog#ate#the#pizza is a word in $L(G_4)$. The terminal symbol # separates English words to make a word in $L(G_4)$ an English sentence. Is the#dog#ate#pizza a word in $L(G_4)$?

No, the#dog#ate#pizza is not a word in $L(G_4)$ because if we try to generate it, we will get:

$$S \Rightarrow_1 NP\#VP \Rightarrow_4 A\#N\#VP \Rightarrow_5 \text{the}\#N\#VP \Rightarrow_6 \text{the\#dog}\#VP \Rightarrow_2 \text{the\#dog}\#V\#NP$$

$$\Rightarrow_8 \text{the\#dog\#ate}\#NP \Rightarrow_4 \text{the\#dog\#ate}\#A\#N$$

But, now we must replace the A by *the*, which doesn't appear in this position in the#dog#ate#pizza.

What production rule could we add to G_4 so that the#dog#ate#pizza would be a word in $L(G_4)$?

___

___

___

(Answer: 9. $NP \rightarrow N$)

PRACTICE

3. Let G_5 be the grammar defined below.

G_5:

$T = \{a, b, c\}$

$NT = \{S, X, Y, Z\}$

Production rules:

1. $S \rightarrow SX$
2. $S \rightarrow cZX$
3. $X \rightarrow aa$
4. $Z \rightarrow bb$

(a) Show that *cbbaaaa* is a word in $L(G_5)$.

(b) Show that *bb* is not a word in $L(G_5)$.

(c) Describe $L(G_5)$.

Computers always check to see that a user's instructions are words in the language being used before attempting to carry out the instructions. A set of instructions must be grammatically perfect before it is executed.

For example, in one version of the high-level language LISP, the instruction for ending a programming session is **(system)**. Somewhere in the production rules for this language are the rules:

$$S \rightarrow (A)$$

and

$$A \rightarrow \text{system}$$

where capital letters are nonterminal symbols and lowercase letters and parentheses are terminal symbols. This computer language will not accept the strings **(quit)** or **(end)** or **(stop)** because there are no production rules that will produce them. Thus, the user will not be able to end a programming session until she inputs **(system)**.

We have seen how to decide whether or not a specific string is a word of a given language. Now let us look briefly at the reverse problem: How can we write a grammar for a given set of words? This is a task faced by the programmers who design computer languages and write compilers.

Suppose we would like to write a grammar for the set of words

$$B = \{aba, aaba, aaaba, aaaaba \ldots\} = \{x\text{: } wba \text{ where } w \text{ is a string of one or more } a\text{'s}\}$$

Our first step will be to define the set of terminals. Since all the words of B consist only of a's and b's, we know that $T = \{a, b\}$. We also know that NT includes at least S, but right now we don't know how many other nonterminals we need. Let's move to the production rules.

Every word in B ends in ba. So, a reasonable first guess at a production rule would be

$$1.\ S \rightarrow Xba$$

Here the nonterminal X means that something will come before the ba. What comes before it?

Right! One or more a's come before the ba. So, we need a production rule (or rules) that allow us to replace X by one or more a's. Here are some that would do just that:

$$X \rightarrow a$$

$$X \rightarrow aa$$

$$X \rightarrow aaa$$

$$X \rightarrow aaaa$$

and so on

The problem with this approach is that we cannot list all the rules we need. (Note that *aaaaaaaaaaaaaaaaaaaaaaaaaaaaaaaaaaaaba* is a word in the language B!) So, instead we write two rules that allow us to replace X by as many a's as we like:

$$2.\ X \rightarrow aX$$
$$3.\ X \rightarrow a$$

Notice that we can repeat rule 2 over and over again to produce as many a's as we like and then finish it off with rule 3.

You should check to see that rules 1–3 will generate *aba*, *aaba*, *aaaaaba*, and *aaaaaaaaaaba*.

We have constructed the grammar G_6:

$T = \{a, b\}$
$NT = \{S, X\}$
Production rules:

1. $S \rightarrow Xba$
2. $X \rightarrow aX$
3. $X \rightarrow a$

and $L(G_6) = B$.

PRACTICE

4. Construct a grammar G_7 such that $L(G_7) = \{bab, baab, baaab, baaaab, baaaaab, \ldots\}$.

5. Construct a grammar G_8 such that $L(G_8) = \{bcabc, bcbcabcbc, bcbcbcabcbcbc, \ldots\}$.

Situation 35

We Got the Account!

Sam and Janet Evenings' small music business is growing. Janet, who is the bookkeeper for the business, decides that it is time they got a computer to help with the everyday tasks of running a business. The salesman at the Evenings' local computer store tells them that regardless of the make of computer they buy, the Evenings must have at least three kinds of software: a word processor, a spreadsheet, and a DBMS.

What are these three types of software and why are they important for a business?

____

____

____

____

Many businesses, large and small, rely on computers and computer software for a variety of needs. In the next four sections, we will look at the major categories of software used by businesses. Because there are many different brands of business-related software with different specific instructions, and these programs are being continuously updated and modified, we will not examine the operation of any specific software package in detail. Instead, we will look only at the common features and purposes of these **software tools**. You are encouraged to learn how to use at least one brand of each of these types of software through a course or self-study.

12.5 WORD PROCESSORS

Word processors are programs that have been designed to make the production of written texts easier and more efficient. Word processors possess a variety of useful features. In this section, we will look at some of the major features of word processing software. I have selected features present in all modern word processors like WordPerfect® or Word®.

1. *Editing features.* Word processors accept written text from the keyboard and then provide the user with the opportunity to make editing changes without extensive retyping. For example, suppose that I was composing a letter to my Aunt Dorothy. I had written the following:

> *I was happy to receive your letter from Floorida. It was very cold in Michigan that day and you letter warmed me up. I certainly would like to come visit you!*

Now suppose that I felt the last sentence should be first. With a word processor, all I need to do is mark the last sentence and then instruct the word processor to move it to a new position. In most cases, a few short key strokes would give me this:

> *I certainly would like to come visit you! I was happy to receive your letter from Floorida. It was very cold in Michigan that day and you letter warmed me up.*

It would be just as easy to move, erase, or copy whole paragraphs or pages.

You may have noticed that in my letter to my Aunt Dorothy, I misspelled *Florida*. With a word processor, I can remove the extra *o* simply by marking it and deleting it. The word processor will close up the gap created and my letter will then look like this:

> *I certainly would like to come visit you! I was happy to receive your letter from Florida. It was very cold in Michigan that day and you letter warmed me up.*

2. *Spelling and grammar checking.* If I hadn't noticed that *Florida* was misspelled, the word processor would have called it to my attention when I activated the **spell checker**. The spell checker is a program within the word processor that tries to match each word in your text with a word in its own dictionary. If it can't find your word in its dictionary, then the word processor suggests an alternative. For example, my word processor suggested *Florida*, *fluoride*, and *fluorite* and alternatives to my *Floorida*.

Here's my letter again.

I certainly would like to come visit you! I was happy to receive your letter from Florida. It was very cold in Michigan that day and you letter warmed me up.

Do you notice any other errors in it?

Right! I wrote "you letter warmed me up" instead of "your letter warmed me up." A spell checker would not notice this error because *you* is spelled correctly. For these kinds of errors, some word processors have **grammar checkers** to point out possible grammatical errors.

In this case, the grammar checker would notice that *you* is neither an adjective nor a possessive pronoun and suggest that it may be an error.

3. *Desktop publishing features.* Many word processors also allow you to incorporate graphs and charts into your text, make newspaper columns, and change the printing fonts. Here are some typical fonts available with word processors:

Baskerville
Garamond
Helvetica
Oxford
Rockland
New York Deco
Courier
Palatino

A quality word processor, computer, and printer will enable you to produce quite professional looking documents—sometimes in color!

Word processors have all but replaced typewriters in businesses. I strongly recommend that you learn how to use a major word processor.

12.6 SPREADSHEETS

Just as word processors have made the preparation of documents a computer activity, **spreadsheets** have helped to computerize the quantitative aspects of business. A spreadsheet (or **worksheet**) is a matrix of labeled rows and columns. The rows are usually labeled with integers and the columns are usually labeled with the letters of the English alphabet. A typical spreadsheet might look like this:

	A	B	C	D	E	F	G
1							
2							
3							
4							
5							
6							
7							
8							
9							
10							

Spreadsheet software, which are usually just called spreadsheets, give the user the ability to do extensive calculations with values entered into the matrix. Most spreadsheet software also will do statistical analysis, produce graphics, and allow the user to access and display this information via a word processor. Two of the most popular spreadsheets are Lotus 1-2-3® and QuattroPro®.

In this section, we will look at some of the more general features shared by all spreadsheet programs. Again, I encourage you to learn how to use one of the major spreadsheet programs.

Let's begin with some elementary calculations. All spreadsheets allow you to enter numerical or nonnumerical information into any location of the worksheet. Suppose we enter the information shown in **bold** print:

	A	B	C	D	E	F	G
1	**HEIGHT (I**	**NCHES) &**	**WEIGHT (P**	**OUNDS) OF**	**SOME PEO**	**PLE WE KN**	**OW**
2							
3	**Height**	**Weight**					
4	**64**	**103**					
5	**72**	**190**					
6	**68**	**150**					
7	**75**	**210**					
8							
9							
10							

The worksheet is divided into **cells** labeled by row number and column letter. For example, the value 64 in the worksheet shown above is in cell A4, the value 210 is in cell B7, and the word *Height* is in cell A3.

The line of text that we entered at the top of the worksheet is also viewed as cells by the spreadsheet program. The string **HEIGHT** (*i* occupies cell A1, the string **NCHES**) & occupies cell B1, and the string **WEIGHT** (**p** occupies cell C1.

PRACTICE

Identify the contents of each cell for the worksheet shown below.

	A	B	C	D	E	F	G
1	HEIGHT (I	NCHES) &	WEIGHT (P	OUNDS) OF	SOME PEO	PLE WE KN	OW
2							
3	Height	Weight					
4	64	103					
5	72	190					
6	68	150					
7	75	210					
8							
9							
10							

1. Cell A5

2. Cell B5

3. Cell D1

4. Cell E1

Suppose that we were interested in the mean height for the data in our worksheet. The spreadsheet program allows us to enter into a cell a command that will instruct the computer to calculate the mean. For example, we could put the following instruction into cell A9:

average(A4–A7)

(*Warning:* This is not an actual command for some specific spreadsheet! It is only an illustration of how a spreadsheet works. Each spreadsheet has its own syntax that must be followed.)

The resulting worksheet would not show the command, but would instead display the average of the values in cells A4 through A7.

	A	B	C	D	E	F	G
1	**HEIGHT (I**	**NCHES) &**	**WEIGHT (P**	**OUNDS) OF**	**SOME PEO**	**PLE WE KN**	**OW**
2							
3	**Height**	**Weight**					
4	**64**	**103**					
5	**72**	**190**					
6	**68**	**150**					
7	**75**	**210**					
8							
9	**69.75**						
10							

Similarly, we could compute the mean of the weights by entering the instruction **average(B4–B7)** into cell B9. The resulting spreadsheet would look like this:

	A	B	C	D	E	F	G
1	**HEIGHT (I**	**NCHES) &**	**WEIGHT (P**	**OUNDS) OF**	**SOME PEO**	**PLE WE KN**	**OW**
2							
3	**Height**	**Weight**					
4	**64**	**103**					
5	**72**	**190**					
6	**68**	**150**					
7	**75**	**210**					
8							
9	**69.75**	**163.25**					
10							

Calculations within a spreadsheet are usually dynamic. This means that if a cell value is changed, then any other cell whose value is computed using the original cell is automatically changed accordingly. For example, suppose that we change the weight in cell B6 to 175. If all we do is type 175 into cell B6, the spreadsheet will look like this:

	A	B	C	D	E	F	G
1	**HEIGHT (I**	**NCHES) &**	**WEIGHT (P**	**OUNDS) OF**	**SOME PEO**	**PLE WE KN**	**OW**
2							
3	**Height**	**Weight**					
4	**64**	**103**					
5	**72**	**190**					
6	**68**	**175**					
7	**75**	**210**					
8							
9	**69.75**	**169.5**					
10							

Notice that the mean weight value in cell B9 has automatically changed to reflect the change to cell B6. This is one of the major strengths of a spreadsheet: Calculations are updated automatically.

Spreadsheets are capable of doing a wide range of calculations and other mathematical tasks quite easily. For example, referring to our sample worksheet, we would calculate the standard deviation of height by entering a command like **std(A4–A7)**, and produce a graph of height versus weight by typing **graph(A4–A7, B4-B7)**. We could also calculate a regression line for height vs. weight, and calculate the correlation coefficient for these data. Furthermore, any change in the data for height or weight that we make would automatically change the statistics and graph!

Spreadsheets are not limited to statistical calculations. All the financial calculations of Chapter 1, all the matrix operations of Chapter 9, all the optimization techniques of Chapter 6, all the formulae used by accountants, and a complete range of bar graphs, pie graphs, and linear graphs are included in a spreadsheet program.

As you can imagine, spreadsheets are a very important tool for businesses. Spreadsheets, like word processors, are essential for the operation of even small businesses today.

12.7 DATABASE MANAGEMENT SYSTEMS

Database management systems (often referred to by the acronym **DBMS**) comprise the third major type of computer programs for business. Like word processors and spreadsheets, the primary purpose of DBMS is to make common business activities easier and more efficient.

Database management systems are programs that help human users manipulate and get information from databases. A **database** is an organized collection of data stored in some location. A telephone directory is a database of people in a community with telephones. The telephone directory is a database that is organized alphabetically by the name of the telephone owner. The Yellow Pages is a database of addresses and telephone numbers that is alphabetically organized by type of business or service.

A computerized database is a collection of **files** stored on a computer disk or magnetic tape. Files, in turn, are collections of lines of information called **records**. Each record is subdivided into sections of information called **fields**. Below is an example of a database that we refer to throughout this section.

DATABASE DATABX

Name	Title	Years	Office	Salary
Abner, L.	Editor	5	Uptown	35
Brown, K.	Writer	10	Downtown	45
Carmelo, M.	Writer	3	Downtown	27
Clemente, R.	Printer	11	Uptown	34
Goren, F.	Editor	7	Downtown	47
Hodge, J.	Trainer	12	Downtown	50
Kwak, T.	Editor	8	Uptown	49
Rose, P.	Writer	3	Uptown	27
Tam, A.	Writer	12	Downtown	50
Zork, F.	Printer	6	Downtown	24

This database, which I will call *DataBX*, consists of one file with 10 records. Each record contains five fields. The first field contains the name of an employee, the second field the job title of that employee, the third field the number of years the employee has worked for this company, the fourth field the location of the office at which the employee works, and the last field the employee's annual salary in thousands of dollars. For example, P. Rose is a writer who has worked at the company for 3 years, works in the uptown office, and earns $27,000 annually.

PRACTICE

All exercises refer to the database DataBX.

DATABASE DATABX

Name	Title	Years	Office	Salary
Abner, L.	Editor	5	Uptown	35
Brown, K.	Writer	10	Downtown	45
Carmelo, M.	Writer	3	Downtown	27
Clemente, R.	Printer	11	Uptown	34
Goren, F.	Editor	7	Downtown	47
Hodge, J.	Trainer	12	Downtown	50
Kwak, T.	Editor	8	Uptown	49
Rose, P.	Writer	3	Uptown	27
Tam, A.	Writer	12	Downtown	50
Zork, F.	Printer	6	Downtown	24

1. What is A. Tam's job title?

2. What is M. Carmelo's annual salary?

3. For how many years has J. Hodge worked for the company?

4. How many people work in the uptown office?

5. How many editors are there?

6. How many people earn more than $48,000 annually?

Database management systems provide the programs necessary for a user to perform important **database operations**. Database operations include **file creation**, **file editing**, **sorting**, and **querying**.

When we **create** a file, we name and specify the order of the fields for each record of the file and then enter each record into the file. In our example, the fields, in order of appearance, are *name*, *title*, *years*, *office*, and *salary*. When I created the database, I first established these fields and then entered each record with information in the proper order.

File editing includes all the necessary facilities for making changes in a database file. A file can be **modified** by changing a specific field's contents. For example, we could change F. Zork's salary to $30,000. We can also add and delete (erase) records. In our example, we would add a new record if a new employee is hired and delete a record if an existing employee retires, resigns, or is dismissed.

Sorting operations give us the ability to arrange the records in a database file in the numerical or alphabetical order of one of the fields. The example database, DataBX, is already sorted alphabetically by the name field. If we sorted DataBX according to the years field from high to low, we would have this database:

DATABASE DATABX

Name	Title	Years	Office	Salary
Hodge, J.	Trainer	12	Downtown	50
Tam, A.	Writer	12	Downtown	50
Clemente, R.	Printer	11	Uptown	34
Brown, K.	Writer	10	Downtown	45
Kwak, T.	Editor	8	Uptown	49
Goren, F.	Editor	7	Downtown	47
Zork, F.	Printer	6	Downtown	24
Abner, L.	Editor	5	Uptown	35
Carmelo, M.	Writer	3	Downtown	27
Rose, P.	Writer	3	Uptown	27

Notice that complete records are moved about in the sorting process. In this way, the information in each record is unaltered.

PRACTICE

7. Sort the database DataBX according to the salary field from low to high.

8. Sort the database DataBX according to the title field.

A **query** operation is one that allows the user to ask (query) specific questions about the database. All database management systems provide for an extensive variety of querying operations. Although some DBMS provide for English language queries like ''How many editors are there?'', most DBMS require that you formulate your queries in a specific querying language. Each DBMS has its own querying language, so to illustrate the basic types of database queries, we will use a general querying language.

The most fundamental query is one that searches the database for a specific value. Suppose that we wanted to find the record for T. Kwak. We would query the DBMS with the command

FIND(Name = Kwak, T.)

The DBMS would then respond with the record:

Kwak, T.	Editor	8	Uptown	49

Similarly, if we wanted to know who the editors were, we would write

FIND(Title = Editor)

and the DBMS would respond with the records:

Abner, L.	Editor	5	Uptown	35
Goren, F.	Editor	7	Downtown	47
Kwak, T.	Editor	8	Uptown	49

We can also ask for comparative information. For example, if we wanted to know how many people earned more than $40,000 a year, we could write the query

FIND(Salary > 40)

and the DBMS would respond with the records:

Brown, K.	Writer	10	Downtown	45
Goren, F.	Editor	7	Downtown	47
Hodge, J.	Trainer	12	Downtown	50
Kwak, T.	Editor	8	Uptown	49
Tam, A.	Writer	12	Downtown	50

PRACTICE

All exercises refer to the database DataBX.

DATABASE DATABX

Name	Title	Years	Office	Salary
Abner, L.	Editor	5	Uptown	35
Brown, K.	Writer	10	Downtown	45
Carmelo, M.	Writer	3	Downtown	27
Clemente, R.	Printer	11	Uptown	34
Goren, F.	Editor	7	Downtown	47
Hodge, J.	Trainer	12	Downtown	50
Kwak, T.	Editor	8	Uptown	49
Rose, P.	Writer	3	Uptown	27
Tam, A.	Writer	12	Downtown	50
Zork, F.	Printer	6	Downtown	24

9. Write the DBMS response to the query

FIND(Years = 7)

10. Write the DBMS response to the query

FIND(Title = Writer)

11. What query would find the records of all the people who worked in the uptown office?

12. Write the DBMS response to the query

FIND(Years ≤ 5)

A second type of query to a DBMS is one that calls for all values of a particular field. For example, an accountant may wish to have a list of all the salaries paid to employees of the company. A DBMS query to accomplish this task might be written like this:

LIST(Salary)

The DBMS response would be

Salary
35
45
27
34
47
50
49
27
50
24

The last basic querying operation we will examine is the **join** query. A join query asks the database for all records that satisfy two or more conditions. For example, if we wanted all the writers who have more than 4 years with the company, we could write the query

FIND(Title = Writer AND Years > 4)

and the DBMS would respond with the records:

Brown, K.	Writer	10	Downtown	45
Tam, A.	Writer	12	Downtown	50

PRACTICE

All exercises refer to the database DataBX.

13. Write the DBMS response to the query

LIST(Title)

14. Write the DBMS response to the query

FIND(Title = Editor AND Salary > 40)

15. Write the DBMS response to the query

FIND(Years > 10 OR Salary > 35)

In addition to the basic editing and querying functions, database management systems also give the user the ability to create statistical graphs from databases, perform some elementary statistical calculations on databases (although not at the level of the spreadsheet), and make business report forms, invoices, and mailing labels.

Modern DBMS, spreadsheets, and word processors have the ability to incorporate information from and pass information to each other. Thus, a company brochure designed and written with a word processor can include tables and graphs from a DBMS and financial projections calculated by a spreadsheet.

Situation 36 Thinking Machines and Imaginary Worlds

Fern walks up to the door of her apartment. ''I'm home,'' she says.

Her robot/computer, whom she has named ''Rob,'' responds to her voice and opens the apartment door for her. Sensing that Fern is thirsty, Rob prepares a cold drink of iced tea for her.

''Thanks, Rob,'' says Fern, ''Do I have any phone messages?''

''Yes,'' responds Rob, ''Your brother called to ask if you would be able to join him for dinner on Thursday. You are scheduled to have dinner with Margaret on Thursday, so I asked your brother if Wednesday would be OK. He said that it would. That was your only message.''

''How about mail?'' asks Fern.

''A couple of bills,'' Rob replies. I checked them for accuracy and paid them. You also got an electronic flyer from Ben's of Boston. You were down to two pairs of socks so I ordered you a pair in red. The rest was junk mail.''

''Thanks, Rob,'' says Fern, ''After I finish this tea, I'd like a small mushroom pizza and a spinach salad for dinner. And please set me in Tuvalu until dinner.''

''OK,'' Rob answers. Fern's living room becomes a south sea island complete with sand, palm trees, and a gentle ocean breeze.

How close to reality is the preceding tale? Can computers speak and understand English? Can computers think and make decisions? Can a computer make your living room look like an island in the Pacific Ocean? Can a robot make a pizza?

____

____

____

In the last three sections of this chapter, we will look at the rapidly advancing fields of computer science that study artificial intelligence and artificial life. The purpose of these sections is to give you an overview of these areas of computer science.

12.8 ARTIFICIAL INTELLIGENCE AND EXPERT SYSTEMS

Artificial intelligence is the field of computer science devoted to the creation of computer software that models intelligent human behavior. Researchers in artificial intelligence have worked on and written programs that plan, learn, understand human language, *speak* human language, carry on conversations, drive cars, play chess, write poetry, summarize newspaper articles, translate Chinese into English, prove theorems in mathematics, and do a variety of other tasks that we would usually call intelligent.

We will not look in detail at the design of any of these programs because that would be beyond the goals of this book. However, we will look at two of the most fundamental aspects of all artificial intelligence software, and then we will examine briefly an area of artificial intelligence that is used commercially right now.

Let's consider one of the activities performed by the fictional computer Rob in Situation 36. Rob told Fern, ''You got an electronic flyer from Ben's of Boston. You were down to two pairs of socks so I ordered you a pair in red.''

If a computer is going to be able to make intelligent decisions on behalf of a person, then that computer should at least know what the human knows.

What do you need to know and be able to do to order clothing from a catalog?

___

___

___

___

___

You probably listed many things in answer to my last question, but among them are probably the following:

What items do I need?
What items do I want?
What styles do I like?
What colors do I like?
What size do I wear?
Can I afford these items?
How soon do I need these items?

When you think about the answers to these questions, you rely on your past experiences and knowledge. You then use this information and the current information to make a decision. Artificial intelligence software operates in a similar fashion. The computer must have an organized body of knowledge to draw from and the ability to retrieve and use that knowledge.

A **knowledge representation system** is a way of organizing information so that it can be processed intelligently. Information organized using some knowledge representation system is called a **knowledge base**. There are many different kinds of knowledge representation systems used by artificial intelligence software, but one of the more popular is the **rule-based** system. The knowledge in a rule-based knowledge representation system is structured as a series of if–then rules.

For example, part of the knowledge base for the software that shops for me might look like this:

RULE 456: IF I have two pairs of socks or less
THEN I need to buy socks

RULE 457: IF I need to buy socks
THEN I should look for the pair of socks in the Sox R Us catalog

RULE 458: IF I should look for the pair of socks in the Sox R Us catalog
AND the pair of socks I see are the pair of socks that I like
THEN I should buy the pair of socks that I like

RULE 459: IF the pair of socks I see are red
AND the pair of socks I see come in size 13
THEN the pair of socks I see are the pair of socks that I like.

Every *rule* consists of a condition, marked by the word *if*, and an action, marked by the word *then*, to be taken if the condition is met. In a carefully designed, rule-based knowledge base, every possible condition has an appropriate action and every reasonable action has a condition.

A **search procedure** is a method of accessing the information in a knowledge base. A search procedure depends on the knowledge representation system. If the knowledge representation system is a rule-based system, then the search procedure is one that follows the links between actions and conditions.

There are two main ways of following the links between actions and conditions in a rule-based knowledge base: **forward chaining** and **backward chaining**. To see how these two search methods work, it is useful to view the rule-based system as a graph (see Section 7.7).

Here is the graph for the rules on buying socks. The arc connecting two edges leading to an action indicates that both conditions must be true before that action can be true.

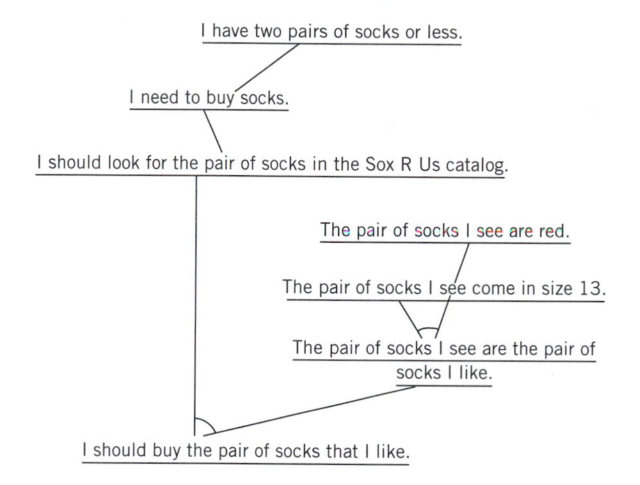

A search procedure moves along the edges of the graph. If I use forward chaining, then I begin with the first condition

I have two pairs of socks or less

and then follow along the edge to the action

I need to buy socks.

This action is itself a condition so the search procedure would direct the computer to its action

I should look for the pair socks in the Sox R Us catalog.

Now this is part of the condition for another rule, so the computer will attempt to satisfy the other condition

The pair of socks I see are the pair of socks that I like.

If both conditions can be satisfied, then the action

I should buy the pair of socks that I like

is taken.

In order to satisfy the condition

The pair of socks I see are the pair of socks that I like

the computer must go back to the conditions that produce this action and see that they are satisfied. Thus, if either

The pair of socks I see are red

or

The pair of socks I see come in size 13

are not satisfied, then we would be unable to reach the action

The pair of socks I see are the pair of socks that I like.

Backward chaining is the process of beginning with an action and then going back to check a condition for an action. If the primary search procedure was backward chaining, then we would begin at the action

I should buy the pair of socks that I like.

We would then move up to conditions

The pair of socks I see are red

and

The pair of socks I see come in size 13

and try to satisfy them. We would continue to move up the graph until we satisfied all conditions or were prevented from moving further.

In summary, forward chaining starts with given rule conditions and attempts to reach a specific action, or goal. Backward chaining begins with a specific action and then attempts to obtain the conditions necessary to cause that action. Here is another example that illustrates the difference between forward chaining and backward chaining.

Suppose that the graph of the rules of a knowledge base looks like this:

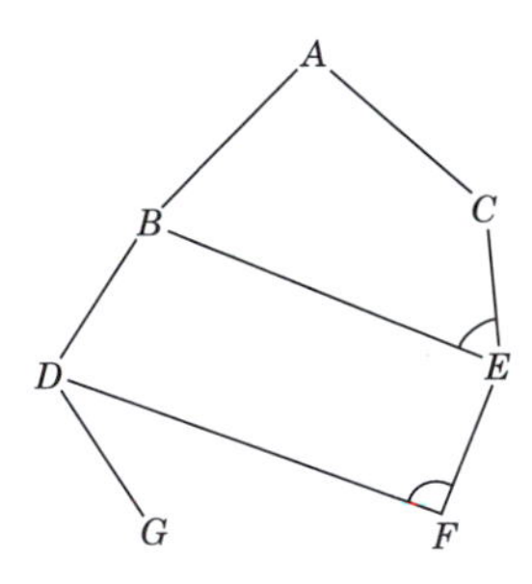

Further suppose that we wish to reach the action labeled F.

If our search procedure is forward chaining, then we would first ask if A were true. If not, then we would stop. If so, then we could conclude B and C. Then we could conclude D and E. Finally, we could conclude F.

On the other hand, if we used backward chaining, we would begin with F. We would move up to D. D could be concluded if we had B and B could be concluded if we had A. Thus, we would ask if A were true. If so, we would then look at E. E could be concluded if B and C were. We already know that B is true, so we need to examine C. We know that A is true, so we can conclude C, then E and consequently F.

PRACTICE

All questions refer to the rule-based knowledge base graph shown below. Assume that the edges move down from conditions to actions.

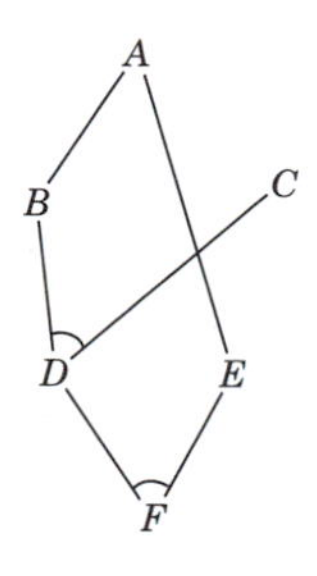

1. Explain how a forward-chaining search procedure would try to conclude *F*.

2. Explain how a backward-chaining search procedure would try to conclude *F*.

Expert systems are widely used commercial applications of the basic artificial intelligence software described above. Expert systems attempt to model the decision-making ability of human experts in particular specialized areas called **knowledge domains**.

Many expert systems use rule-based knowledge representation systems with forward and backward chaining. Expert systems are used by doctors to help diagnose diseases, computer manufacturers to help configure systems for customers, financial planners, auditors, and hundreds of other people in businesses in a large variety of tasks. To read more about expert systems, I encourage you to consult the book *Expert Systems* by Paul Harmon and David King (Wiley, New York: 1985).

12.9 ARTIFICIAL LIFE AND GENETIC ALGORITHMS

Rule-based knowledge representation and search procedures are at the heart of what is referred to as classical or traditional artificial intelligence. Recently, however, researchers have begun to question this traditional approach in which the programmer designs the structure of the knowledge base and the method of retrieving information from it. Many current artificial intelligence researchers are experimenting with ways in which the computer can design its own knowledge structures and write its own programs. These researchers feel that this approach, which includes neural networks, cellular automata, genetic algorithms, and genetic programming, is more likely to produce true artificial intelligence. **Neural network** research seeks to design a computer version of the human brain that can then learn and organize on its own. **Genetic algorithms** and **genetic programming** use the model of human biological genetics to allow computer programs to breed and evolve. As fascinating as each of these areas is, it is well beyond the scope of this book to investigate them. Instead, in this section, we will look at only cellular automata, the basis of a field of computer science known as artificial life.

Cellular automata are systems of dots (cells) that, through the application of simple rules, are self-generating (automata). Cellular automata have important applications in the simulation of some physical phenomena (e.g., wind turbulence), design of super computers, and simulation of biological systems. The simulation of biological systems is known as **artificial life.**

Here we will look at one of the most widely known *worlds* of artificial life involving two-dimensional cellular automata. The world, called Life, consists of a two-dimensional grid whose occupants are dots. A dot lives and dies within a single cell of the grid. Here is an example of the Life world with dots living in cells C5, D5, and E5.

	A	B	C	D	E	F	G	H	I	J	K	L	M	N	O
1															
2															
3															
4															
5			●	●	●										
6															
7															
8															
9															
10															
11															
12															
13															

Each cell has eight neighboring cells. Consequently, each dot in the Life world has eight potential neighbors. In the picture below, each of the neighboring cells of the dot in F5 is marked with an X.

	A	B	C	D	E	F	G	H	I	J	K	L	M	N	O
1															
2															
3															
4					X	X	X								
5					X	●	X								
6					X	X	X								
7															
8															
9															
10															
11															
12															
13															

The Life world, like all cellular automata worlds, is governed by a set of simple rules of birth, life, and death. In order to give our world a temporal dimension, we define the passage of time in terms of these rules. We can view the life history of a dot or group of dots in terms of its births and deaths. Here are the rules of Life.

For each dot and empty cell, sum up the number of neighboring adult dots.

1. *Birth.* If an empty cell has exactly three neighboring adult dots, then a baby dot, denoted by ❍, is born in that cell. That baby will become an adult, denoted by ●, in the next generation.
2. *Life.* If an adult dot has two or three neighboring adult dots, then that dot lives to the next generation.
3. *Death.* If an adult dot has less than two neighboring adult dots or more than three neighboring adult dots, then that dot does not survive to the next generation.

Let's see how the rules of Life affect the three adult dots in the world, called *Blinker*, shown below.

BLINKER: FIRST GENERATION

	A	B	C	D	E	F	G	H	I	J	K	L	M	N	O
1															
2															
3															
4															
5						●	●	●							
6															
7															
8															
9															
10															
11															
12															
13															

Let's look at births first. A birth will occur in any empty cell that has exactly three neighboring dots. We see that cell G4 has three neighboring dots in cells F5, G5, and H5. So, a new dot will be born in G4. Similarly, a new dot will be born in G6.

BLINKER: SHOWING NEW BIRTHS

	A	B	C	D	E	F	G	H	I	J	K	L	M	N	O
1															
2															
3															
4							○								
5						●	●	●							
6							○								
7															
8															
9															
10															
11															
12															
13															

Now let's first determine the fate of the three adult dots. The dot in cell F5 (which I will refer to as F5 for convenience) has only one neighboring adult dot (G5). Thus, F5 will not survive. G5 has two neighboring adult dots (F5 and H5), so G5 will survive. H5 has only one neighboring adult dot and will not survive.

So, in the next generation, F5 and H5 will vanish, G5 will persist, and G4 and G6 will become adults. The resulting picture will be

BLINKER: SECOND GENERATION

	A	B	C	D	E	F	G	H	I	J	K	L	M	N	O
1															
2															
3															
4							●								
5							●								
6							●								
7															
8															
9															
10															
11															
12															
13															

In the next generation, there will be births in cells H5 and F5 because both are empty cells with exactly three adult neighbors. Draw these births as circles in the picture above.

Also, G5 will survive to the next generation, but G4 and G6 will not. Explain why.

____

____

____

Thus, in the third generation, we will have

BLINKER: THIRD GENERATION

	A	B	C	D	E	F	G	H	I	J	K	L	M	N	O
1															
2															
3															
4															
5						●	●	●							
6															
7															
8															
9															
10															
11															
12															
13															

This is the same arrangement with which we began! We know, then, that the fourth generation will look like the second, and the fifth will look like the first, and the sixth will look like the second, and so on. Our original arrangement is self-perpetuating as it alternates between F5–G5–H5 and G4–G5–G6.

Some arrangements, also called **colonies**, are not so fortunate. Consider this next colony, which is called *Star4*.

STAR4: FIRST GENERATION

	A	B	C	D	E	F	G	H	I	J	K	L	M	N	O
1															
2															
3															
4															
5									●						
6									●						
7							●	●	●	●	●				
8									●						
9									●						
10															
11															
12															
13															

Are there any births in Star4?

No, there are not any births in Star4, because no empty cell has exactly three adult neighbors. What about the fate of the adults?

Correct! A quick look at the number of adult neighbors of the adults in Star4 reveals that each adult in Star4 has either more than three adult neighbors or less than two adult neighbors.

With no survivors and no births, this colony is extinct in the second generation.

As a final example, let's look at the colony L, shown below.

L: FIRST GENERATION

	A	B	C	D	E	F	G	H	I	J	K	L	M	N	O
1															
2															
3															
4															
5				●											
6				●											
7				●	●	●									
8															
9															
10															
11															
12															
13															

Here are the births in L. Make sure you see that each birth occurred in an empty cell with exactly three adult neighbors.

L: BIRTHS IN FIRST GENERATION

	A	B	C	D	E	F	G	H	I	J	K	L	M	N	O
1															
2															
3															
4															
5				●											
6			❍	●											
7				●	●	●									
8					❍										
9															
10															
11															
12															
13															

The adults that survive to the next generation are D6, D7, and E7. So, we have

L: SECOND GENERATION

	A	B	C	D	E	F	G	H	I	J	K	L	M	N	O
1															
2															
3															
4															
5															
6			●	●											
7				●	●										
8					●										
9															
10															
11															
12															
13															

The births in the second generation are in cells C7, D8, and E6, giving:

L: BIRTHS IN SECOND GENERATION

	A	B	C	D	E	F	G	H	I	J	K	L	M	N	O
1															
2															
3															
4															
5															
6			●	●	❍										
7			❍	●	●										
8				❍	●										
9															
10															
11															
12															
13															

The adults C6, D6, E7, and E8 survive to the next generation, giving:

L: THIRD GENERATION

	A	B	C	D	E	F	G	H	I	J	K	L	M	N	O
1															
2															
3															
4															
5															
6			●	●	●										
7			●		●										
8				●	●										
9															
10															
11															
12															
13															

The succeeding generations are shown below. Make sure that you agree with them!

L: BIRTHS IN THIRD GENERATION

	A	B	C	D	E	F	G	H	I	J	K	L	M	N	O
1															
2															
3															
4															
5				❍											
6			●	●	●										
7			●		●	❍									
8				●	●										
9															
10															
11															
12															
13															

L: FOURTH GENERATION

	A	B	C	D	E	F	G	H	I	J	K	L	M	N	O
1															
2															
3															
4															
5				●											
6			●		●										
7			●			●									
8				●	●										
9															
10															
11															
12															
13															

L: FIFTH GENERATION

	A	B	C	D	E	F	G	H	I	J	K	L	M	N	O
1															
2															
3															
4															
5				●											
6			●		●										
7			●			●									
8				●	●										
9															
10															
11															
12															
13															

L: SIXTH GENERATION AND ON

	A	B	C	D	E	F	G	H	I	J	K	L	M	N	O
1															
2															
3															
4															
5				●											
6			●		●										
7			●			●									
8				●	●										
9															
10															
11															
12															
13															

The colony L becomes immortal with no births or deaths after the fourth generation.

PRACTICE

Trace the evolution of each of the following colonies in the cellular automata world of Life:

1. **RAMP: FIRST GENERATION**

	A	B	C	D	E	F	G	H	I	J	K	L	M	N	O
1															
2															
3															
4															
5				●											
6				●	●										
7				●	●	●									
8															
9															
10															
11															
12															
13															

2. **GLIDER: FIRST GENERATION**

	A	B	C	D	E	F	G	H	I	J	K	L	M	N	O
1															
2															
3															
4															
5															
6				●	●										
7			●	●											
8					●										
9															
10															
11															
12															
13															

3.

T: FIRST GENERATION

	A	B	C	D	E	F	G	H	I	J	K	L	M	N	O
1															
2															
3															
4															
5															
6				●	●	●									
7					●										
8					●										
9															
10															
11															
12															
13															

Cellular automata, even the very simple two-dimensional automata of the Life world, mimic the biological processes of birth, death, and evolution. When the ideas of evolution and survival are extended to complete programs, we have what are known as **genetic programs.**

In genetic programming, the computer is allowed to create its own programs that are then judged as to their correctness for a given task. The programs that are most correct are allowed to survive and breed new programs. The process continues until a satisfactory program is created.

Genetic programming, while fascinating, is well beyond the goals of this book, so I will end here and hope that you will be interested enough to explore some of the topics of artificial life on your own.

12.10 WHAT DO YOU KNOW?

If you have worked carefully through this chapter, then you know the basics of computer architecture, languages, and programs, can determine the language generated by a formal grammar, construct formal grammars for specified languages, know the basic ideas behind fundamental applications software, and have some familiarity with the concepts of artificial intelligence. The exercises in this section are designed to test and refine your skills in the mathematics of computers.

Identify each of the following:

1. CPU
2. Output device
3. Arithmetic and logic unit
4. Internal memory
5. External memory
6. Hardware
7. Software
8. Hard disk
9. Floppy disk
10. Compiler
11. Assembler
12. Your home computer is an IBM PC with 640K RAM, a 2-megabyte hard disk, and one floppy disk drive. You would like to purchase the computer game ''Bouncing Baby Beetles.'' The game's package says that the game requires 512K RAM and a 1-megabyte hard disk. Can you run this game on your computer?
13. Your home computer is a Tandy 1000EX with 256K RAM and two floppy disk drives. You would like to purchase the computer game ''Bouncing Baby Beetles.'' The game's package says that the game requires 512K RAM and a 1-megabyte hard disk. Can you run this game on your computer?
14. Your home computer is a Zenith PC with 1056K RAM and two floppy disk drives. You would like to purchase the computer game ''Bouncing Baby Beetles.'' The game's package says that the game requires 512K RAM and one floppy disk drive. Can you run this game on your computer?
15. Write an essay in which you compare and contrast the Macintosh operating system with that of Windows.
16. Go to a local store that sells computer software. Randomly select 20 software packages. Find out what the memory and operating system requirements are for each of the 20. What size and type of machine would you need to run all these programs?
17. Find out how to write the instruction *add 2 to 5* in an assembly language.
18. Find out how to write the instruction *add 2 to 5* in a machine language.

Suppose that the postage on a letter in some country costs 25¢ per ounce for letters under 1 pound and 20¢ per ounce for letters weighing more than 1 pound.

19. Write the instructions in Pascal for a computer to figure the postage given the weight of the letter.
20. Write the instructions in FORTRAN for a computer to figure the postage given the weight of the letter.
21. Write the instructions in C for a computer to figure the postage given the weight of the letter.
22. Write the instructions in LISP for a computer to figure the postage given the weight of the letter.

Suppose that a music store is selling Strawberrry Alarm Clock CDs for $2 a piece if you buy three or less and $1.75 a piece if you buy more than three.

23. Write the instructions in Pascal for a computer to figure the cost given the number of CDs.
24. Write the instructions in FORTRAN for a computer to figure the cost given the number of CDs.
25. Write the instructions in C for a computer to figure the cost given the number of CDs.
26. Write the instructions in LISP for a computer to figure the cost given the number of CDs.

Let G_9 be the grammar defined below.

$T = \{a, b\}$
$NT = \{S, X, Y\}$
Production rules:

1. $S \rightarrow XY$
2. $X \rightarrow aY$
3. $Y \rightarrow bS$
4. $Y \rightarrow a$

27. Show that *abaaaa* is a word in $L(G_9)$.
28. Show that *ababaaaaa* is a word in $L(G_9)$.
29. Show that *abbb* is not a word in $L(G_9)$.
30. Describe $L(G_9)$.

Let G_{10} be the grammar defined below.

$T = \{a, b, c\}$
$NT = \{S, X, Y\}$
Production rules:

1. $S \rightarrow XcY$
2. $X \rightarrow aX$
3. $X \rightarrow c$
4. $Y \rightarrow bY$
5. $Y \rightarrow b$

31. Show that *aaccb* is a word in $L(G_{10})$.
32. Show that *aaaaaccbbb* is a word in $L(G_{10})$.

33. Show that *accabbb* is not a word in $L(G_{10})$.

34. Describe $L(G_{10})$.

Let G_{11} be the grammar defined below.

T = {yabba, dabba, doo, -}
$NT = \{S, X, Y\}$
Production rules:

1. $S \rightarrow$ yabba-X
2. $X \rightarrow$ dabba-X
3. $X \rightarrow$ dabba-Y
4. $Y \rightarrow$ doo

35. Show that yabba-dabba-doo is a word in $L(G_{11})$.

36. Show that yabba-dabba-dabba-dabba-doo is a word in $L(G_{11})$.

37. Show that yabba-yabba-doo is not a word in $L(G_{11})$.

38. Describe $L(G_{11})$.

Let G_{12} be the grammar defined below.

T = {see, spot, run, look, Dick, Jane, and, ·, .}
$NT = \{S, V, NP, M, N\}$
Production rules:

1. $S \rightarrow V \cdot NP.$
2. $S \rightarrow V \cdot NP.S$
3. $V \rightarrow$ see
4. $V \rightarrow$ look
5. $NP \rightarrow N \cdot$ run
6. $NP \rightarrow N$
7. $N \rightarrow M \cdot$ and $\cdot M$
8. $M \rightarrow$ Dick
9. $M \rightarrow$ Jane
10. $N \rightarrow$ Spot

39. Show that see · Spot · run. is a word in $L(G_{12})$.

40. Show that look · Dick · and · Jane.see · Spot · run. is a word in $L(G_{12})$.

41. Show that run · Spot · run. is not a word in $L(G_{12})$.

42. Describe $L(G_{12})$.

43. Write a grammar G_{13} such that

$$L(G_{13}) = \{ab, aabb, aaabbb, \ldots\}$$

44. Write a grammar G_{14} such that

$$L(G_{14}) = \{\text{boo, boohoo, boohoohoo, boohoohoohoo} \ldots\}$$

45. Write a grammar G_{15} such that

$$L(G_{15}) = \{aabc, aabbcc, aabbbccc, aabbbbcccc \ldots\}$$

46. Write a grammar G_{16} such that

$$L(G_{16}) = \{101, 1001, 10001, 100001 \ldots\}$$

47. Write a grammar G_{17} such that

$$L(G_{17}) = \{acb, accb, acccb, accccb, acccccb, \ldots\}$$

48. Write a grammar G_{18} such that

$$L(G_{18}) = \{abba, abaaba, abaabbaaba, \ldots\} = \{x\colon x = ww^R \text{ where } w \text{ is a string of } a\text{'s and } b\text{'s and } w^R \text{ is the reverse of } w\}$$

49. Write a grammar G_{19} such that

$$L(G_{19}) = \{x\colon x = aawbb \text{ where } w \text{ is a string of an odd number of } c\text{'s}\}$$

50. Write a grammar G_{20} such that

$$L(G_{20}) = \{x\colon x \text{ is an even number expressed as a binary numeral}\}$$

51. Write a grammar G_{20} such that

$$L(G_{20}) = \{x\colon x \text{ is an odd number expressed as a binary numeral}\}$$

52. Obtain some advertising information for each of the word processors, MicroSoft Word and WordPerfect. What features do they have in common? What features are unique?

53. Obtain some advertising information for each of the spreadsheets, Lotus 1-2-3 and QuattroPro. What features do they have in common? What features are unique?

54. Find a user's guide or handbook for some word processor in your library or borrow one from a friend. Find out how to

(a) Run the spell checker.

(b) Move a paragraph from one place to another.

(c) Delete a paragraph.

(d) Change from single spacing to double spacing.

55. Find a user's guide or handbook for some spreadsheet program in your library or borrow one from a friend. Find out how to

(a) Find the mean of a column of values.

(b) Draw a bar graph for a column of values.

(c) Find the standard deviation of a column of values.

56. Find a user's guide or handbook for some DBMS in your library or borrow one from a friend. Find out how to

(a) Sort by a field.

(b) Query a database for a specific value.

(c) Produce a list of field values.

Exercises 57–62 refer to the database *Dogs* shown below.

DATABASE DOGS

Name	Breed	Age	Owner	State
Snoopy	Beagle	3	C. Brown	NY
Fido	Beagle	6	T. Fern	WY
Balzac	Poodle	7	G. Whiz	FL
Big Red	Greyhound	2	C. Bunga	NY
Alfie	Terrier	5	C. Diego	NY
Toto	Terrier	3	D. Smith	KS

57. Write the DBMS response to the query

FIND(Age > 5)

58. Write the DBMS response to the query

FIND(Age > 5 AND Breed = Beagle)

59. Write the DBMS response to the query

FIND(State = NY OR Age < 5)

60. Write the DBMS response to the query

FIND(Breed = Terrier OR Breed
= Beagle OR State = NY)

61. Sort the Dogs database according to the age field from high to low.

62. Sort the Dogs database according to the state field.

Exercises 63–68 refer to the database *Cats* shown below.

DATABASE CATS

Name	Breed	Age	Owner	State
Smiles	Persian	13	C. Bono	CA
Claws	Tiger	6	T. Frenzy	WY
Balzac	Tiger	7	G. Whiz	FL
Boozer	Persian	7	F. Zorka	CA
Aleph0	Tiger	15	C. Mendes	NY
Tom	Tiger	7	D. Smith	CA

63. Write the DBMS response to the query

FIND(Age = 7)

64. Write the DBMS response to the query

FIND(Age > 6 AND Breed = Tiger)

65. Write the DBMS response to the query

FIND(State = CA OR Name = Balzac)

66. Write the DBMS response to the query

FIND(Breed = Persian OR Age = 7 OR State = CA)

67. Sort the Cats database according to the age field from high to low.

68. Sort the Cats database according to the name field.

69. Write a series of rules for an expert system in a knowledge domain of your choice.

70. Find a beginning text on artificial intelligence and write an essay in which you explain the fundamentals of the knowledge representation systems called *frames* and *scripts*.

Exercises 71–72 refer to the rule graph shown below:

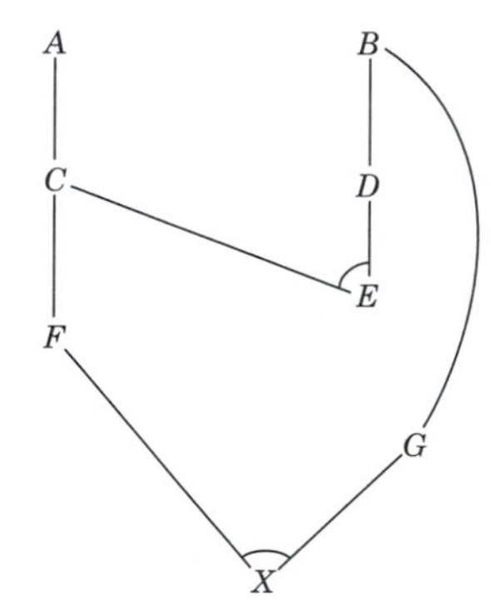

71. Explain how a forward-chaining search would try to reach the action *X*.

72. Explain how a backward-chaining search would try to reach the action *X*.

Exercises 73–74 refer to the rule graph shown below:

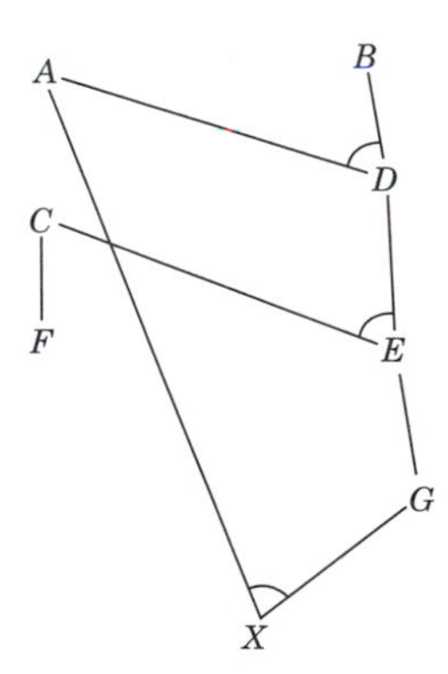

73. Explain how a forward-chaining search would try to reach the action X.

74. Explain how a backward-chaining search would try to reach the action X.

75. Do some research and prepare a paper on one of the following topics in artificial intelligence or artificial life:
   (a) Neural networks
   (b) Fuzzy logic
   (c) Genetic programming
   (d) Natural language processing
   (e) Machine translation

76. Trace the evolution of the following colony in the cellular automata world of Life:

4LINE: FIRST GENERATION

	A	B	C	D	E	F	G	H	I	J	K	L	M	N	O
1															
2															
3															
4															
5				●	●	●	●								
6															
7															
8															
9															
10															
11															
12															
13															

77. Trace the evolution of the following colony in the cellular automata world of Life:

ARROW: FIRST GENERATION

	A	B	C	D	E	F	G	H	I	J	K	L	M	N	O
1															
2															
3															
4						●									
5				●	●	●	●								
6						●									
7															
8															
9															
10															
11															
12															
13															

78. Trace the evolution of the following colony in the cellular automata world of Life:

V: FIRST GENERATION

	A	B	C	D	E	F	G	H	I	J	K	L	M	N	O
1															
2															
3															
4				●				●							
5					●		●								
6						●									
7															
8															
9															
10															
11															
12															
13															

79. Trace the evolution of the following colony in the cellular automata world of Life:

ELEVEN: FIRST GENERATION

	A	B	C	D	E	F	G	H	I	J	K	L	M	N	O
1															
2															
3															
4															
5					●		●								
6					●		●								
7					●		●								
8															
9															
10															
11															
12															
13															

80. Trace the evolution of the following colony in the cellular automata world of Life:

GLIDER CRASH: FIRST GENERATION

	A	B	C	D	E	F	G	H	I	J	K	L	M	N	O
1															
2															
3															
4															
5															
6					●	●									
7					●	●									
8															
9										●	●				
10									●	●					
11											●				
12															
13															

ANSWERS TO PRACTICE EXERCISES

Section 12.1

1. Processing unit, internal memory, arithmetic and logic unit.
2. Internal memory is within the CPU; external memory is outside of the CPU
3. Its internal memory can hold 640,000 bytes of information and its external memory can store 3,000,000 bytes of information.

Section 12.3

Pascal
```
IF BOXES < 7 THEN
  COST := BOXES*3.00
ELSE
  COST := BOXES*2.80;
```

FORTRAN
```
IF (BOXES.LT.7) THEN
  COST = BOXES*3.00
ELSE
  COST = BOXES*2.80
```

C
```
IF (BOXES < 7) THEN
  COST = BOXES*3.00;
ELSE
  COST = BOXES*2.80;
```

COBOL
```
IF BOXES IS LESS THAN 7
  COMPUTE COST = BOXES*3.00
ELSE
  COMPUTE COST = BOXES*2.80.
```

LISP
```
(IF (< BOXES 7)
                 (SETQ COST (* BOXES 3.00))
                 (SETQ COST (* BOXES 2.80)))
```

Section 12.4

1. $S \Rightarrow_2 aS \Rightarrow_2 aaS \Rightarrow_2 aaaS \Rightarrow_2 aaaaS \Rightarrow_1 aaaaaC \Rightarrow_3 aaaaab$
2. (a) $S \Rightarrow_1 aDb \Rightarrow_2 acDb \Rightarrow_2 accDb \Rightarrow_3 acccb$
   (b) $S \Rightarrow_1 aDb \Rightarrow_3 acb$
   (c) The words of $L(G_3)$ are strings that begin with a, end with b, and have one or more c in between.
3. (a) $S \Rightarrow_1 SX \Rightarrow_2 cZXX \Rightarrow_4 cbbXX \Rightarrow_3 cbbaaX \Rightarrow_3 cbbaaaa$
   (b) No production rule that has S on its left-hand side can result in the current string beginning with b.
   (c) The words of $L(G_5)$ are strings that begin with cbb followed by an even number of a's.
4. G_7:
   $T = \{a, b\}$
   $NT = \{S, X\}$
   Production rules:
   1. $S \rightarrow bXb$
   2. $X \rightarrow aX$
   3. $X \rightarrow a$
5. G_8:
   $T = \{a, b, c\}$
   $NT = \{S, X\}$
   Production rules:
   1. $S \rightarrow bcXbc$
   2. $S \rightarrow bcSbc$
   3. $X \rightarrow a$

Section 12.6

1. 72
2. 190
3. OUNDS) OF
4. SOME PEO

Section 12.7

1. Writer
2. $27,000
3. 12
4. 4
5. 3
6. 3
7. **DATABASE DATABX**

Name	Title	Years	Office	Salary
Zork, F.	Printer	6	Downtown	24
Carmelo, M.	Writer	3	Downtown	27
Rose, P.	Writer	3	Uptown	27
Clemente, R.	Printer	11	Uptown	34
Abner, L.	Editor	5	Uptown	35
Brown, K.	Writer	10	Downtown	45
Goren, F.	Editor	7	Downtown	47
Kwak, T.	Editor	8	Uptown	49
Hodge, J.	Trainer	12	Downtown	50
Tam, A.	Writer	12	Downtown	50

8. **DATABASE DATABX**

Name	Title	Years	Office	Salary
Abner, L.	Editor	5	Uptown	35
Goren, F.	Editor	7	Downtown	47
Kwak, T.	Editor	8	Uptown	49
Clemente, R.	Printer	11	Uptown	34
Zork, F.	Printer	6	Downtown	24
Hodge, J.	Trainer	12	Downtown	50
Brown, K.	Writer	10	Downtown	45
Carmelo, M.	Writer	3	Downtown	27
Rose, P.	Writer	3	Uptown	27
Tam, A.	Writer	12	Downtown	50

9. Goren, F. Editor 7 Downtown 47
10. Brown, K. Writer 10 Downtown 45
    Carmelo, M. Writer 3 Downtown 27

Rose, P.	Writer	3	Uptown	27
Tam, A.	Writer	12	Downtown	50

11. FIND(Office = Uptown)

12.

Abner, L.	Editor	5	Uptown	35
Carmelo, M.	Writer	3	Downtown	27
Rose, P.	Writer	3	Uptown	27

13.

	Title	
Editor	Editor	Writer
Writer	Trainer	Printer
Writer	Editor	
Printer	Writer	

14.

Goren, F.	Editor	7	Downtown	47
Kwak, T.	Editor	8	Uptown	49

15.

Brown, K.	Writer	10	Downtown	45
Clemente, R.	Printer	11	Uptown	34
Goren, F.	Editor	7	Downtown	47
Hodge, J.	Trainer	12	Downtown	50
Kwak, T.	Editor	8	Uptown	49
Tam, A.	Writer	12	Downtown	50

Section 12.8

1. Begin at *A*. If *A* is true, conclude *B* and *E*. Examine *C*. If *C* is true, conclude *D*. If *D* has been concluded, then conclude *F*.

2. Begin at *F*. Move up to *D*. *D* is an action so move up to *B*. *B* is an action, so move up to *A*. If *A* is true, examine *C*. If *C* is true, conclude *D*. Now look at *E*. *E* is true if *A* is. If *E* and *D* are true, then conclude *F*.

Section 12.9

1. **RAMP: FIRST GENERATION**

	A	B	C	D	E	F	G	H	I	J	K	L	M	N	O
1															
2															
3															
4															
5				●											
6				●	●										
7				●	●	●									
8															
9															
10															
11															
12															
13															

RAMP: SECOND GENERATION

	A	B	C	D	E	F	G	H	I	J	K	L	M	N	O
1															
2															
3															
4															
5				●	●										
6			●			●									
7				●		●									
8					●										
9															
10															
11															
12															
13															

RAMP: THIRD GENERATION AND ON

	A	B	C	D	E	F	G	H	I	J	K	L	M	N	O
1															
2															
3															
4															
5				●	●										
6			●			●									
7				●		●									
8					●										
9															
10															
11															
12															
13															

2.

GLIDER: FIRST GENERATION

	A	B	C	D	E	F	G	H	I	J	K	L	M	N	O
1															
2															
3															
4															
5															
6				●	●										
7			●	●											
8					●										
9															
10															
11															
12															
13															

GLIDER: SECOND GENERATION

	A	B	C	D	E	F	G	H	I	J	K	L	M	N	O
1															
2															
3															
4															
5															
6			●	●	●										
7			●												
8				●											
9															
10															
11															
12															
13															

GLIDER: THIRD GENERATION

	A	B	C	D	E	F	G	H	I	J	K	L	M	N	O
1															
2															
3															
4															
5				●											
6			●	●											
7			●		●										
8															
9															
10															
11															
12															
13															

GLIDER: FOURTH GENERATION

	A	B	C	D	E	F	G	H	I	J	K	L	M	N	O
1															
2															
3															
4															
5			●	●											
6			●		●										
7			●												
8															
9															
10															
11															
12															
13															

GLIDER: FIFTH GENERATION

	A	B	C	D	E	F	G	H	I	J	K	L	M	N	O
1															
2															
3															
4															
5			●	●											
6		●	●												
7				●											
8															
9															
10															
11															
12															
13															

In the fifth generation, the Glider colony has retaken its original form in a position that is one cell up and one cell to the left of its original position. In succeeding generations, Glider will move left and up through the life world grid.

3.

T: FIRST GENERATION

	A	B	C	D	E	F	G	H	I	J	K	L	M	N	O
1															
2															
3															
4															
5															
6				●	●	●									
7					●										
8					●										
9															
10															
11															
12															
13															

T: SECOND GENERATION

	A	B	C	D	E	F	G	H	I	J	K	L	M	N	O
1															
2															
3															
4															
5					●										
6				●	●	●									
7					●										
8															
9															
10															
11															
12															
13															

T: THIRD GENERATION

	A	B	C	D	E	F	G	H	I	J	K	L	M	N	O
1															
2															
3															
4															
5				●	●	●									
6				●		●									
7				●	●	●									
8															
9															
10															
11															
12															
13															

T: FOURTH GENERATION

	A	B	C	D	E	F	G	H	I	J	K	L	M	N	O
1															
2															
3															
4					●										
5				●		●									
6			●				●								
7				●		●									
8					●										
9															
10															
11															
12															
13															

T: FIFTH GENERATION

	A	B	C	D	E	F	G	H	I	J	K	L	M	N	O
1															
2															
3															
4					●										
5				●	●	●									
6			●	●		●	●								
7				●	●	●									
8					●										
9															
10															
11															
12															
13															

T: SIXTH GENERATION

	A	B	C	D	E	F	G	H	I	J	K	L	M	N	O
1															
2															
3															
4				●	●	●									
5			●				●								
6			●				●								
7			●				●								
8				●	●	●									
9															
10															
11															
12															
13															

T: SEVENTH GENERATION

	A	B	C	D	E	F	G	H	I	J	K	L	M	N	O
1															
2															
3					●										
4				●	●	●									
5			●		●		●								
6		●	●	●		●	●	●							
7			●		●		●								
8				●	●	●									
9					●										
10															
11															
12															
13															

T: EIGHTH GENERATION

	A	B	C	D	E	F	G	H	I	J	K	L	M	N	O
1															
2															
3				●	●	●									
4															
5		●						●							
6		●						●							
7		●						●							
8															
9				●	●	●									
10															
11															
12															
13															

T: NINTH GENERATION

	A	B	C	D	E	F	G	H	I	J	K	L	M	N	O
1															
2					●										
3					●										
4					●										
5															
6	●	●	●				●	●	●						
7															
8					●										
9					●										
10					●										
11															
12															
13															

T: TENTH GENERATION

	A	B	C	D	E	F	G	H	I	J	K	L	M	N	O
1															
2															
3				●	●	●									
4															
5		●						●							
6		●						●							
7		●						●							
8															
9				●	●	●									
10															
11															
12															
13															

T: ELEVENTH GENERATION

	A	B	C	D	E	F	G	H	I	J	K	L	M	N	O
1															
2					●										
3					●										
4					●										
5															
6	●	●	●				●	●	●						
7															
8					●										
9					●										
10					●										
11															
12															
13															

The colony persists as a set of four Blinkers.

Answers to Selected Exercises

Chapter One

1. (b) 2/1; (d) 9/7
3. 30 eggs
5. 16 minutes
7. 18 ounces
9. 103.5
11. .0072
13. 58.99%
15. .988%
17. 176.47
19. 225
21. 30%
23. (a) 20.69%; (c) 57.89%
25. 523.33 acres
27. (a) $800; (b) 6.06%
29. (a) $29.40; (c) $2940
31. (a) $188.80; (b) −5.6%
33. $23.40
35. $5166.67
37. (a) 6%; (c) 72%; (e) 2160%
39. $237.46
41. $8.60
43. (a) 11 months; (b) $87.61; (c) $55.31
45. (a) $382.33; (b) 9.5%
47. (a) 6.69%; (c) 3.24%
49. (a) Monthly payment = $692.02,
    total cost = $260,927.20.
    (c) Monthly payment = $745.71,
    total cost = $199.770.40.

Chapter Two

5. Mean = 250
   Median = 250
   Mode = 100 200 300 400
   sd = 111.80
7. Mean = 23.75
   Median = 20
   Mode = 20
   sd = 12.18
9. Mean = median = mode = 5
   sd = 0
11. The mode
13. This year
15. 88
17. (a) Mean = 93.17, sd = 10.63
18.

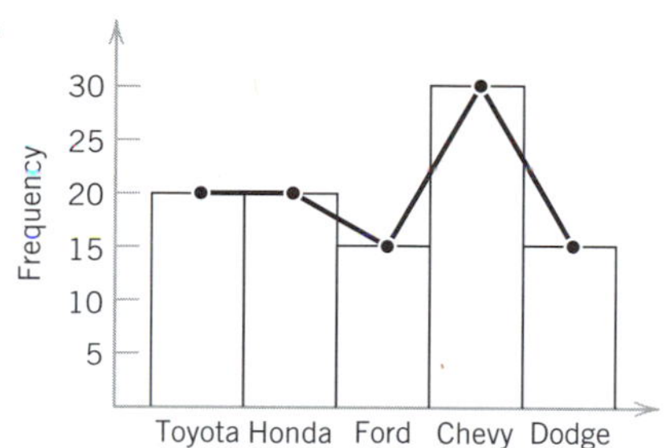

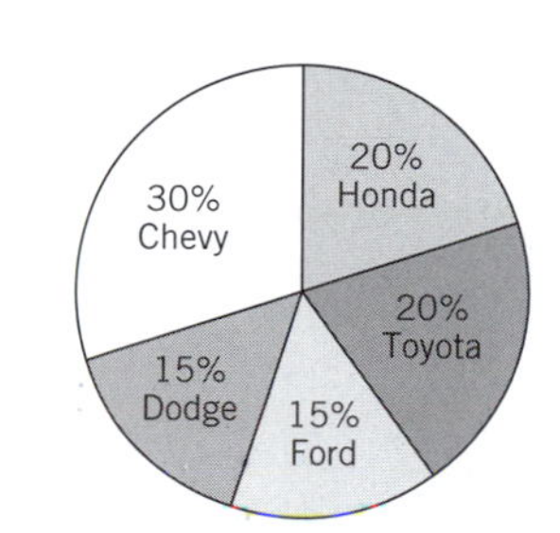

20.

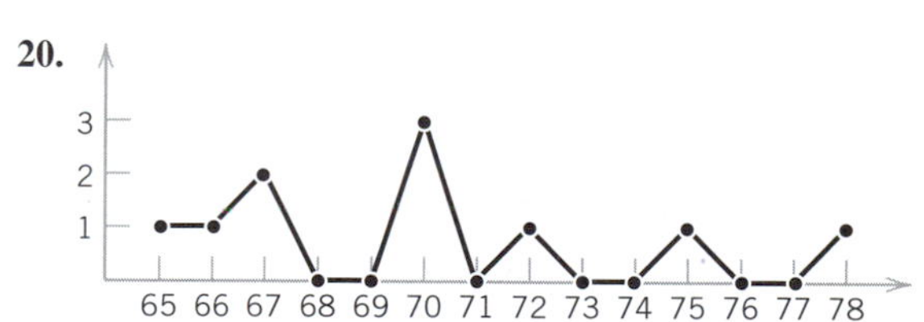

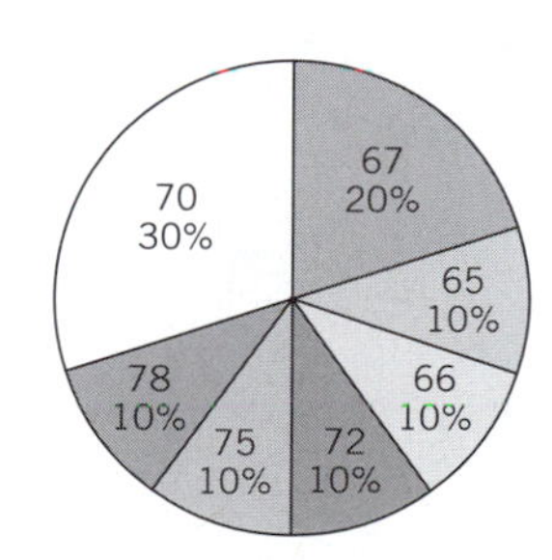

23. 70
25. July
27. 50%
29. 25%
31. 99.88%
33. (a) about 3; (c) about 1547

Chapter Three

1. Yes
3. No
5. Yes
7. Yes
9. {Santiago}
11. {5, 6, 7, 8, 9, 10, 11}
13. {Lions, Colts, Bears, etc.}
15. {x: x is a nonhuman character in ''Winnie the Pooh''}
17. {x: x is a city in California}
19. {x: x is continent that is also a country}
21. True
23. False
25. False
27. True
29. True
31. True
33. True
35. False
37.

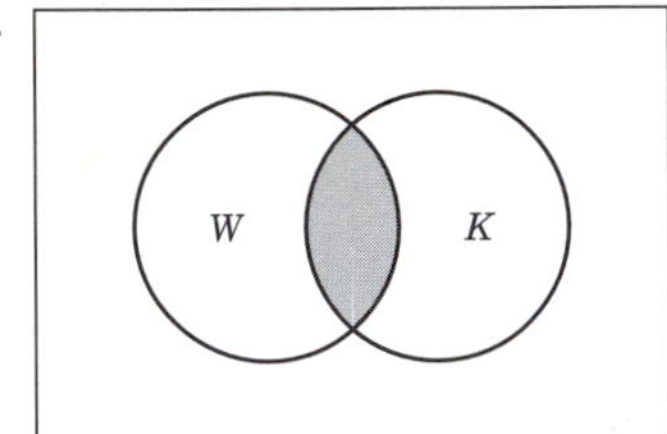

39.

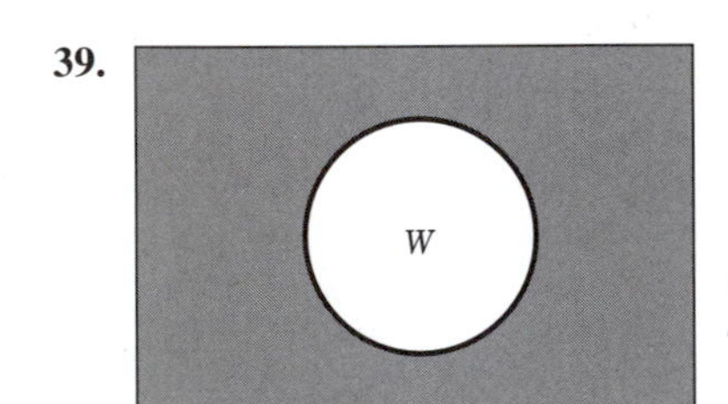

41.

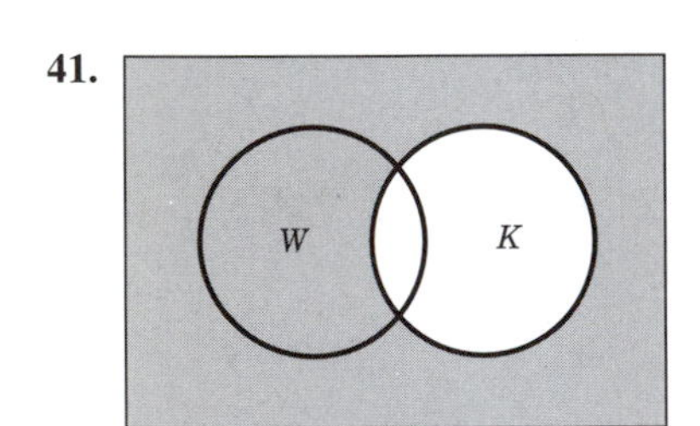

43.

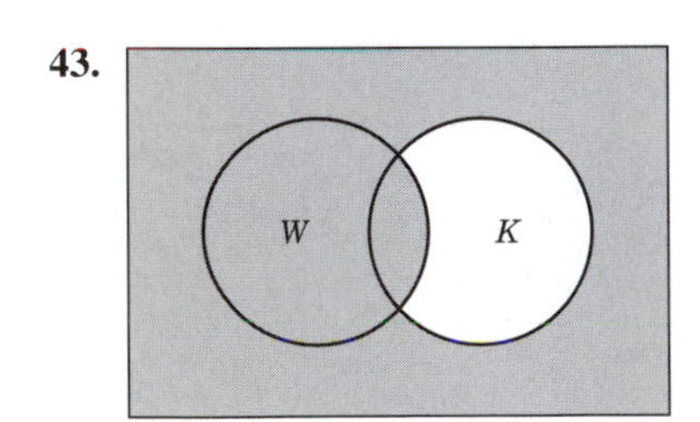

45. (a) If there is an element of H that isn't an element of G, then H is not a subset of G.
47. {s}, {a}, {d}, {i}, {s, a}, {s, d}, {s, i}, {a, d}, {a, i}, {d, i}, {s, a, d}, {s, a, i}, {s, d, i}, {a, d, i}, {s, a, d, i}, empty set
49. (c) They are the same. This result is called De Morgan's law.
51. 3
53. 1
55. 1
57. 4
59. 6
61.

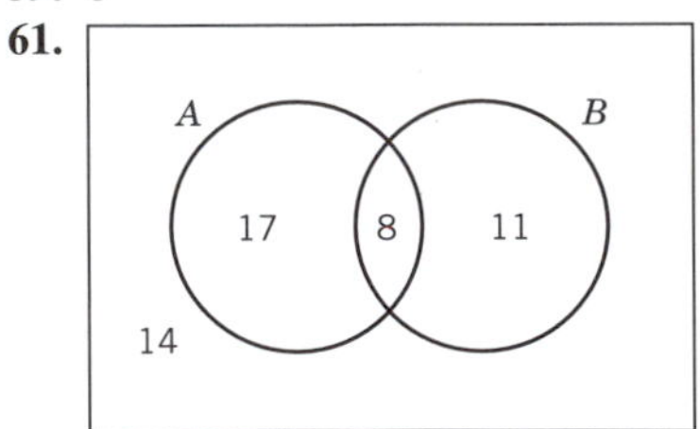

63.

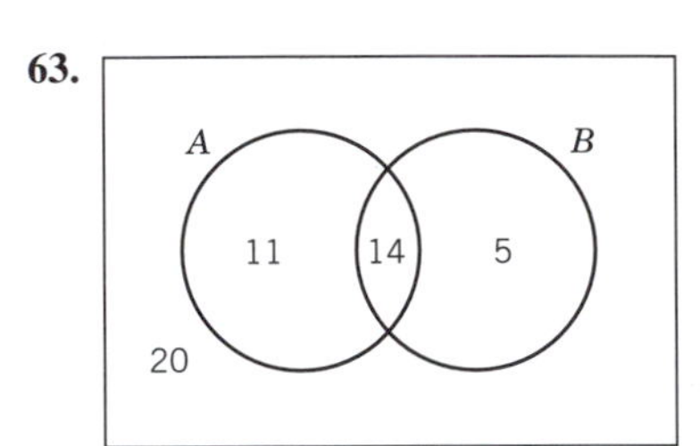

65. 20
67. 138
69. 119
71. 53
73. 236
75. 166
77. 22
79. 130
81. 200
83. 65
85. 206
87. 146
89. {x: x is an odd number between 0 and 25}
91. 15
93. 12
95. {x: x is an even number between 0 and 25}
97. 13
99. n{x: x is an odd number between 0 and 25 other than 3 or 9} = 11
101. Function, domain = {1, 2, 3}, range = {4, 5, 6} = codomain
103. Not a function
105. Function, domain = {1, 3, 5, 7, . . .}, range = {2, 4, 6, 8, . . .} = codomain
107. One-to-one
109. One-to-one and onto
111. One-to-one and onto
113. Function, not onto, not one-to-one
115. Not a function

Chapter Four

1. {Fred, Mel, Polly, Kate}
3. 3/4
5. {HHH, HHT, HTH, THH, HTT, THT, TTH, TTT}

7. 1/8
9. 1/2
11. 1/6
13. {Apple, pear}
15. 2 to 3
17. 3/23, 20/23
19. 1/13
21. 36
23. 1/6
25. $7
27. 120
29. 676
31. 26
33. 1976
35. 24
37. 1/15
39. 2/3
41. 6 to 1
43. 1/10
45. 4/10
47. $37
49. 455/2300
51. $0
53. 3/11
55. 1/52
57. 1/4
59. 1/13
61. 0.4878
63. 0.077
65. 0.00204
67. 0.00000178
69. 0.967
71. 0.425
73. 0.298
75. 12/204
77. −$14.12
79. 1/1296
81. 1/37
83. 18/37
85. −$0.03
87. 0.55 (assuming a 50% chance of rain)
89. Chi-square = 9, *df* = 1, yes
91. Chi-square = 6.43 *df* = 1, no
93. Chi-square = 9, *df* = 1, improvement is not due to something other than chance if she took 10 shots.

Chapter Five

1.

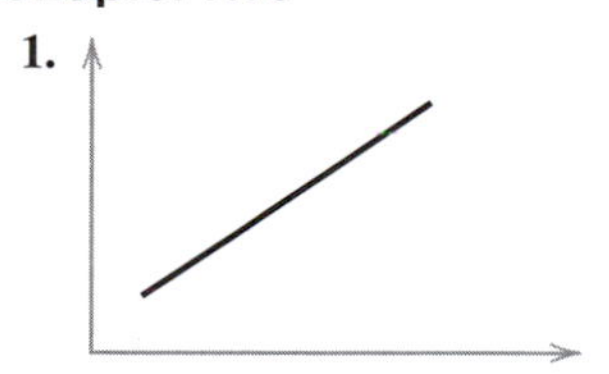

3. $A(-2, 1)$, $B(2, 3)$, $C(2, 2.5)$, $D(4, 0)$ $E(9, -1)$
5. (a) Linear
(b) Quadratic
(c) None
(d) Exponential
(e) Linear
(f) Quadratic
7. Negative
9. (a) $y = 4x + 210$
(b)

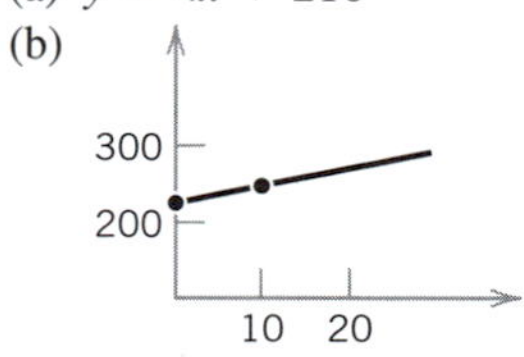

(c) $410
(d) 197
11. (a) $y = 11{,}384.62x - 22{,}529{,}547.66$
(b)

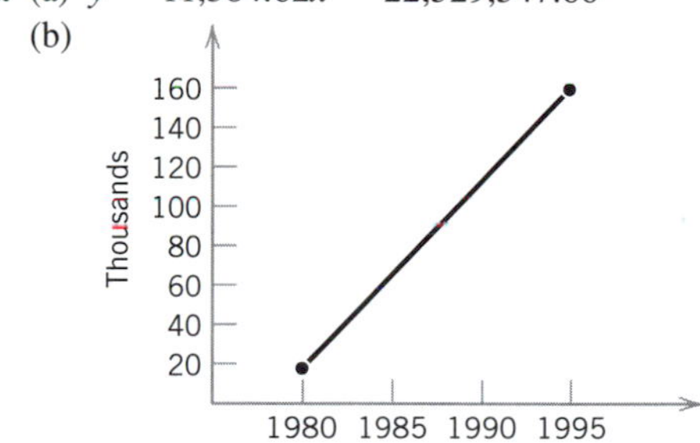

(c) $228,307.72
(d) 2023
13. (a) 0.982
(b) 0.122
(c) −0.944
17. (a) 1126.69
(b) 1,087,125.70
(c) 41.21
(d) 12,488.73
19. (a) 500,000,000
(b) 250,000,000
(c)

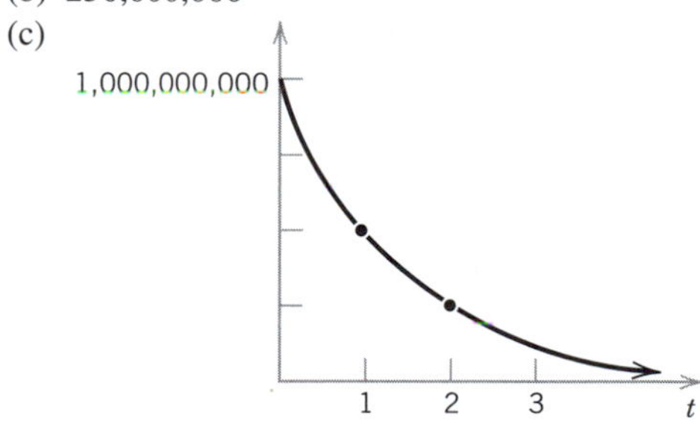

(d) 61035
21. (a) 2^{63}
(b) about 7,278,544,852,320 miles
23. (a) 4.22 feet
(b) 7.97 feet
(c) 0 feet
(d) 128 feet

Chapter Six

1. (−13, −34)
3. (21, 12)
5. (−3.9, −4.3)
7. No solution
9. (7, 16)

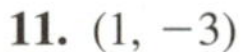

11. (1, −3)

13.

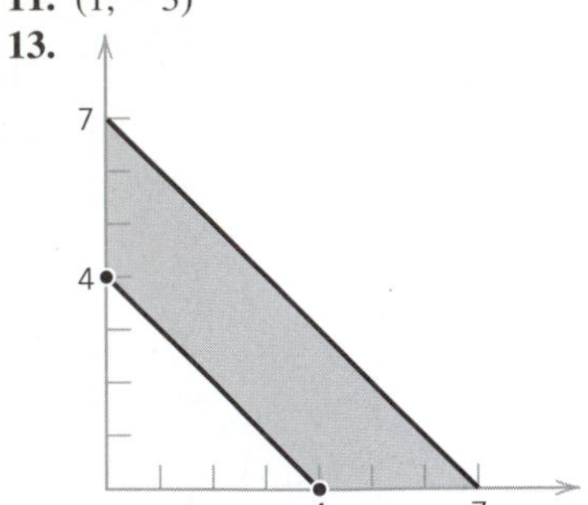

15.

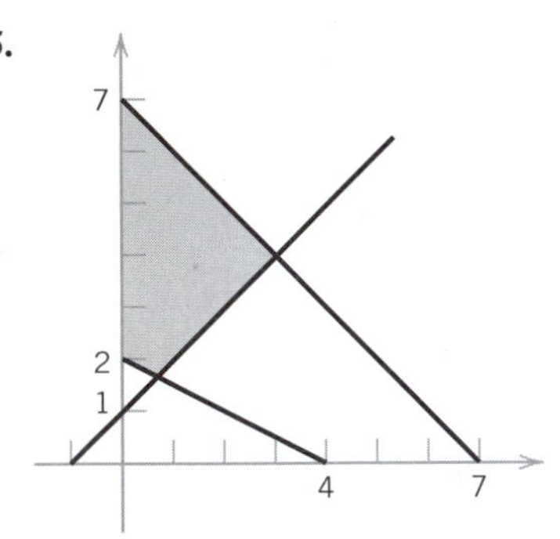

17.

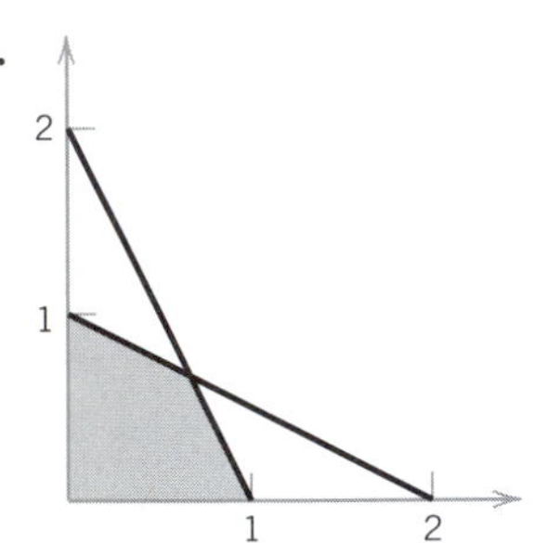

19. (0, 4), (0, 7), (4, 0), (7, 0)
21. (0, 2), (0, 7), (3, 4), (2/3, 5/3)
23. (0, 0), (0, 1), (1, 0), (2/3, 2/3)
25. (a) 28; (b) 29; (c) $10
27. (a) $y = 175x + 100$; (b) 0.57; (c) 1; (d) $25
29. (a) 6 small, 0 large; (b) $42
31. (a) 3 dulcimers, 3 sitars; (b) $1365
33. (a) 0 gallons of Triple Hitter, 200 gallons of Major League; (b) $300

Chapter Seven

1. (a) 150 square miles
(b) 196 square inches
(c) 5 square feet
(d) 500 square feet
(e) 452.39 square inches
(f) 12 square feet
3. (a) $172.55
(b) $238.00
(c) $85.94
5. No, they need to cover 236.55 sq ft twice.
7. (a) 0.13 liters
(b) 200 liters
(c) 6,000,000 liters
(d) 494.8 liters
9. (a) 24,000
(b) No
11. (a) 5 gallons
(b) 450 miles
(c) 2 quarts
(d) 100 cubic inches
13. (a) $3.45 \cdot 10^{-8}$
(b) $1.23567 \cdot 10^{2}$
(c) $1.4 \cdot 10^{28}$
(d) $-8.8 \cdot 10^{38}$
15. 7927.45 years
19. (a) Parent: *A*
Child: *D*
Sibling: *C*
(b) Parent: *A*
Children: *C*, *D*
Siblings: none
(c) Parent: *R*
Children: *F*, *E*
Siblings: *A*, *B*
(d) Parent: B
Children: none
Siblings: C
21. (a)

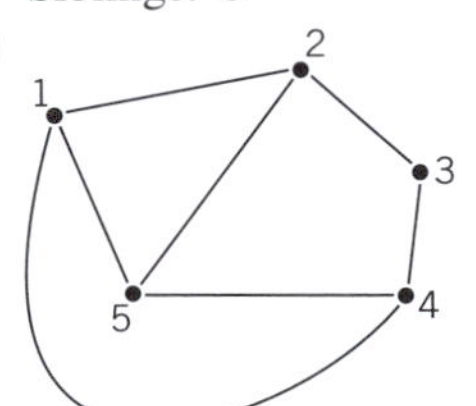

(b)

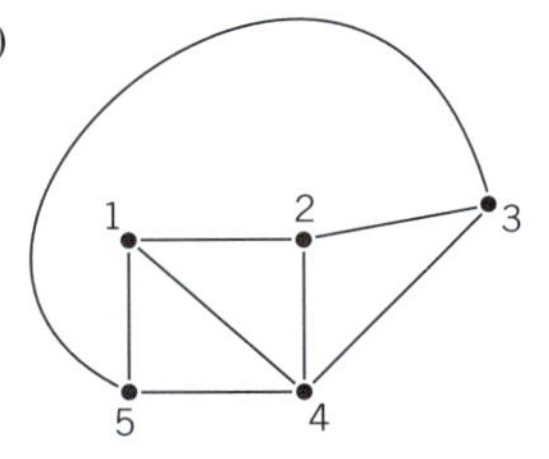

(c)

23. (a) 1-2-6, weight = 5
(b) 1-2-3-6, weight = 5
(c) 1-2-5-4-6, weight = 8

Chapter Eight

1. A numeral is a symbol that names a number.
3. (a) *R*, *Q*, *Z*, *W*
(b) *R*, *Q*, *Z*, *W*, *N*
(c) *R*, *Q*
(d) *R*, *Q*, *Z*, *W*

(e) R, Q
(f) R, Q, Z, W, N
(g) R, Q
(h) R, H
(i) R, H
(j) R, Q

5. Suppose that n is the numerator of the rational number. Divide n by 9 and find the quotient and remainder. The decimal representation of $n/9$ is that quotient followed by a decimal point and then the remainder repeated forever.

7.

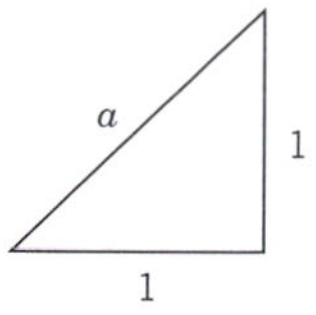

$a = \sqrt{2}$

9. (a) 198; 132; 495; 297
(b) The middle digit of the product is the sum of the digits in the number being multiplied by 11.

11. Primes: 2, 3, 5, 7, 11, 13, 17, 19, 23, 29, 31, 37, 41, 43, 47, 53, 59, 61, 67, 71, 73, 79, 83, 89, 97, 101, 103, 107, 109, 113, 127, 131, 137, 139, 149, 151, 157, 163, 167, 173, 179, 181, 191, 193, 197, 199, 211, 223, 227, 229, 233, 239, 241, 251, 257, 263, 269, 271, 277, 281, 283, 293

13. (a) 61 + 37
(b) 31 + 7
(c) 557 + 563

15. 1

17. 01 04 05 01 06 05 14 09 14 13 06 03 01 13 07 01 14

19. 18 02 17 16 10 17 02 15 09 01 17 03

21. BAG MAKE

23. (a) LHQLB THQ KALB
(b) $d = 2$

25. (a) Rectangular, triangular, hexagonal
(b) Rectangular
(c) Rectangular
(d) Rectangular, triangular, pentagonal

27. 16, 19, 22, 25, 28

29.

13	1	deficient
14	10	deficient
15	9	deficient
16	15	deficient
17	1	deficient
18	21	abundant
19	1	deficient
20	22	abundant

31. Suppose that p is a perfect number. Then the sum of the proper divisors of p is p. Suppose that k is a proper divisor of p. Then $2k$ is a proper divisor of $2p$. Thus, the sum of the proper divisors of $2p$ is at least $2p$. But p is a proper divisor of $2p$. So, the sum of the proper divisors of $2p$ is at least $2p + p = 3p$, which is greater than $2p$. Hence, $2p$ is abundant.

33. (a) 2002
(b) 1993, 1997, 1999
(c) Yes, 1997 and 1999

Chapter Nine

1. 1×3

3. 2×5

5.

	Sam	Janet
June	6	12
July	12	14
August	7	10

7. [10 15 20]

9. [20 19]

11. Impossible. B is 2×4 but D is 3×2.

13. $\begin{bmatrix} 10 & 16 \\ 10 & 26 \end{bmatrix}$

15. Impossible. C and E are not the same order.

17. (a)

	A	B	C
Apples	20	25	40
Peaches	40	35	20

(b)

	A	B	C
Apples	20	20	40
Peaches	40	25	20

19. Strictly determined, 2

21. Strictly determined, 8

23. Not strictly determined

25. Strictly determined, −8

27. 0.275

29. 0.5

31. Row: [0.7 0.3]; column: [0.3 0.7]; $E = -1.7$

33. Row: [0.5 0.5]; column: [0.5 0.5]; $E = 2$

37. (a)

		Pitcher	
		Fastball	Curveball
Batter	Fastball	10	−5
	Curveball	−5	3

(b) Batter: [0.348 0.652]
Pitcher: [0.348 0.652]
(c) 0.217; the batter has a slight advantage.

Chapter Ten

1. (a) $4 \cdot 10^4 + 5 \cdot 10^3 + 6 \cdot 10^2 + 7 \cdot 10 + 8$
(c) $2 \cdot 10^{12}$

2. (b)

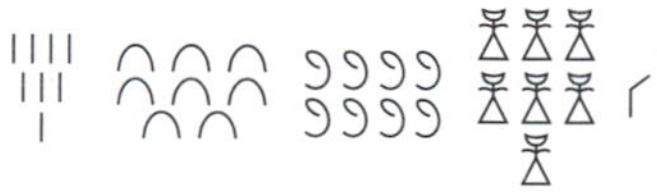

(d)

3. (b) 10,601
(d) 101,260

4. (b) , ι ζ

(d) ͵αξϛ

5. (b) 40,521
(d) $\overline{100},001,001$
6. (b) $\overline{\text{XVII}}$
(d) MLXVI
7. (b) 1746
(d) 90,020
8. (b) 萬
七
千

(d) 千
六
十
六

9. (b) 5046
(d) 1529
10. (b) ··
$\overset{..}{—}$
····
◎

(d) ··
$\overset{..}{≡}$
$\overset{.}{—}$

11. (b) 73,130,402
(d) 5,765,782
12. (b) ▾▾▾▾ ≺≺≺≺▾▾▾ ≺≺

(d) ≺▾▾▾▾▾▾▾ ≺≺≺≺▾▾▾▾▾▾

13. (b) 129,636,012
(d) 15,415,201
14. (b) 100001001101000
(d) 10000101010
15. (b) 31
(d) 87
16. (b) *ome-tzuntli-on-matlactli-poualli-on-yey-poualli-on-chicace*
(d) *chicuey-tzuntli-on-matlactli-poualli*
17. (b) 2714
(d) 32,866
18. (b) *pai-erh-shih-si*
(d) *chiu-pai-pa-shih-chi*
20. (a) *kumi na nne*
(c) *mia mojo thelathini na sita*
21. (a) *tri ar dec*
(c) *tri ar ugeint*
(e) *petwar ar petwar-ugeint*
(g) *pimp ar cant*
22. (b) *uaclahun*
(d) *canlahun kal*
23. (b) 2 less than 20
(d) *metadilogota*

Chapter Eleven

1. Proposition
3. Not a proposition
5. Proposition
7. Not a proposition
9. Proposition
11. $s \wedge e$
13. $s \wedge b$
15. $\neg p \rightarrow b$
17. $d \vee s$
19. $\neg u \rightarrow \neg b$
21. If there are bear tracks around our campsite, then the bear ate our pizza.
23. If there are not bear tracks around our campsite, then the bear ate our pizza.
25. The raccoons stole our shoes or there are bear tracks around our campsite.
27. There are not bear tracks around our campsite and the raccoons stole our shoes.
29. If the raccoons didn't steal our shoes, then the bear ate our pizza or there are bear tracks around our campsite.
31. If it is the case that if the bear ate our pizza then the raccoons stole our shoes, then there are bear tracks around our campsite.
33. False
35. True
37. True
39. True
41. True
43.

p	q	p	$\rightarrow$	$(p \vee q)$
T	T		**T**	T
T	F		**T**	T
F	T		**T**	T
F	F		**T**	F

45.

p	q	$\neg q$	$\vee$	$(p \wedge q)$
T	T	F	**T**	T
T	F	T	**T**	F
F	T	F	**F**	F
F	F	T	**T**	F

47.

p	q	r	$(p \wedge q)$	$\rightarrow$	$(r \vee p)$
T	T	T	T	**T**	T
T	T	F	T	**T**	T
T	F	T	F	**T**	T
T	F	F	F	**T**	T
F	T	T	F	**T**	T
F	T	F	F	**T**	F
F	F	T	F	**T**	T
F	F	F	F	**T**	F

49.

p	q	r	$(\neg p \to r)$	$\to$	$\neg q$
T	T	T	T	**F**	F
T	T	F	T	**F**	F
T	F	T	T	**T**	T
T	F	F	T	**T**	T
F	T	T	T	**F**	F
F	T	F	F	**T**	F
F	F	T	T	**T**	T
F	F	F	F	**T**	T

51.

p	q	$(p$	$\to$	$(p \to q))$	$\to$	q
T	T		T	T	**T**	
T	F		F	F	**T**	
F	T		T	T	**T**	
F	F		T	T	**F**	

53. 42, 43, 47, and 50 are tautologies.

55.

p	q	$p \to \neg q$	q	$\neg p$
T	T	F	T	F
T	F	T	F	F
F	T	T	T	T
F	F	T	F	T

Valid

57.

p	q	r	$\neg(p \to q)$	q	r
T	T	T	F		
T	T	F	F		
T	F	T	T	F	
T	F	F	T	F	
F	T	T	F		
F	T	F	F		
F	F	T	F		
F	F	F	F		

Valid

59.

p	q	r	$(r \to p)$	$\to$	q	$\neg p$	$r \vee q$
T	T	T		T		F	
T	T	F		T		F	
T	F	T		F			
T	F	F		F			
F	T	T		T		T	T
F	T	F		T		T	T
F	F	T		T		T	T
F	F	F		F			

Valid

61.

p	q	r	$p \to (r \vee q)$	$\neg r$	$\neg q$	p
T	T	T	T	F		
T	T	F	T	T	F	
T	F	T	T	F		
T	F	F	F			
F	T	T	T	F		
F	T	F	T	T	F	
F	F	T	T	F		
F	F	F	T	T	T	F

Invalid

63. b = I went to the beach
a = I saw my friend Al
g = Al was playing golf

b
$\neg a \to g$
$\neg b \vee \neg g$
$\therefore a$

a	b	g	b	$\neg a \to g$	$\neg b \vee \neg g$	a
T	T	T	T	T	F	
T	T	F	T	T	T	T
T	F	T	F			
T	F	F	F			
F	T	T	T	T	F	
F	T	F	T	F		
F	F	T	F			
F	F	F	F			

Valid

65. s = Sam went to Chicago
y = Yohei went to Austin
m = Myrna went to Lahore

$s \vee y$
$y \to m$
$\neg m$
$\therefore s$

s	y	m	$s \vee y$	$y \to m$	$\neg m$	s
T	T	T	T	T	F	
T	T	F	T	F		
T	F	T	T	T	F	
T	F	F	T	T	T	T
F	T	T	T	T	F	
F	T	F	T	F		
F	F	T	F			
F	F	F	F			

Valid

67. **69.**

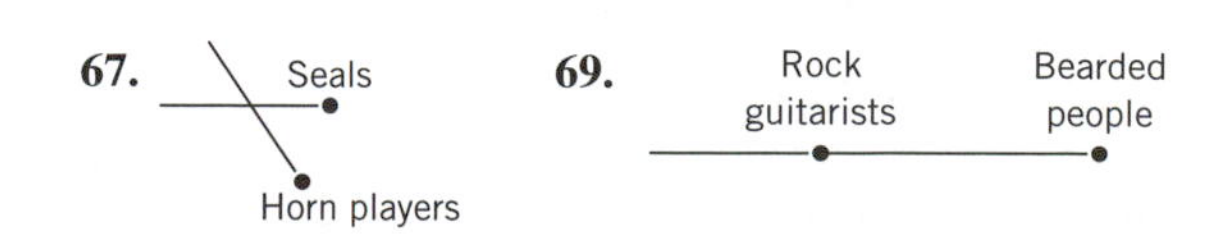

71.

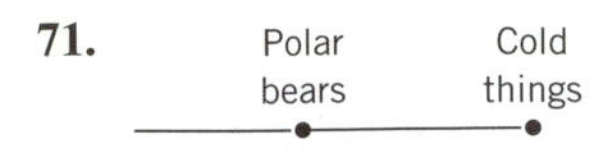

73. ¬Readers
Gorillas

75. Aztecs
Writers

77. No seals are horn players.

79. Some rock guitarists do not have beards.

81. Some polar bears are not cold.

83. All gorillas can read.

85. No Aztecs can write.

87.

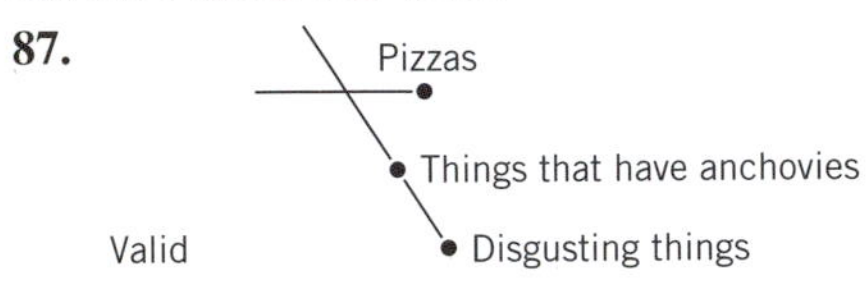

Valid

89.

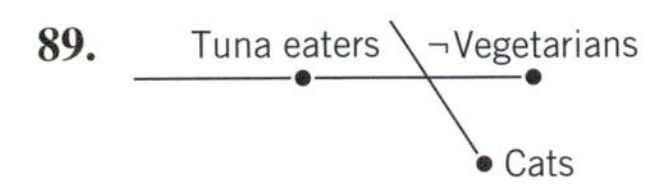

or

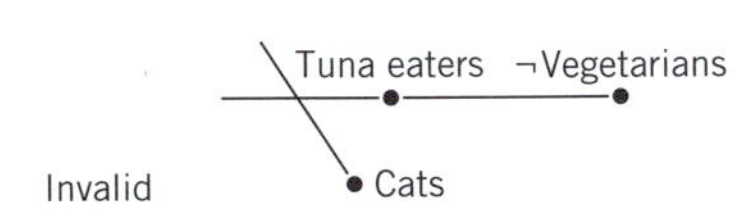

Invalid

91. Fat books Hard to read books
Math books

Invalid

93.

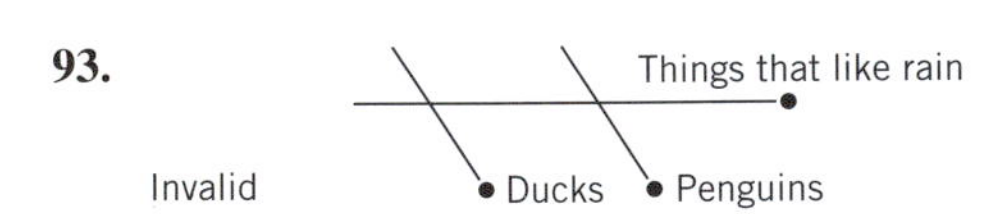

Invalid

95.

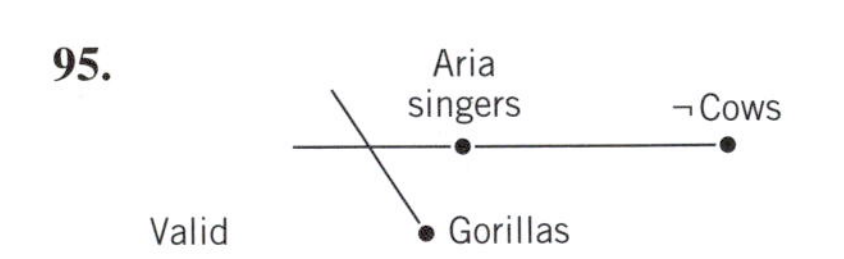

Valid

97. (a) $f \to b \wedge \neg a$
$p \to a \wedge b$
(b) No

Chapter Twelve

13. No.

19. If weight < 16 then
cost := 0.25 * weight
else
cost := 0.20 * weight;

21. If (weight < 16) then
cost = 0.25 * weight;
else
cost = 0.20 * weight;

23. If CD < 4 then
cost := CD * 2.00
else
cost := CD * 1.25;

25. If (CD < 4) then
cost = CD * 2.00;
else
cost = CD * 1.25;

27. $S \Rightarrow_1 XY \Rightarrow_2 aYY \Rightarrow_3 abSY \Rightarrow_1 abXYY \Rightarrow_2 abaYYY \Rightarrow_4$ (three times) *abaaaa*

29. $S \Rightarrow_1 XY \Rightarrow_2 aYY \Rightarrow_3 abSY$ generation is blocked because we cannot get another b from S.

31. $S \Rightarrow_1 XcY \Rightarrow_2 aXcY \Rightarrow_2 aaXcY \Rightarrow_3 aaccY \Rightarrow_5 aaccb$

33. $S \Rightarrow_1 XcY \Rightarrow_2 aXcY \Rightarrow_3 accY$ generation is blocked because we cannot get an a from Y.

35. $S \Rightarrow_1$ *yabba-X* $\Rightarrow_3$ *yabba-dabba-Y* $\Rightarrow_4$ *yabba-dabba-doo*

37. $S \Rightarrow_1$ *yabba-X* generation is blocked because we cannot get a *yabba-* from X.

39. $S \Rightarrow_1 V \cdot NP. \Rightarrow_3$ see $\cdot$ $NP. \Rightarrow_5$ see $\cdot$ N $\cdot$ run. $\Rightarrow_{10}$ see $\cdot$ Spot $\cdot$ run.

41. $S \Rightarrow_1 V \cdot NP.$ or $S \Rightarrow_2 V \cdot NP.S$ generation is blocked in both cases because no rule gives *run* from V.

43. $T = \{a, b\}$
$NT = \{S\}$
1. $S \to$ ab
2. $S \to$ aSb

45. $T = \{a, b, c\}$
$NT = \{S, X\}$
1. $S \to aaX$
2. $X \to bc$
3. $X \to bXc$

47. $T = \{a, b, c\}$
$NT = \{S, X\}$
1. $S \to aXb$
2. $X \to c$
3. $X \to cX$

49. $T = \{a, b, c\}$
$NT = \{S, B, X\}$
1. $S \to aaXbb$
2. $X \to c$
3. $X \to cB$
4. $B \to cX$

51. $T = \{0, 1\}$
$NT = \{S, X, Y\}$
1. $S \to X1$
2. $S \to 1$
3. $X \to 1Y$
4. $X \to 1$
5. $Y \to 0Y$
6. $Y \to 0$
7. $Y \to 1Y$
8. $Y \to 1$

57.

Fido	Beagle	6	T. Fern	WY
Balzac	Poodle	7	G. Whiz	FL

59.

Snoopy	Beagle	3	C. Brown	NY
Big Red	Greyhound	2	C. Bunga	NY
Alfie	Terrier	5	C. Diego	NY
Toto	Terrier	3	D. Smith	KS

61.

Balzac	Poodle	7	G. Whiz	FL
Fido	Beagle	6	T. Fern	WY
Alfie	Terrier	5	C. Diego	NY
Snoopy	Beagle	3	C. Brown	NY
Toto	Terrier	3	D. Smith	KS
Big Red	Greyhound	2	C. Bunga	NY

63.

Balzac	Tiger	7	G. Whiz	FL
Boozer	Persian	7	F. Zorka	CA
Tom	Tiger	7	D. Smith	CA

65.

Smiles	Persian	13	C. Bono	CA
Balzac	Tiger	7	G. Whiz	FL
Boozer	Persian	7	F. Zorka	CA
Tom	Tiger	7	D. Smith	CA

67.

Aleph0	Tiger	15	C. Mendes	NY
Smiles	Persian	13	C. Bono	CA
Balzac	Tiger	7	G. Whiz	FL
Boozer	Persian	7	F. Zorka	CA
Tom	Tiger	7	D. Smith	CA
Claws	Tiger	6	T. Frenzy	WY

71. If A then C and F. If B then D and G. F and G gives X.

73. If A and B then D. If C then F and E (with D). E gives G. G and A gives X.

77. **ARROW: FIRST GENERATION**

	A	B	C	D	E	F	G	H	I	J	K	L	M	N	O
1															
2															
3															
4						●									
5				●	●	●	●								
6						●									
7															
8															
9															
10															
11															
12															
13															

ARROW: SECOND GENERATION

	A	B	C	D	E	F	G	H	I	J	K	L	M	N	O
1															
2															
3															
4						●	●								
5							●								
6						●	●								
7															
8															
9															
10															
11															
12															
13															

ARROW: THIRD GENERATION

	A	B	C	D	E	F	G	H	I	J	K	L	M	N	O
1															
2															
3															
4						●	●								
5								●							
6						●	●								
7															
8															
9															
10															
11															
12															
13															

ARROW: FOURTH GENERATION

	A	B	C	D	E	F	G	H	I	J	K	L	M	N	O
1															
2															
3															
4							●								
5								●							
6							●								
7															
8															
9															
10															
11															
12															
13															

ARROW: FIFTH GENERATION

	A	B	C	D	E	F	G	H	I	J	K	L	M	N	O
1															
2															
3															
4															
5							●	●							
6															
7															
8															
9															
10															
11															
12															
13															

Extinct in the sixth generation.

79.

ELEVEN: FIRST GENERATION

	A	B	C	D	E	F	G	H	I	J	K	L	M	N	O
1															
2															
3															
4															
5					●		●								
6					●		●								
7					●		●								
8															
9															
10															
11															
12															
13															

ELEVEN: SECOND GENERATION

	A	B	C	D	E	F	G	H	I	J	K	L	M	N	O
1															
2															
3															
4															
5															
6				●	●		●	●							
7															
8															
9															
10															
11															
12															
13															

Extinct in the third generation

Appendix A

VALUES OF $(1 + R/12)^N$

	N			
R	**180**	**240**	**300**	**360**
0.070	2.848947	4.038739	5.725418	8.116497
0.075	3.069452	4.460817	6.482880	9.421534
0.080	3.306921	4.926803	7.340176	10.935730
0.085	3.562653	5.441243	**8.310413**	12.692499
0.090	3.838043	6.009152	9.408415	14.730576
0.095	4.134593	6.636061	10.650941	17.094862
0.100	4.453920	7.328074	12.056945	19.837399
0.105	4.797761	8.091918	13.647852	23.018509
0.110	5.167988	8.935015	15.447889	26.708098
0.115	5.566613	9.865552	17.484440	30.987181
0.120	5.995802	10.892554	19.788466	35.949641
0.125	6.457884	12.025975	22.394964	41.704262
0.130	6.955364	13.276792	25.343491	48.377089
0.135	7.490939	14.657109	28.678761	56.114160
0.140	8.067507	16.180270	32.451308	65.084661
0.145	8.688187	17.860991	36.718246	75.484592
0.150	9.356334	19.715494	41.544120	87.540995

Example: $(1 + 0.085/12)^{300}$ is shown in **bold print.**

Appendix B

z = z-score

An entry in the table is the area under the curve between $z = 0$ and a positive value of z. Areas for negative values of z are obtained by symmetry.

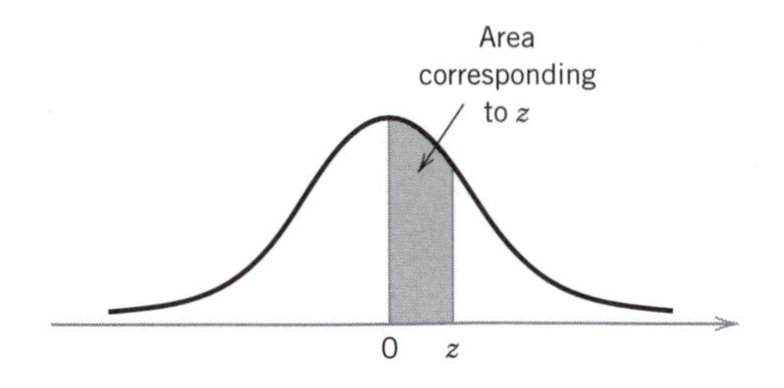

NORMAL CURVE TABLE*

z	**0.00**	**0.01**	**0.02**	**0.03**	**0.04**	**0.05**	**0.06**	**0.07**	**0.08**	**0.09**
0.0	0.0000	0.0040	0.0080	0.0120	0.0160	0.0199	0.0239	0.0279	0.0319	0.0359
0.1	0.0398	0.0438	0.0478	0.0517	0.0557	0.0596	0.0636	0.0675	0.0714	0.0753
0.2	0.0793	0.0832	0.0871	0.0910	0.0948	0.0987	0.1026	0.1064	0.1103	0.1141
0.3	0.1179	0.1217	0.1255	0.1293	0.1331	0.1368	0.1406	0.1433	0.1480	0.1517
0.4	0.1554	0.1591	0.1628	0.1664	0.1700	0.1736	0.1772	0.1808	0.1844	0.1879
0.5	0.1915	0.1950	0.1985	0.2019	0.2054	0.2088	0.2123	0.2157	0.2190	0.2224
0.6	0.2257	0.2291	0.2324	0.2357	0.2389	0.2422	0.2454	0.2486	0.2517	0.2549
0.7	0.2580	0.2611	0.2642	0.2673	0.2703	0.2734	0.2764	0.2794	0.2823	0.2852
0.8	0.2881	0.2910	0.2939	0.2967	0.2995	0.3023	0.3051	0.3078	0.3106	0.3133
0.9	0.3159	0.3186	0.3212	0.3238	0.3264	0.3289	0.3315	0.3340	0.3365	0.3389
1.0	0.3413	0.3438	0.3461	0.3485	0.3508	0.3531	0.3554	0.3577	0.3599	0.3621
1.1	0.3642	0.3665	0.3686	0.3708	0.3729	0.3749	0.3770	0.3790	0.3810	0.3830
1.2	0.3849	0.3869	0.3888	0.3907	0.3925	0.3944	0.3962	0.3980	0.3997	0.4015
1.3	0.4032	0.4049	0.4066	0.4082	0.4099	0.4115	0.4131	0.4147	0.4162	0.4177
1.4	0.4192	0.4207	0.4222	0.4236	0.4251	0.4265	0.4279	0.4292	0.4306	0.4319
1.5	0.4332	0.4345	0.4357	0.4370	0.4382	0.4394	0.4406	0.4418	0.4429	0.4441
1.6	0.4452	0.4463	0.4474	0.4484	0.4495	0.4505	0.4515	0.4525	0.4535	0.4545
1.7	0.4554	0.4564	0.4573	0.4582	0.4591	0.4599	0.4608	0.4616	0.4625	0.4633
1.8	0.4641	0.4649	0.4656	0.4664	0.4671	0.4678	0.4686	0.4693	0.4699	0.4706
1.9	0.4713	0.4719	0.4726	0.4732	0.4738	0.4744	0.4750	0.4756	0.4761	0.4767
2.0	0.4772	0.4778	0.4783	0.4788	0.4793	0.4798	0.4803	0.4808	0.4812	0.4817
2.1	0.4821	0.4826	0.4830	0.4834	0.4838	0.4842	0.4846	0.4850	0.4854	0.4857
2.2	0.4861	0.4864	0.4868	0.4871	0.4875	0.4878	0.4881	0.4884	0.4887	0.4890
2.3	0.4893	0.4896	0.4898	0.4901	0.4904	0.4906	0.4909	0.4911	0.4913	0.4916
2.4	0.4918	0.4920	0.4922	0.4925	0.4927	0.4929	0.4931	0.4932	0.4934	0.4936
2.5	0.4938	0.4940	0.4941	0.4943	0.4945	0.4946	0.4948	0.4949	0.4951	0.4952
2.6	0.4953	0.4955	0.4956	0.4957	0.4959	0.4960	0.4961	0.4962	0.4963	0.4964
2.7	0.4965	0.4966	0.4967	0.4968	0.4969	0.4970	0.4971	0.4972	0.4973	0.4974
2.8	0.4974	0.4975	0.4976	0.4977	0.4977	0.4978	0.4979	0.4979	0.4980	0.4981
2.9	0.4981	0.4982	0.4982	0.4983	0.4984	0.4984	0.4985	0.4985	0.4986	0.4986
3.0	0.4987	0.4987	0.4987	0.4988	0.4988	0.4989	0.4989	0.4989	0.4990	0.4990

* This table is reprinted from *Finite Mathematics with Applications, Sixth Edition* by A. Mizrahi (NY: John Wiley & Sons, 1992), Table 2, p. 287.

$z = \frac{x - \bar{x}}{s}$; $\bar{x}$ = mean; s = standard deviation

Appendix C

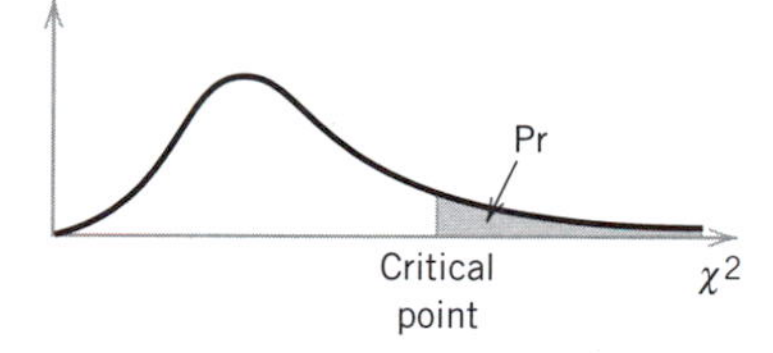

CHI-SQUARE TABLE*

d.f. \ Pr	.250	.100	.050	.025	.010	.005	.001
1	1.32	2.71	3.84	5.02	6.63	7.88	10.8
2	2.77	4.61	5.99	7.38	9.21	10.6	13.8
3	4.11	6.25	7.81	9.35	11.3	12.8	16.3
4	5.39	7.78	9.49	11.1	13.3	14.9	18.5
5	6.63	9.24	11.1	12.8	15.1	16.7	20.5
6	7.84	10.6	12.6	14.4	16.8	18.5	22.5
7	9.04	12.0	14.1	16.0	18.5	20.3	24.3
8	10.2	13.4	15.5	17.5	20.1	22.0	26.1
9	11.4	14.7	16.9	19.0	21.7	23.6	27.9
10	12.5	16.0	18.3	20.5	23.2	25.2	29.6
11	13.7	17.3	19.7	21.9	24.7	26.8	31.3
12	14.8	18.5	21.0	23.3	26.2	28.3	32.9
13	16.0	19.8	22.4	24.7	27.7	29.8	34.5
14	17.1	21.1	23.7	26.1	29.1	31.3	36.1
15	18.2	22.3	25.0	27.5	30.6	32.8	37.7
16	19.4	23.5	26.3	28.8	32.0	34.3	39.3
17	20.5	24.8	27.6	30.2	33.4	35.7	40.8
18	21.6	26.0	28.9	31.5	34.8	37.2	42.3
19	22.7	27.2	30.1	32.9	36.2	38.6	43.8
20	23.8	28.4	31.4	34.2	37.6	40.0	45.3
21	24.9	29.6	32.7	35.5	38.9	41.4	46.8
22	26.0	30.8	33.9	36.8	40.3	42.8	48.3
23	27.1	32.0	35.2	38.1	41.6	44.2	49.7
24	28.2	33.2	36.4	39.4	43.0	45.6	51.2
25	29.3	34.4	37.7	40.6	44.3	46.9	52.6
26	30.4	35.6	38.9	41.9	45.6	48.3	54.1
27	31.5	36.7	40.1	43.2	47.0	49.6	55.5
28	32.6	37.9	41.3	44.5	48.3	51.0	56.9
29	33.7	39.1	42.6	45.7	49.6	52.3	58.3
30	34.8	40.3	43.8	47.0	50.9	53.7	59.7
40	45.6	51.8	55.8	59.3	63.7	66.8	73.4
50	56.3	63.2	67.5	71.4	76.2	79.5	86.7
60	67.0	74.4	79.1	83.3	88.4	92.0	99.6
70	77.6	85.5	90.5	95.0	100	104	112
80	88.1	96.6	102	107	112	116	125
90	98.6	108	113	118	124	128	137
100	109	118	124	130	136	140	149

* Reprinted from T. H. Wonnacott and R. J. Wonnacott, *Introductory Statistics for Business & Economics* (New York, John Wiley & Sons: 1972), Table IIa, p. 592.

Index